D0706767

INDUSTRIAL MECHANICS AND MAINTENANCE

THIRD EDITION

LARRY CHASTAIN
Athens Technical College

Upper Saddle River, New Jersey 07458

Library of Congress Cataloging-in-Publication Data

Chastain, Larry
Industrial mechanics and maintenance/Larry Chastain.—3rd ed.
p. cm.
Includes index.
ISBN-13: 978-0-13-515096-2 (alk. paper)
ISBN-10: 0-13-515096-5 (alk. paper)
1. Industrial equipment—Maintenance and repair. 2. Machinery—Maintenance and repair.
I. Title.
TS191.C45 2009
621.8'16—dc22 2008013928

Vice President and Executive Publisher: Vernon R. Anthony
Acquisitions Editor: Eric Krassow
Editorial Assistant: Sonya Kottcamp
Project Manager: Wanda Rockwell
Creative Director: Jayne Conte
Cover Designer: Margaret Kenselaar
Cover Image: Getty Images Inc.
Director of Marketing: David Gesell
Marketing Manager: Derril Trakalo
Marketing Coordinator: Alicia Dysert

This book was set in 10/12 Melior by Integra. It was printed and bound by Courier Companies. The cover was printed by Phoenix Color Corp./Hagerstown.

Pearson Education Ltd., London
Pearson Education Singapore, Pte. Ltd
Pearson Education Canada, Inc.
Pearson Education–Japan
Pearson Education Australia PTY, Limited

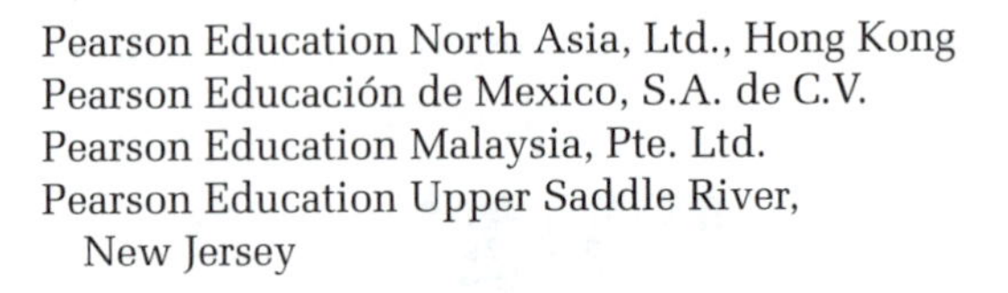
Pearson Education North Asia, Ltd., Hong Kong
Pearson Educación de Mexico, S.A. de C.V.
Pearson Education Malaysia, Pte. Ltd.
Pearson Education Upper Saddle River, New Jersey

10 9 8 7 6 5 4
ISBN-13: 978-0-13-515096-2
ISBN-10: 0-13-515096-5

PREFACE

While in engineering college, I noticed many textbooks dealing with the theoretical side of mechanics but few dealing with actual industrial mechanical equipment. Individuals seeking a solid background in the basic principles and real-life equipment used in industrial mechanics today need a modern, advanced, industrial textbook with a practical hands-on approach. The text should also be useful for those already working in industry who want to enrich and expand their knowledge of mechanical equipment used in industry.

FOR WHOM THIS BOOK WAS WRITTEN

This book was written to fill the need for an up-to-date, practical teaching resource that focuses on the needs of industrial mechanics, technicians, and engineers working with industrial mechanical and power-transmission products. It is based on feedback from students and instructors along with ideas from different industries with which I have worked over the years.

This text is designed to be used as the principal text for training the industrial mechanic in postsecondary vocational schools and colleges, on-the-job training, and apprenticeship programs. It can also serve as a real-world text for mechanical engineering or engineering technology.

ORGANIZATION OF THIS BOOK

Readability is my prime concern, so I have used a direct, straightforward approach along with detailed illustrations. Instead of separate chapters on safety and troubleshooting, I have integrated safety and troubleshooting tips in each chapter. This comprehensive approach encourages students to learn about the safety and troubleshooting components along with the basic information associated with a topic. Instructors can teach the material in the 13 chapters in sequential order or change the order of chapters to accomplish specific goals.

ACKNOWLEDGMENTS

I would like to thank the following for their helpful reviews of this new edition: Michael Case, Middle Georgia Technical College; Jason Kefover, York Technical College; Michael E. Brumbach, York Technical College; and Edward T. Moore, York Technical College.

I would also like to thank the Dodge Division of Rockwell Automation and the Browning/Morse Division of Emerson Electric for supplying many of the photos used in this text.

I would like to especially thank Dr. Tom Joiner who has helped in more ways than I could possibly list.

ONLINE INSTRUCTOR'S RESOURCES

To access supplementary materials online, instructors need to request an instructor access code. Go to **www.pearsonhighered.com/irc,** where you can register for an instructor access code. Within 48 hours after registering, you will receive a confirming e-mail, including an instructor access code. Once you have received your code, go to the site and log on for full instructions on downloading the materials you wish to use.

CONTENTS

CHAPTER 1

Hand Tools

1-1 INTRODUCTION

It has been said that the ability to use tools is a measure of the intelligence of a species. Many hand tools still used today have a history as old as people themselves. From the beginning of civilization when people started to build things, they began to design and use hand tools. Industry today requires the use of many tools, a majority of which are very specialized for particular jobs (see Figure 1.1). Industrial technicians must be proficient in the use of basic hand tools as well as some specialized tools. They should use the correct tool for the job being performed and use it correctly. They should always be sure that their tools are well maintained and in good working condition. Lastly, and most importantly, industrial technicians should always consider safety first when using any tool and concentrate on the task at hand.

This chapter deals with the basic types of hand tools, correct terms, and how to use the tool safely.

1-2 SCREWDRIVERS

The screwdriver is one of the most basic hand tools. It is probably the most frequently misused and abused of any type of hand tool.

A screwdriver is designed to install and remove all types of screws and fasteners. Some people also use it as a pry bar, as a chisel, and for other things for which it was not designed. Don't be guilty of misusing this tool. Always get the correct tool designed for your application.

A screwdriver has three basic parts: a tip, a shank, and a handle. The oldest style tip is the *flat blade* (see Figure 1.2). The blade tip should be flat on the end and should fit squarely into the screw slot. If it is rounded, it will try to rise out of the screw slot when being used. A worn flat-blade tip should be reground

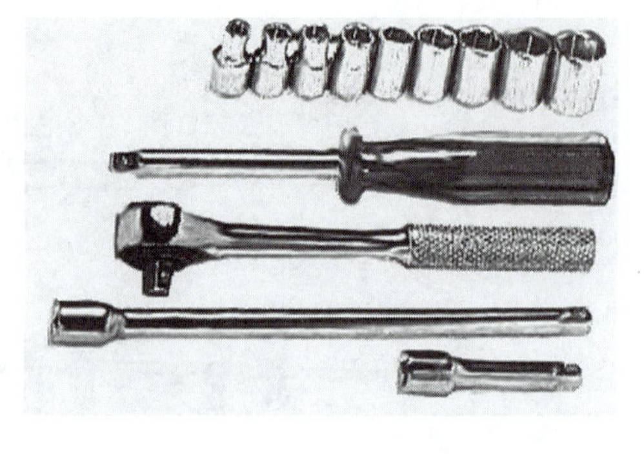

FIGURE 1.1
Basic hand tools.

FIGURE 1.2
Flat-blade screwdriver.

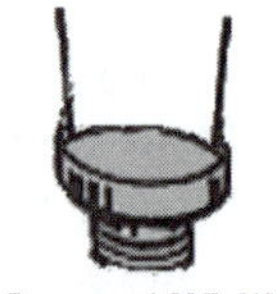

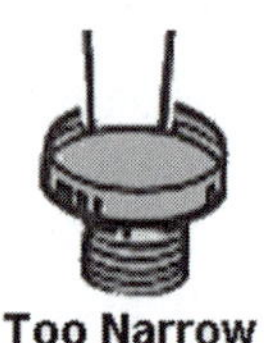

Blade Too Thick

FIGURE 1.3
Always use the correct size of screwdriver blade.

or replaced. The correct size flat-blade tip should be used for different fasteners (see Figure 1.3). The flat-blade tip is the most misused and misapplied of all the different types of tips. If the flat-blade tip is correctly selected, it should fill the slot almost completely and it should contact the bottom of the slot. The width of the blade should extend to the edges of the fastener. Table 1.1 gives the correct blade tip width for different screw sizes.

The *Phillips* tip is another common type of screwdriver tip (see Figure 1.4). This *cross-slot* tip has two slots at right angles. This tip is less likely to slip out of the slots because of its self-centering action. The Phillips tip was one of the first of many different tips developed primarily for production work. There are four standard sizes of Phillips screwdriver tips. Always use the correct size tip for the fastener to be driven. Incorrect matching of any tip with its fastener can cause fastener damage, screwdriver damage, or personal injury. Table 1.2 shows the correct Phillips tip number for different screw sizes.

TABLE 1.1
Screwdriver flat-blade tip widths for different screw sizes.

Flat-Blade Tip Width	Screw Size
1/4 in.	#6 to #8
5/16 in.	#8 to #12
3/8 in.	#12 to #16
7/16 in.	#16 to #20
1/2 in.	#20 to #24

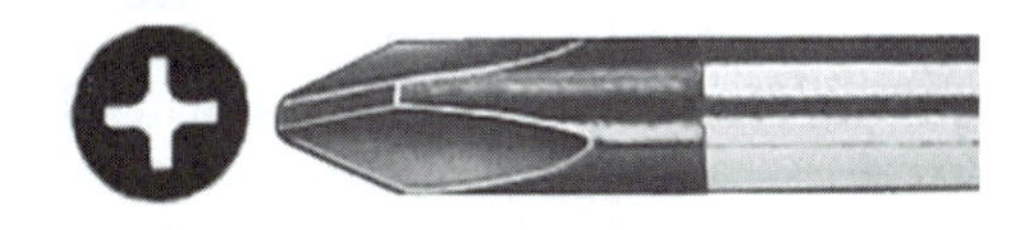

FIGURE 1.4
Phillips screwdriver tip.

TABLE 1.2
Phillips screwdriver tips for different screw sizes.

Phillips Tip Number	Screw Size
No. 1	#4 and smaller
No. 2	#5 to #9
No. 3	#10 to #16
No. 4	#18 and larger

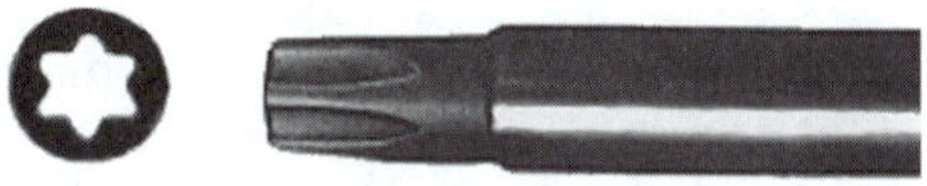

FIGURE 1.5
Torx screwdriver tip.

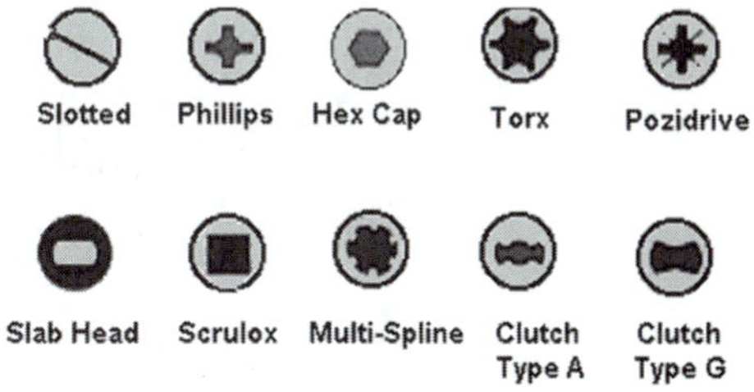

FIGURE 1.6
Common screwdriver tip types.

Torx screwdriver tips are somewhat similar to Phillips tips (see Figure 1.5). The main difference is that these tips have six driving surfaces and are flat across the end. Torx fasteners are usually hardened because of their tendency to strip out. The thinness of the internal webs of the fastener causes this type of damage.

There are many more different types of tips used in industry today. Most are used in production instead of maintenance. Figure 1.6 shows some of the more common types of screwdriver tips.

FIGURE 1.7
Electrically insulated screwdriver.

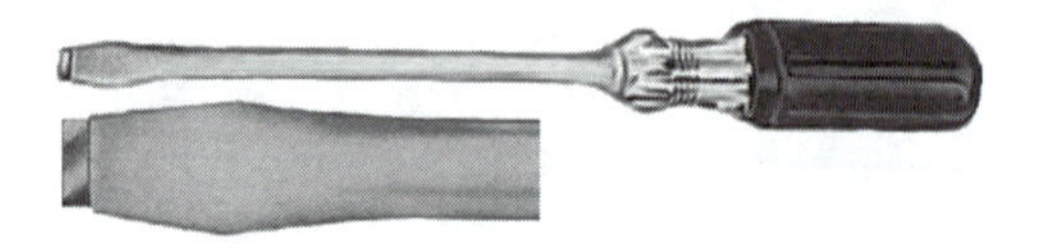

Screwdrivers are specified by the length of their shank as well as the type of tip or blade. The shank length is the distance from the handle (not including the handle length) to the end of the tip. Common screwdriver sizes range from 2 1/2 to 12 in. However, there are many special-purpose screwdrivers that may be smaller or larger.

Screwdriver handles may be smooth, but they usually have grooves to improve grip of the screwdriver. Handles may be wood, metal, or hard rubber, but today most manufacturers use plastic. Plastic handles are durable and inexpensive to manufacture. Rubber-cushioned handles are also common with many screwdrivers. Rubber-cushioned handles are *not* safe for live electrical circuits. You should *never* use a screwdriver on a live electrical circuit. Always check electrical circuits, turn off the power, and use approved lockout/tagout procedures. If you must work on a "live" circuit, use a screwdriver designed for this purpose (see Figure 1.7). Such screwdrivers have an insulated shank and are marked "electrically insulated for live circuits." If at all possible, always remove power before working on any electrical circuit.

When using a screwdriver, always follow these safety precautions:

- Always use the correct type and size screwdriver for the fastener.
- Make sure the screwdriver is in good condition. Repair or replace it if necessary.
- Keep screwdrivers clean and free of oil or grease.
- When using a screwdriver, keep the blade pointed away from you.
- Never hold the part you are working on while using a screwdriver. If the screwdriver slips, you could stab yourself.
- Never use a screwdriver as a chisel, punch, or pry bar.
- Never use a hammer on a screwdriver.
- Never use a screwdriver on a live electrical circuit unless it is specially designed for that purpose.

1-3 PLIERS

Pliers are used to hold, bend, rotate, or cut wire parts or fasteners. Pliers are available in two major configurations: *solid-joint* pliers and *adjustable-joint* pliers. Solid-joint pliers are nonadjustable and have a set range of gripping sizes. Examples of solid-joint pliers are lineman's pliers, needle-nose pliers, and diagonal cutting pliers. Adjustable-joint pliers are more versatile but are usually not as rugged as solid-joint pliers. The adjustability of this type of

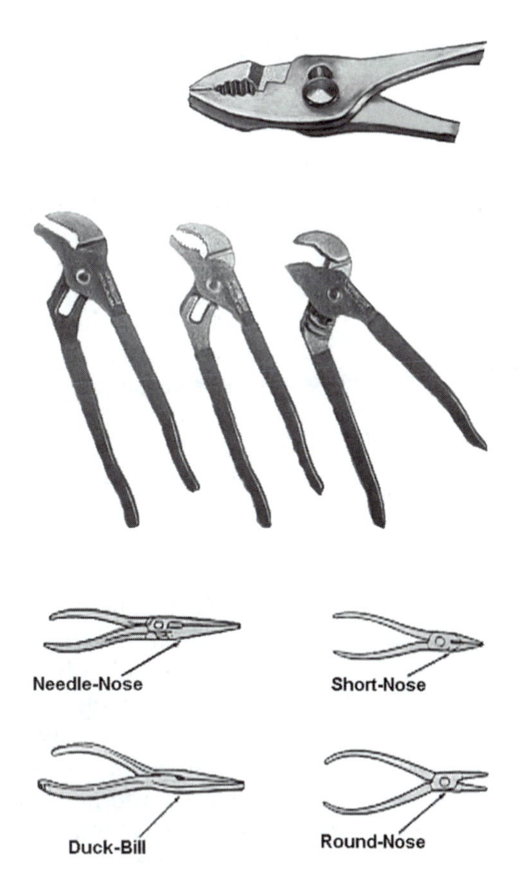

FIGURE 1.8
Slip-joint pliers.

FIGURE 1.9
Common arc-joint pliers.

FIGURE 1.10
Different types of solid-joint pliers.

pliers allows the user to hold various sizes of parts with maximum gripping power. Examples of adjustable-joint pliers include slip-joint pliers (combination), arc-joint pliers (channel locks), and locking pliers (vise grips) (see Figures 1.8 and 1.9).

Pliers are also described by their type of nose and their intended purpose (see Figure 1.10). For example, needle-nose, duck-bill, square-nose, and flat-nose are all different types of pliers. Water-pump, snap-ring, and lineman's pliers are also descriptive names for pliers.

Specifications for the pliers' size are given in inches of overall length. Of course, the longer the length, the larger the pliers are. Larger pliers mean more gripping power and larger jaw openings. The disadvantage of the larger size is that they can become awkward when used with smaller parts.

Pliers should never be used to take the place of a wrench. They should not be used on nuts or bolt heads. They will round off the edges of the hex and ruin the flats so a wrench can no longer be used to hold the fastener properly. Pliers can also slip off, causing injury to your hand.

Some pliers come with plastic or rubber-coated handles, which is done for cushioning and *not* for electrical insulation. *Never* use pliers on a live electrical circuit unless the pliers are rated for live electrical work. Electricity can kill!

1-4 WRENCHES

Wrenches are tools designed to install and remove nuts and bolts (see Figure 1.11). One or both ends of the wrench will have an opening that fits snugly over the fastener. Most industrial nuts and bolts are hexagonal (six sides). The bolt head or nut is measured on the outside of the flat part of the fastener, and this measurement determines the correct wrench size. This dimension is not the same as the diameter or size of the fastener itself. The wrench size is usually stamped on the head of the tool. The correct-size wrench should always be used to prevent damage to the fastener or personal injury to the user. The wrench size may be given in English or metric units. English and metric tools cannot be used interchangeably; the wrench may fit loosely and damage the fastener or cause user injury. When using any wrench, you should *pull* rather than push on it for safety reasons. If you push and the nut or bolt breaks loose suddenly, you can injure your hand. This is less likely to happen when you pull the wrench, but you should always be ready for the fastener to break loose suddenly.

A wrench with an opening at one or both ends is called an *open-end* wrench (see Figure 1.12). The heads in most of these wrenches have an offset at 15°, which makes them more versatile for operating in tight places. To use an open-end wrench, place the open end over the flats of the bolt head or nut. The proper size wrench should fit snugly (see Figure 1.13). *Pull* (do not push) to turn the nut or bolt. Wrenches with smaller openings usually have shorter handles to reduce the chance of applying too much torque on a fastener. Open-end wrenches should not be used to break a fastener loose because they have a tendency to slip off when too much force is applied.

An adjustable open-end wrench, usually called an *adjustable wrench*, has an adjustable jaw that can be moved to narrow or widen the distance between

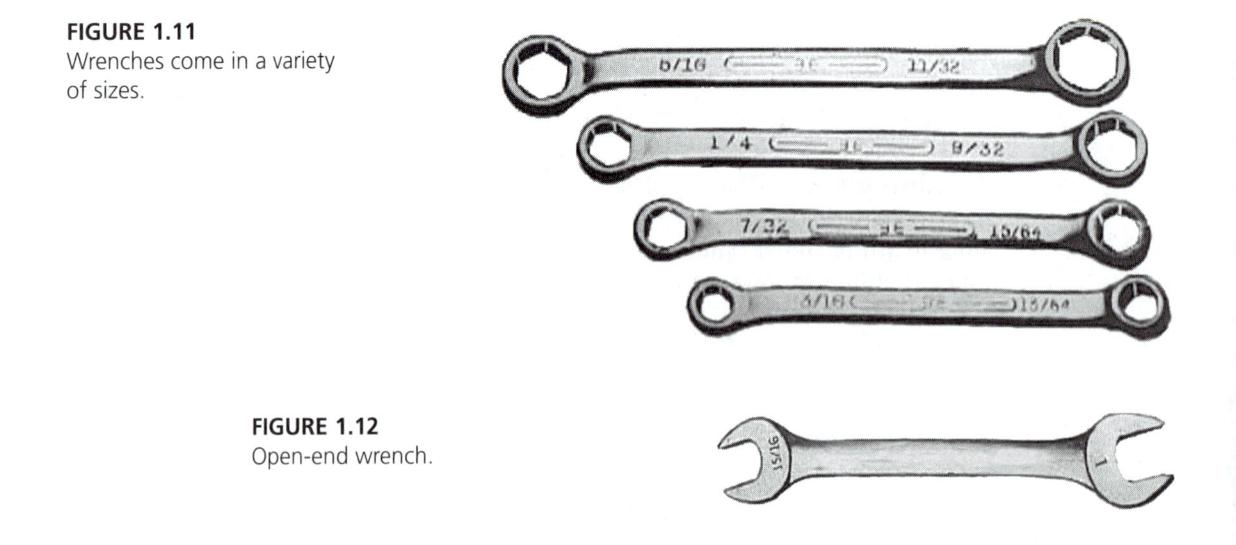

FIGURE 1.11
Wrenches come in a variety of sizes.

FIGURE 1.12
Open-end wrench.

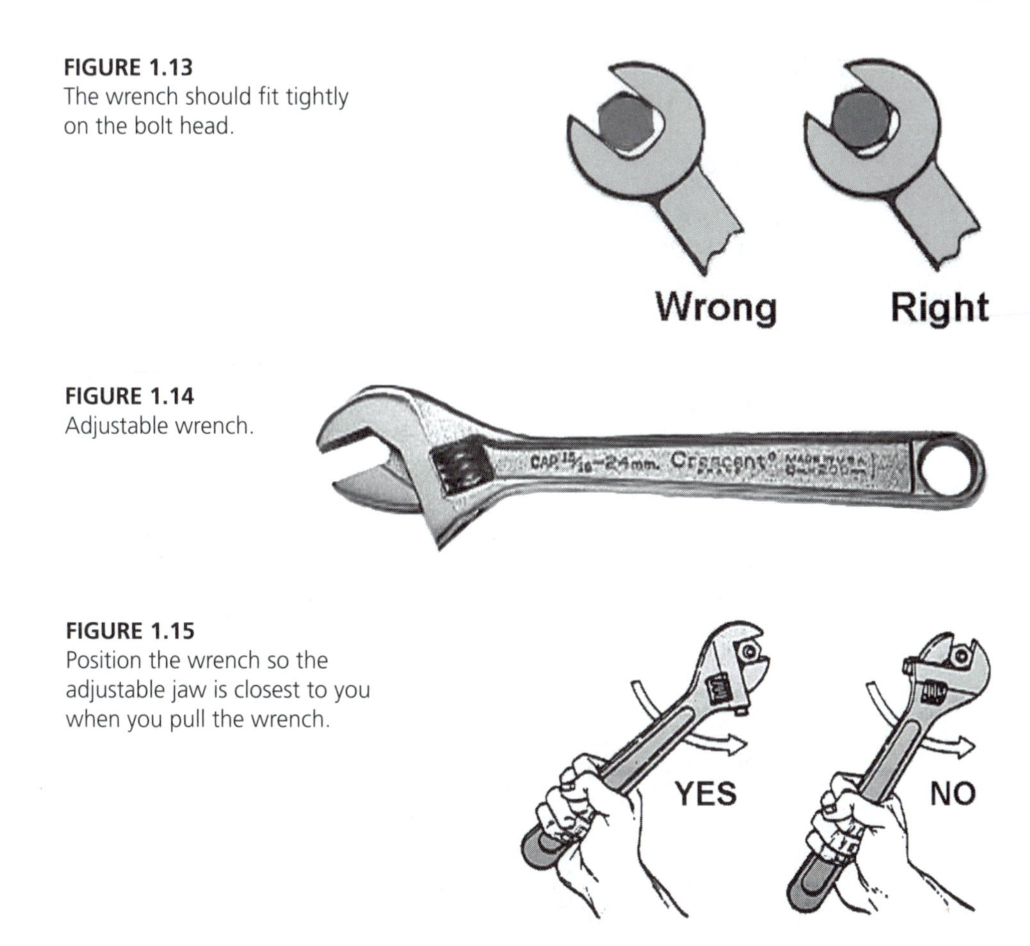

FIGURE 1.13
The wrench should fit tightly on the bolt head.

FIGURE 1.14
Adjustable wrench.

FIGURE 1.15
Position the wrench so the adjustable jaw is closest to you when you pull the wrench.

the jaws (see Figure 1.14). This wrench can fit a range of English or metric sizes. This multisize wrench is *not* intended to replace standard wrenches. When using an adjustable wrench, make sure to adjust the jaws *tightly* on the nut or bolt. To do this, place the wrench on the fastener and then tighten the jaws. Adjusting the wrench before placing on the fastener does not ensure the tightest fit. Always position the adjustable wrench so the adjustable jaw is closest to you when you pull the wrench. This position causes the pulling force to keep the adjustable jaw tight against the nut or bolt (see Figure 1.15). In general, adjustable wrenches should be used only when a fixed-size wrench is not available or does not fit the fastener.

A *box-end* wrench has a closed-end head (see Figure 1.16). The head fits snugly over the fastener and will not slip off. The head opening usually has 6 or 12 corners to grip the nut or bolt. The 6-corner wrench is stronger, but the 12-corner wrench works better in tight quarters. The 12-corner wrench needs to be rotated only half as far as the 6-corner wrench before it slides on a 6-sided nut or bolt. Most box-end wrench heads are offset 10° to allow clearance for the user's hand.

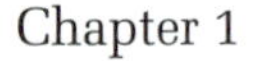

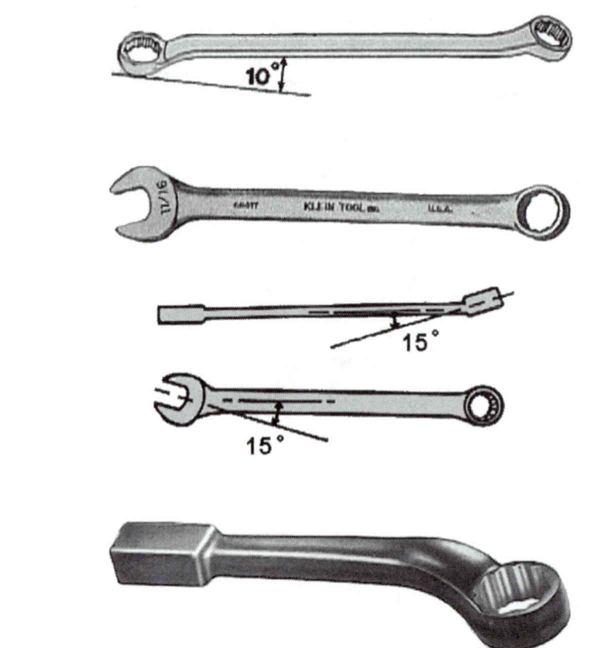

FIGURE 1.16
Box-end wrench.

FIGURE 1.17
Combination wrench.

FIGURE 1.18
Striking-face box wrench.

The combination wrench has both open-end and box-end heads (see Figure 1.17), which are the same size. The box end of the combination wrench may also have a 15° offset that allows clearance for the user's hand. The box end is used to break loose the nut or bolt, and then the open end is used to remove the fastener quickly.

The *striking-face box wrench* (also called an *impact wrench*) is a special type of box-end wrench designed for use on stubborn nuts or bolts and larger fasteners (see Figure 1.18). It is the *only* wrench you should *ever* strike with a hammer. You should fit the striking-face wrench over the nut or bolt and then strike the end of the wrench with a hammer. This impact multiplies your strength many times and can break loose even the most stubborn fastener.

When using any type of wrench, you should follow these rules:

- Always select the proper wrench for the job.
- Be sure to use the correct-size wrench so that it fits snugly on the nut or bolt.
- Always pull rather than push a wrench if possible. If you must push, then push the wrench with an open palm.
- Never use a "cheater bar," a pipe that extends a wrench handle for extra leverage.
- Never hammer a wrench to break loose a stubborn nut or bolt unless you are using a striking-face wrench.
- Frequently check wrenches for cracks or wear. Good wrenches carry a lifetime guarantee; replace them *before* an injury occurs.

1-5 SOCKET TOOLS

A socket tool is made up of a handle and drive that turns a socket. The socket slides over the nut or bolt and uses corners to drive it. The socket wrench is somewhat similar to the box-end wrench, but because of the detachable sockets, a single handle can be used for many sizes of fasteners (see Figure 1.19).

Sockets are classified by drive size. The *drive* is the square lug attached to the handle that couples it to the socket. The most common drive sizes include 1/4 in., 3/8 in., and 1/2 in. The 1/4 in. size is lightweight and meant for light-duty jobs, whereas the 1/2 in. size is of heavier construction for larger-size fasteners. Larger-size drives of 5/8 in., 3/4 in., 1 in., and 1 1/2 in. are not as common but may be used on larger nuts and bolts and heavy equipment. Metric sockets also use inch-size lugs for driving different size sockets.

Most sockets have either 6 or 12 corners, or points. The 12-point socket needs to be rotated only half as far as the 6-point socket before it locks on a 6-sided nut or bolt. This feature can be useful in tight quarters. The 6-point socket is stronger and is better suited for nut and bolts that have worn corners. An 8-point socket is also available. This special socket is used on square-head set screws, bolts, and pipe plugs.

Sometimes a standard socket cannot reach a nut used with a larger stud. An extended-length socket, called a *deep* socket, is useful for this type of application. Figure 1.20 shows standard and deep sockets.

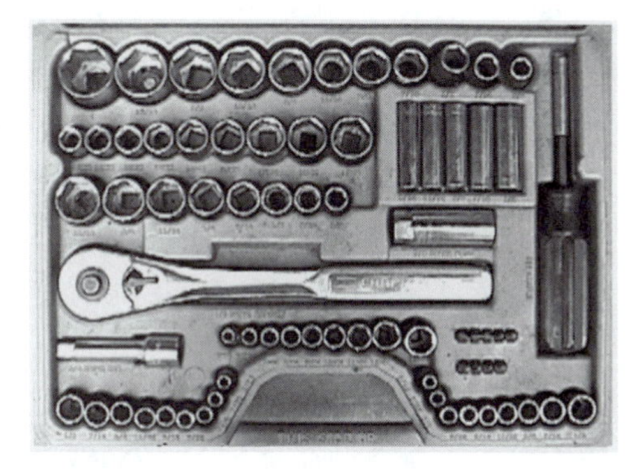

FIGURE 1.19
Socket wrench set.

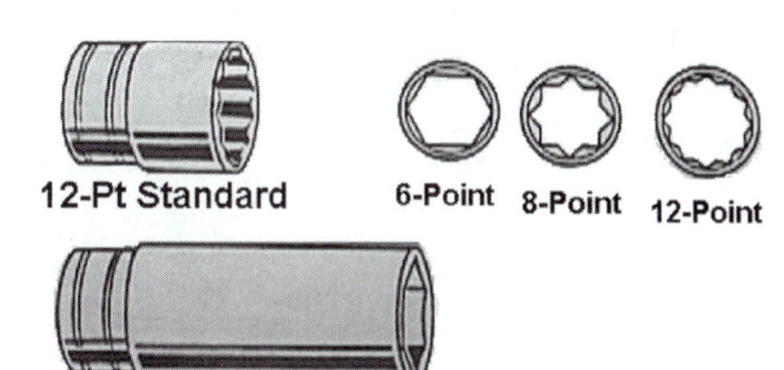

FIGURE 1.20
Standard and deep sockets.

1-6 TORQUE WRENCHES

In most modern equipment, engineers have determined the exact amount of torque needed to tighten nuts or bolts. Fasteners that are not tight enough may vibrate loose and fall out. Nuts and bolts that are tightened too much may be overstressed or broken. The torque wrench is a special tool with an indicating device that tells the user how much **torque** is being applied to a fastener. Figure 1.21 shows the three basic types of torque wrenches.

The **deflecting-beam torque wrench** is the most basic type of torque wrench (see Figure 1.22). The amount of torque applied is indicated by a pointer on a scale as the handle is pulled. The **dial-indicating torque wrench** works in much the same way as a deflecting-beam torque wrench. It is used where more accurate torque settings are required. A precision dial-indicator is built into this type of torque wrench. The **micrometer torque wrench** (sometimes called a *clicker* torque wrench) is replacing most other types of torque wrenches because it is accurate and easy to use (see Figure 1.23). A micrometer-type handle is used to preset a specified torque. Rather than reading the torque on a dial or gauge as it is applied, the head breaks away

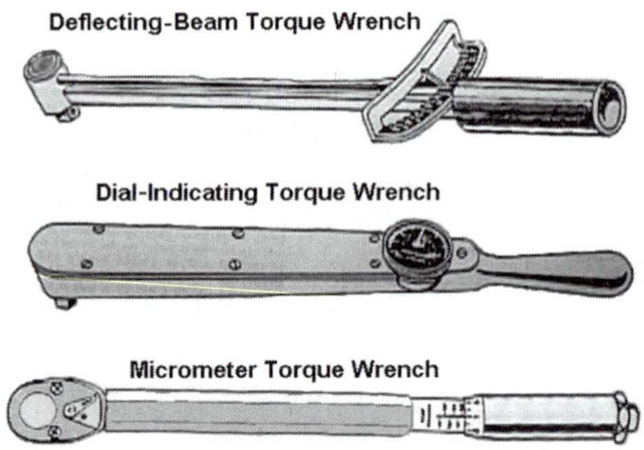

FIGURE 1.21
Types of torque wrenches.

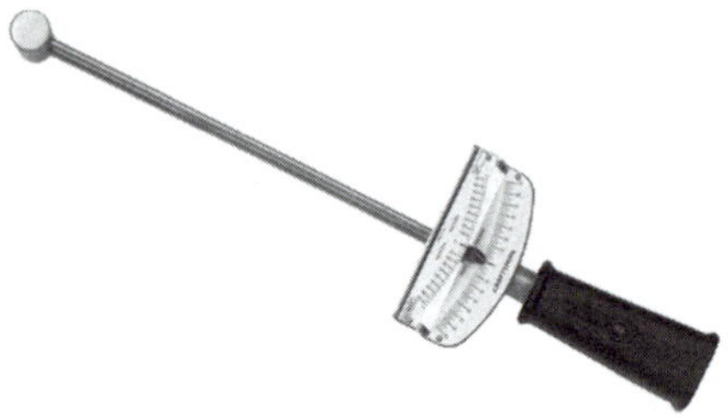

FIGURE 1.22
Deflecting-beam torque wrench.

FIGURE 1.23
Micrometer torque wrench.

and produces an audible click. Micrometer-style torque wrenches should always be turned down and stored at the wrench's lowest possible setting. Storing micrometer-style torque wrenches in any setting besides the lowest one will cause wear of main spring and, possibly, failure over a period of time.

To use the micrometer torque wrench you should do the following:

1. Rotate locking ring to the *unlock* position.
2. Set the appropriate torque value by rotating the knurled handle; tens are marked along the fixed portion of the handle, and ones are marked along the rotating portion of the handle.
3. Rotate locking ring to the *lock* position.
4. Install the fastener and tighten until you hear or feel a click. (If desired, after the initial click, back the handle off and tighten once more until you hear a second click; this ensures that the fastener is properly torqued.)

Most torque wrenches used in the United States measure torque in pound-feet (lb-ft). Smaller fasteners may need a torque wrench that measures pound-inches (lb-in.). One pound-foot equals 12 lb-in.

Metric torque wrenches measure torque in kilogram-meters (kg·m) and kilogram-centimeters (kg·cm). One kilogram-meter equals 100 kg·cm. To convert kilogram-meters to pound-feet, multiply kilogram-meters by 7.233.

Fasteners to be torqued should be clean and lightly lubricated. Any blind holes should be clean and free of excess oil. Multiple fasteners to be torqued (head bolts of an air compressor, for example) should be torqued evenly in increments not to exceed one-third of final torque. Always follow the manufacturer's recommendations for torque patterns and procedures.

When working with any type of torque wrench, you should follow these guidelines:

- Do a visual inspection of the torque wrench before and after every use. Look for any signs for wear or deterioration.
- A torque wrench is a precision measuring instrument. Handle it with care and do not drop it.
- As with any precision measuring instrument, you need to recalibrate a torque wrench periodically to ensure accuracy.
- Grip the handle securely in the proper position and brace yourself for possible breakage of the tool.
- Always use a smooth, steady pulling motion when using a torque wrench. A jerky or quick motion will not give you an accurate torque reading.
- Do not use torque wrenches to do jobs for which they were not intended.

1-7 HAMMERS

Hammers are used for driving, striking, and shaping things. A hammer is a simple tool and was probably one of the first tools used. It is also probably one of the most-used tools. Almost everyone has used a hammer at one time or another. A hammer should be used only when it is the correct tool for the job.

Hammers are classified as hard- or soft-faced. Most hard-faced hammers have steel heads. One of the most common hard-faced hammers used in industry is the *ball-peen* hammer. As shown in Figure 1.24, this hammer has a flat-faced head used for driving and a rounded, ball-shaped head originally designed for flattening rivets. Hammers are specified by the weight of the head. Ball-peen hammers usually range in weight from 2 oz to 3 lb. They are hardened and tempered. The smaller sizes are used for layout work, and the larger ones are used for general benchwork. A well-used hammer will eventually develop burrs on the hammer face. Left unrepaired, slivers of metal break off, which can cause personal injury or equipment damage. You should always inspect hammers before using them and regrind them on a bench grinder when necessary. Don't use heavy force when regrinding because it can cause the hammer face to turn blue, indicating the metal has become too hot and has lost its temper. If this happens, you should replace the hammer.

Other hard-faced hammers include the cross-peen hammer shown in Figure 1.25 and a blacksmith hammer, or maul, shown in Figure 1.26.

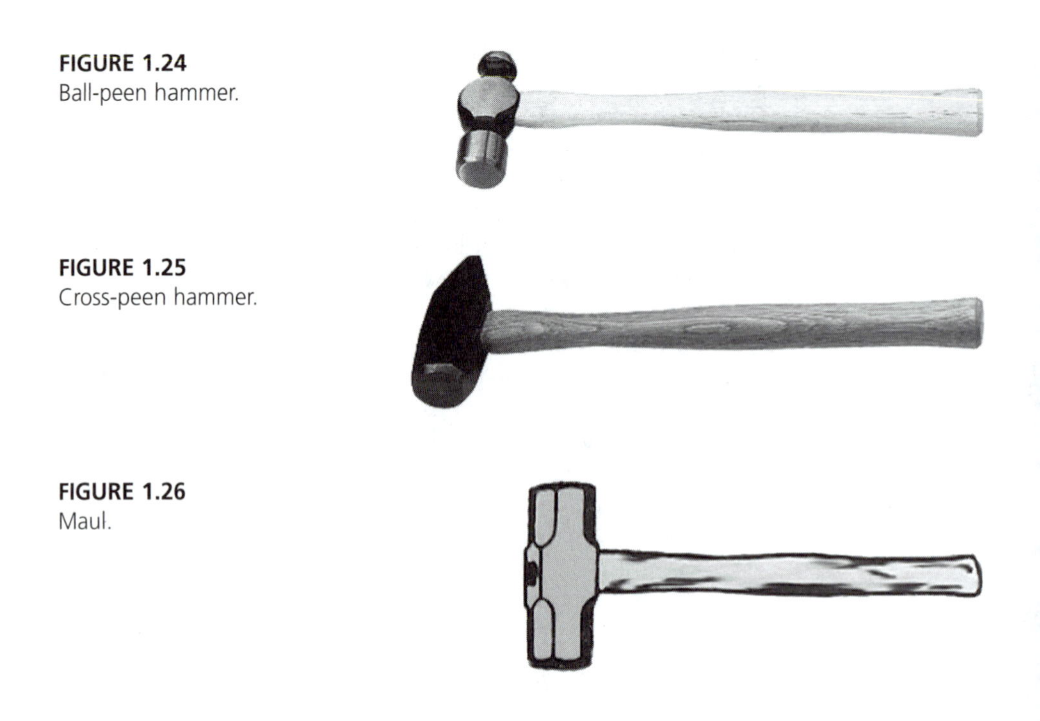

FIGURE 1.24
Ball-peen hammer.

FIGURE 1.25
Cross-peen hammer.

FIGURE 1.26
Maul.

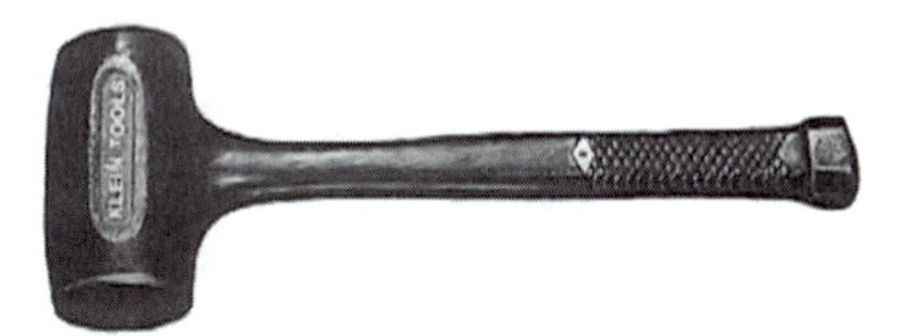

FIGURE 1.27
Dead-blow hammer.

You should use a soft-faced hammer, or **mallet**, where a hard-faced hammer may mar or damage the object being hammered. Soft faces may be made of rawhide, rubber, plastic, brass, copper, or lead (see Figure 1.27). Lead used to be a popular choice because it did not rebound when striking, as a rubber mallet would. Most soft-faced hammers are now modern **dead-blow hammers**. The face of this hammer is made from tough, space-age plastics that can be used even on sharp edges without damage. The hammer head is hollow and loaded with lead shot that absorbs most bounce, or rebound. As with any hammer, you should always inspect it for damage before using. Make sure the handle is secure and free from oil or grease that could cause it to slip in your hand.

1-8 CHISELS AND PUNCHES

Chisels and punches are tools designed to be used with a hammer (see Figure 1.28). Chisels are used to cut metal, whereas punches are used primarily to move or mark an object. Chisels and punches are heat-treated on the cutting end, whereas the end being hammered is relatively soft. This is done to prevent small pieces of metal from breaking off when you strike the chisel with a hard-faced hammer. After moderate use, the chisel or punch will begin to mushroom where it has been struck by the hammer. You can easily repair it by grinding it back to its original shape. The cutting, or working, edge of the chisel or punch will also become dull or deformed. Because this end is hardened, you must be careful when returning it to its original shape. A fine-grit grinding wheel or file works best. Be careful not to overheat the

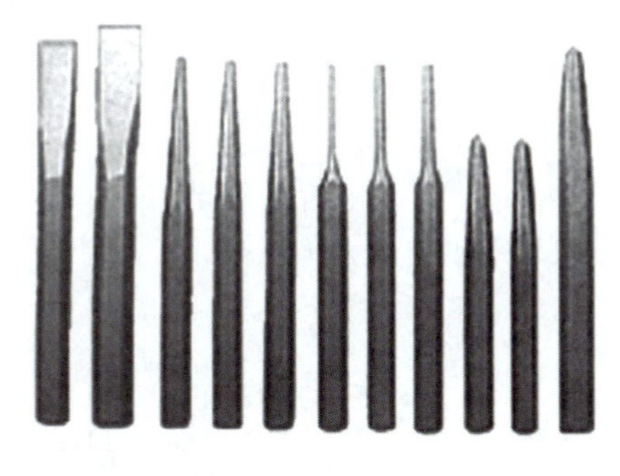

FIGURE 1.28
Chisels and punches.

metal and destroy the temper. Any bluish color is an indication of overheating that may damage the temper of the tool. Keeping a chisel or punch in good repair makes it work better and safer.

The most common chisel is called the **cold chisel** (see Figure 1.29). This general-purpose chisel has a straight edge used to cut metal that has not been heated. You can use it to cut metals of all kinds except hardened steel. A chisel should always be harder than the metal it cuts. You should generally choose a cold chisel to cut rusted or frozen parts or fasteners. You can also use it for cutting small rods or bars, cutting shapes from sheet metal, and cleaning up rough castings.

The **cape chisel** has a narrow pointed edge (see Figure 1.30). The area behind the edge is ground smaller to avoid binding when working in a slot. You can use a cape chisel for cleaning keyway slots or in any application that requires a sharp inside corner.

The **round-nose chisel** has a round cross section at the cutting edge. The edge is ground at an angle of 60° with the chisel axis. You can use round-nose chisels to clean up round slots, damaged bolt holes, and rounded inside corners or to cut oil grooves in plain bearings or bushings.

The **diamond-point chisel** is ground at an angle across the corners of the chisel that gives the cutting edge a diamond-shaped face (see Figure 1.31). This sharp-pointed chisel can cut V-grooves, slot a joint for welding, and start slots for cutting with a cold chisel.

You should never do any of the following:

- Use a cold chisel on concrete.
- Use a chisel on anything harder than the chisel.
- Use a dull chisel.
- Use a damaged chisel.
- Use a chisel with a mushroomed head.

FIGURE 1.29
Common cold chisel.

FIGURE 1.30
Cape chisel.

FIGURE 1.31
Diamond-point chisel.

You should always do the following:

- Wear safety goggles.
- Wear clean and dry gloves.
- Keep working surfaces free from oil and grease.
- Make sure the chisel is sharp and in good condition.
- Resharpen a chisel carefully; do not overheat it when grinding.
- Discard any chisel if it is bent, is cracked, or shows excessive wear.

Punches, like chisels, have a hardened tool steel point and a soft-steel striking end. Punches are designed to mark metal, drive pins and rivets, and align holes.

The *center* punch is probably the most frequently used type of punch (see Figure 1.32). You use the center punch primarily to mark the center of a hole to be drilled. This keeps the drill bit on center when starting a hole. You can also use the center punch to mark parts before disassembly so they can be reassembled in the same positions. For example, it might be possible to assemble a gear box cover plate in more than one position. You can use a center punch to mark the mating edges of the cover and the housing. Aligning the punch marks during assembly ensures correct reassembly.

Drift, or *starter*, punches have a long, continuous taper from the point to the shank (see Figure 1.33). You use a drift punch to start to remove a frozen bolt or pin. You then use a *pin* punch to finish the job. The drift punch is stronger because of the added material in the taper. That is why you use it initially to loosen the bolt or pin (see Figure 1.34). Pin punches are not tapered; the shank

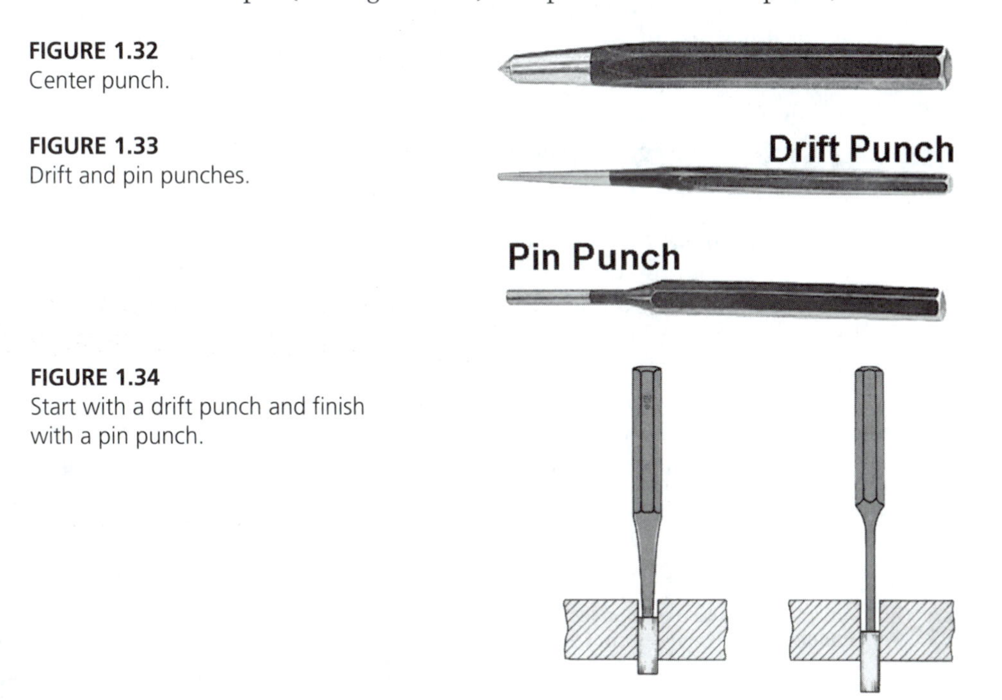

FIGURE 1.32
Center punch.

FIGURE 1.33
Drift and pin punches.

FIGURE 1.34
Start with a drift punch and finish with a pin punch.

FIGURE 1.35
Transfer punch set.

is ground to the same diameter as the tip for an extended length. They have a much longer reach than the drift punches. Both drift punches and pin punches should be ground with a circular, flat point.

The transfer punch looks like a pin punch with a point on its end similar to a center punch. The transfer punch is used to transfer the exact location of the center of a hole on one part to another part. These punches usually come in sets with various sizes to be used with different hole sizes. Figure 1.35 shows a typical transfer punch set.

The *aligning* punch is usually about 12 in. long. It has a long, continuing taper from the point to the punch body. The aligning punch is useful for lining up holes in mating parts during assembly. Sometimes, you can successfully use drift punches for alignment. However, you should never use an aligning punch to remove drive-fit pins or fasteners. The taper of the aligning punch can destroy the drive-fit hole if it is driven too far. Always use the proper tool for the job.

1-9 HACKSAWS AND FILES

The *hacksaw* is probably the most common saw an industrial technician uses (see Figure 1.36). Hacksaws may have adjustable frames or solid frames. Most adjustable frames will hold blades from 8 to 12 in. long. Solid frames are more rugged and tend to hold blades more rigidly, but they can hold a blade of only one length (usually 12 in. long). Hacksaw blades may be mounted vertically or horizontally, depending on the application.

Blades for hacksaws are made of hardened tool steel. Some blades have the same hardened material throughout the blade. Other blades have a flexible back with hardened, tempered teeth. Both types cut equally well, but the all-hardened

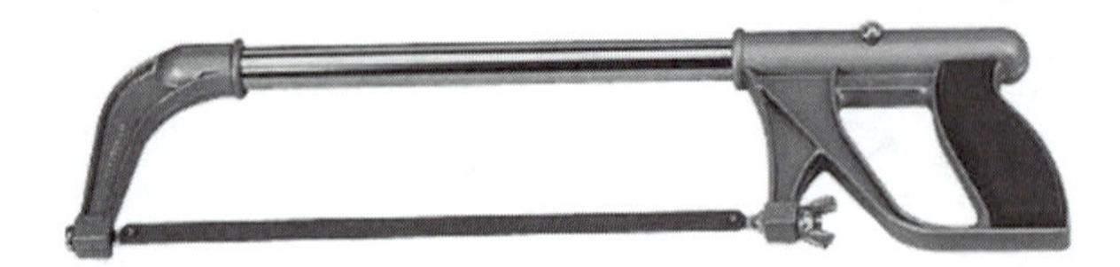

FIGURE 1.36
Hacksaw.

blades tend to break and shatter if abused. This can be a serious safety hazard, causing personal injury. Flexible-back blades are more expensive, but they are safer and usually last longer.

Hacksaw blades are specified by their blade lengths and the number of teeth per inch. The higher the number of teeth per inch, the smaller the teeth. You should always have a minimum of two to three teeth in contact with the material at all times. Therefore, the blade selected for thinner materials should have more teeth per inch. For material 1/8 in. thick or thinner, you should use a 32-teeth-per-inch blade. For 1/8-to 1/4-in. material, a blade with 24 teeth per inch would be a good selection. It is not a good idea to use a blade with 32 teeth per inch for everything. For thicker material, a coarse tooth blade cuts faster and lasts longer. In addition, fine-tooth blades do not have enough chip clearance for thicker material and will become clogged and tend to overheat, possibly destroying the blade.

When using a hacksaw, you should hold it firmly with both hands. The material to be cut should be well secured. A vise works well for this purpose (see Figure 1.37). Keep uniform pressure on the forward stroke. Do not allow the teeth to slide over the material; this dulls the teeth and may cause tooth breakage. Lift the hacksaw slightly on the return stroke to avoid dulling the teeth. Keep a steady rate of about 50 strokes per minute. Cutting too fast overheats the blade and dulls the teeth. When you have almost cut through the material, the blade makes a different sound, warning you to reduce the pressure that you are applying. This prevents an abrupt breakthrough at the end of the cut, which could cause you to lose your balance or damage the hacksaw blade or material being cut.

Files are usually used to remove burrs or sharp edges or to fit a part to be assembled. Files come in many sizes and shapes and have many uses. Figure 1.38 shows a common file and the names of its parts. The length of a file is measured from the heel to the point, not including the tang. The most common file sizes range from 4 to 18 in. Other lengths are available for special application.

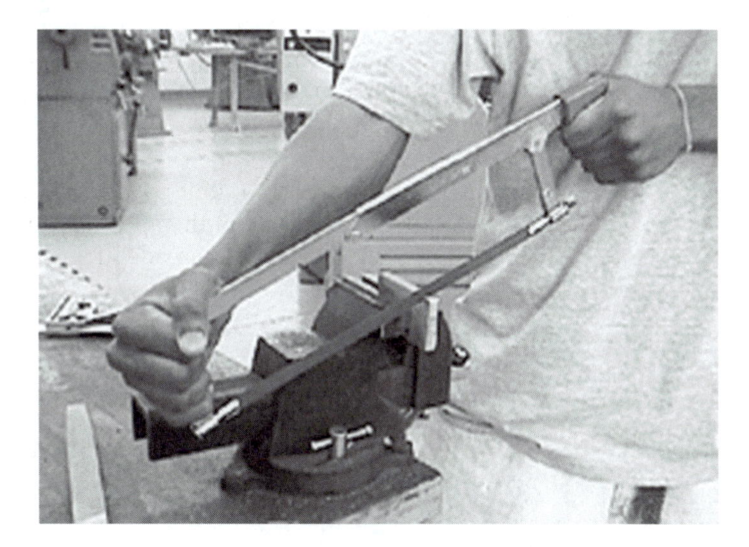

FIGURE 1.37
Using a hacksaw safely.

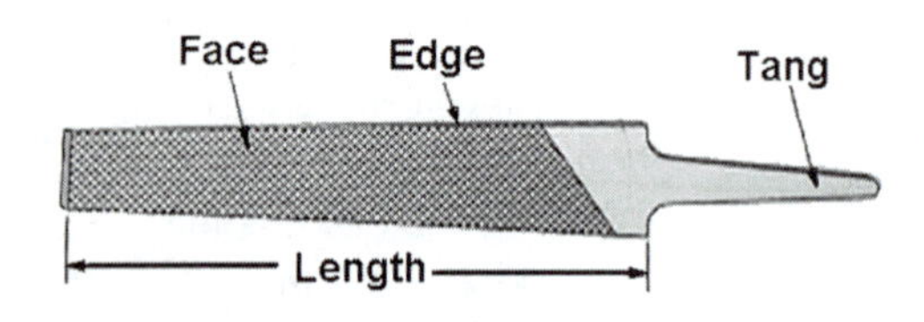

FIGURE 1.38
Common file.

The *tang* is the narrow, tapered section where the handle is attached. You should never use a file without a handle. The tang is sharp enough to cause severe injuries if the file hangs up while you are using it.

The *edge* of a file is its narrowest surface. The teeth on the edge are usually coarser. The edge of the file is used for filing notches, corners, special shapes, or any tight areas. Some files have *safe edges*, which means that the file has one or both edges with no teeth. This allows you to file close to other objects without accidentally filing and damaging them.

The *face* of the file is its main metal-removing part. It looks like a series of tiny chisels, each with a sharp cutting edge. As the face of the file is moved across the material, the teeth remove tiny shavings of metal. The coarseness of the file determines how much metal is removed per pass. The term *cut* is used to designate the coarseness of the face.

Files are manufactured in four types of cuts (see Figure 1.39):

1. *Single-cut files* have a single series of parallel teeth across the file. They are used to produce a smooth surface finish. These require only light pressure to cut. Single-cut files are also called *Mill files.*
2. *Double-cut files* have two series of diagonal rows of teeth. They will remove metal faster, but they will also produce a rougher surface finish. They require heavier pressure than single-cut files do.

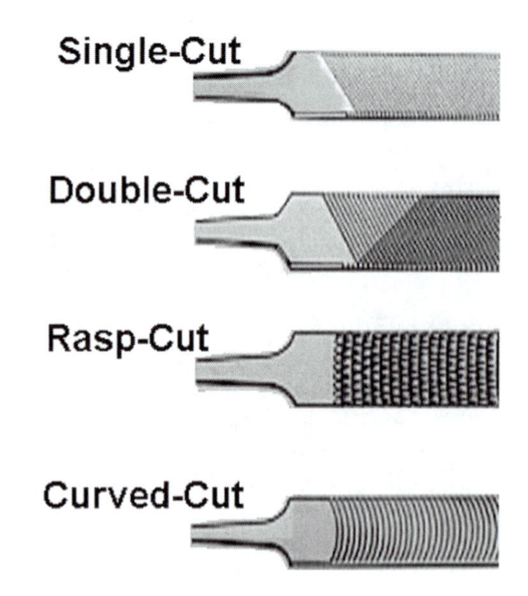

FIGURE 1.39
Files have different types of cuts.

3. *Rasps* are usually used on wood or other soft materials where a large amount of material needs to be removed.
4. *Curved-tooth files* are typically used on soft metals such as aluminum, brass, or lead. They can also be used on wood or plastic. Their self-cleaning design works well on soft material that tends to clog the teeth of other types of files.

The different cuts listed here can also vary in their coarseness: listed from the coarsest to the smoothest cut, these are categorized as rough, coarse, bastard, second cut, smooth, and dead smooth. The files most often used in industry are bastard, second cut, and smooth. The length of the file also determines its coarseness. The longer the file is, the coarser the cut will be. For example, a 12 in. single-cut bastard file is coarser than a 6 in. single-cut bastard file.

Files come in many different shapes (see Figure 1.40). They can be either blunt or tapered. A blunt file has the same size from the heel to the point. A tapered file becomes narrower near the point. Some of the more common shapes are as follows:

1. A *flat* file is a rectangular-shaped file used as a general-purpose file.
2. A *mill* file is a thinner version of the flat file. Mill files are single-cut and were originally used to sharpen large saws in lumber mills.
3. A *pillar* file is a thicker version of the flat file. This file is almost square and usually has one or two safe edges.
4. A *square* file is a square-shaped file. It is usually double-cut and is used to file keyseats, slots, recesses, and holes.
5. A *warding* file is a very thin rectangular file used for narrow work, often by locksmiths.
6. A *knife* file is a rectangular knife-shaped file. The included angle between the faces is about 10°. This file is also used for narrow work where a more substantial file is required.
7. A *three-square* file is a triangular-shaped file used for filing sharp angles, corners, grooves, and notches.
8. A *round* file is a circular-shaped file used to dress and enlarge holes. These files are available in many different diameters.
9. A *half-round* file is a file with a curved face on one side and flat on the other. It is used for larger diameter holes and curves.

FIGURE 1.40
Some common file shapes.

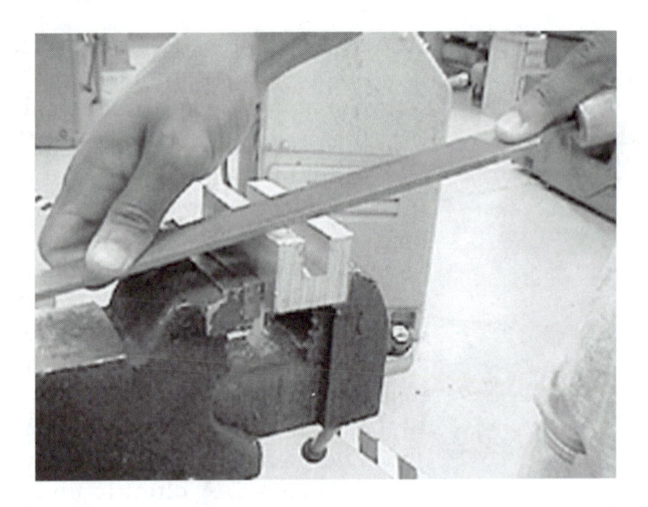

FIGURE 1.41
Apply just enough pressure to cut.

Using too much or too little pressure on the cutting stroke can damage files. You should apply just enough pressure to cut through the entire forward stroke (see Figure 1.41). Too much pressure overloads the file and causes the teeth to clog and break. Too little pressure allows the teeth to slide over the material and become dull. You should lift the file from the work on the reverse stroke. The exception to this rule is when filing soft metal. The return stroke is used to clear the teeth of removed metal, and you should use no more pressure than the weight of the file. If the material is too hard to file, you should not attempt to cut it with a hacksaw.

To care for any type of file, you should do the following:

- Keep the file clean.
- Keep the file and object free of oil.
- Clean the surface being filed with a cloth or brush—never with your hand.
- Do not use files for pry bars or hammer on them. Remember that files are *very* brittle.
- Keep files stored separately from each other in a clean storage area.
- Use the correct file for the application.
- Always use a handle on the tang.

1-10 INDUSTRIAL MEASURING TOOLS

Two basic linear measuring systems are commonly used: A basic unit of length in the SI (Systeme International), or metric, system of measurement is the meter. In the English, or customary, system of measurement, which is still widely used in the United States, feet and inches are used to measure length.

The primary divisions on a metric ruler or tape are centimeters (cm). One hundred centimeters equals 1 m. On most rulers centimeters are divided into millimeters (mm): 1 cm = 10 mm. Metric calipers and micrometers are divided into even smaller divisions.

On most English system rulers or tapes, the main divisions are inches. Inches are then divided into halves, quarters, eighths, and sixteenths of an inch. Precision rulers are further divided into thirty-seconds and sixty-fourths of an inch. Machinist's rulers are divided into inches, tenths, and hundredths of an inch. Caliper rulers and micrometers are used for more precise measurement (see Figure 1.42).

Calipers are instruments used to transfer measurements from a work piece to a measuring tool or from a measuring tool to a work piece. Calipers have two straight or curved legs connected by a hinge and can be adjusted to different sizes. There are two basic types of calipers: inside calipers, used to transfer inside measurements, and outside calipers, used to transfer outside measurements. Calipers should be handled with care. They should always be kept clean and should be lightly oiled to prevent rust.

Caliper rules or slide calipers are used to take inside or outside measurements directly read from the caliper itself. They have two index lines marked for inside and outside use. To measure the outside diameter of pipe or other round objects, move the jaws into tight contact with the surface of the pipe, as shown in Figure 1.43. Then read the measurement at the index line labeled *out*.

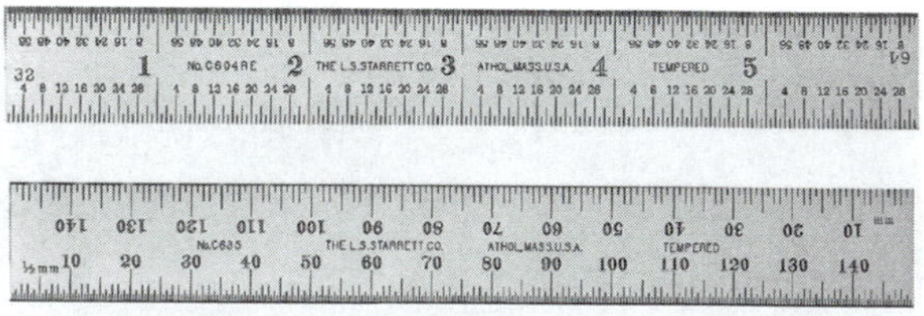

FIGURE 1.42
English and metric 6 in. rules.

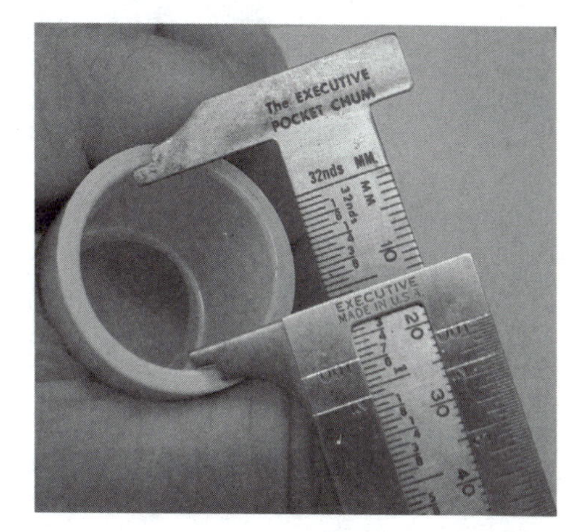

FIGURE 1.43
Measuring inside diameter.

To measure the inside diameter of a pipe or hole, insert the rounded tips of the slide caliper inside the pipe and expand them until they fit tightly against the walls of the pipe. Read the measurement indicated by the index line labeled *in*.

A *vernier* caliper is much more accurate than a basic slide caliper. This greater accuracy is made possible because of a special *vernier scale* attached to the index line. The scale is made so that, for any measurement, only one of its graduations lines up with a graduation on the main scale.

The vernier scale subdivides the smallest graduations on the main scale. To read the measurement, you must add the value of the lined-up graduation on the vernier scale to the value of the main scale indicated by the index line.

Although a vernier caliper is highly reliable and accurate, the nature of its function makes it somewhat difficult to read, requiring more time and practice to become proficient. In recent years, vernier calipers have been widely replaced by direct-reading electronic digital calipers (see Figure 1.44). As the sliding jaw moves along the beam, the position is shown on the digital display. The display may be set to zero at any point and may be switched from inch to metric measurement.

A micrometer caliper, commonly referred to as a *micrometer*, is one of the most widely used precision measuring tools found in industry. It uses a fine-pitch screw to keep track of the measurement. Figure 1.45 shows the parts of a micrometer. The *spindle* is the main part. It is attached to a very accurate inside-the-gage barrel called the *thimble*. The thread turns through a fixed nut. As you turn the thimble, the distance between the end of the spindle and the anvil changes.

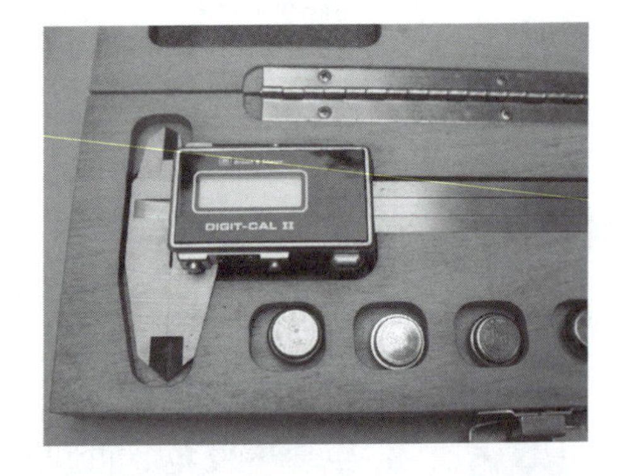

FIGURE 1.44
Direct reading digital calipers.

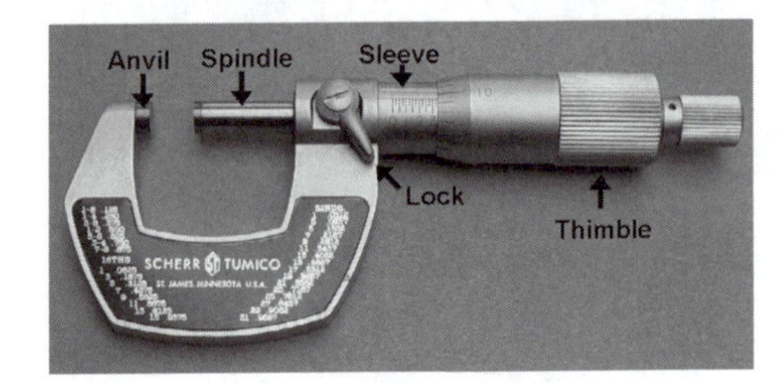

FIGURE 1.45
Parts of a micrometer.

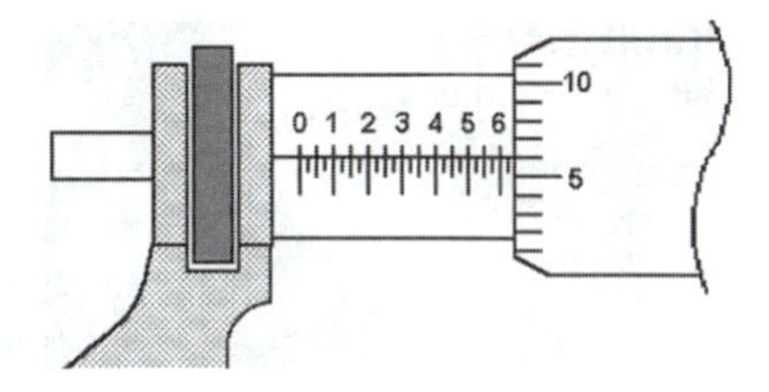

FIGURE 1.46
Reading is 0.631 in.

Dimensions requiring the use of micrometers are generally expressed to three decimal places. In the case of the inch micrometer, the reading is to thousandths of an inch. For example, 0.056 in. is read as fifty-six thousandths inch, and 0.163 in. is read as one hundred sixty-three thousandths inch.

On the sleeve of the micrometer is a graduated scale with 10 numbered divisions, each one being 1/10 in., or 0.100 in. (one hundred thousandths) in. apart. Each of these divisions is further divided into four equal parts, which makes the distance between these graduations 0.025 (25 thousandths) in. When the spindle is turned one complete revolution, it has moved 0.025 in. The thimble has 25 evenly spaced divisions around its circumference. Each mark represents 0.001 in. Rotating the thimble from one line to the next moves the spindle 0.001 in.

Metric micrometers are similar to inch micrometers in operation but are graduated in millimeters. On the sleeve of the micrometer is a graduated scale with 25 equal divisions; every fifth division is numbered, each one being 1 mm (1.0 mm) apart. When the spindle is turned one complete revolution, it moves 0.5 mm. The thimble has 50 evenly spaced divisions around its circumference, with every fifth division numbered. Each mark represents 0.01 mm. Rotating the thimble from one line to the next moves the spindle 0.5 mm.

To read the micrometer, first determine the value indicated by the lines exposed on the sleeve (see Figure 1.46). The edge of the thimble shows six major divisions, which represents 0.600 (600 thousandths) in. There is also one minor division showing on the sleeve. The value of this is 0.025 (25 thousandths) in. The thimble reading is 6, which indicates 0.006 (6 thousandths) in. The micrometer reading is determined by adding all the readings. In this example, the reading is 0.600 in. + 0.025 in. + 0.006 in., or 0.631 in. to the nearest thousandth.

A micrometer should be held by the frame (see Figure 1.47), leaving the thumb and forefinger free to operate the thimble. You should hold the part being measured in your other hand. Turn the thimble to tighten the spindle on the part. The tightening force should be great enough to make a snug fit but loose enough to be able to move the micrometer over the part. If the micrometer has a ratchet stop, you will hear it click when it is tight enough. Then turn the locking screw and take the reading.

There are many different sizes of micrometers. The total range of most micrometers is only 1 in. The smallest size measures from 0 to 1 in.; the next sizes are from 1 to 2 in., 2 to 3 in., and so forth. Some of the more common types of micrometers are outside, inside, and depth. There are also special micrometers for different applications.

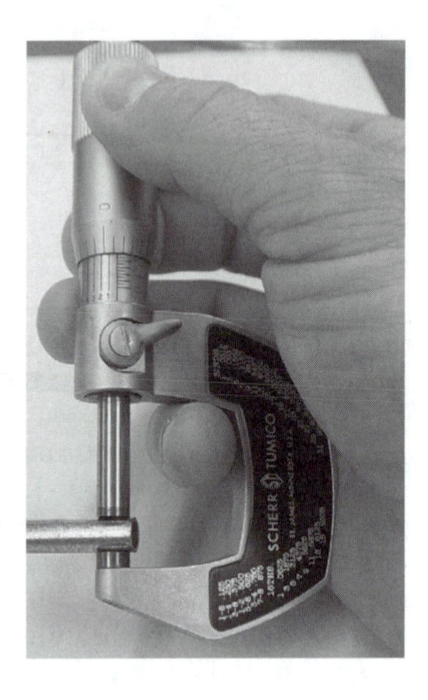

FIGURE 1.47
Using a micrometer.

QUESTIONS

1. What are the three basic parts of a screwdriver?
2. When is it acceptable to use a screwdriver as a pry bar?
3. What are the four sizes of Phillips screwdriver tips?
4. What types of screwdrivers are safe for live electrical work?
5. What are the two major configurations of pliers?
6. What determines the correct wrench size to use for a particular bolt?
7. When using a wrench, should you pull or push on the wrench?
8. Should open-end adjustable wrenches be used to replace all fixed open-end wrenches?
9. Why does the combination wrench have the same size on both the box and the open end?
10. Which is stronger, a 6-point or a 12-point socket?
11. Why is the micrometer torque wrench sometimes called a clicker torque wrench?
12. What is a dead-blow hammer?
13. When regrinding a chisel, what does a bluish color indicate?
14. What type of chisel should be used for cleaning a keyway slot?
15. What tool should be used to mark the center of a hole to be drilled?
16. Why should you first use a drift punch instead of a pin punch when removing a frozen bolt?
17. What type of hacksaw blade should be chosen for safety?
18. When choosing a hacksaw blade, how many teeth should be in contact with the material?
19. What are the four types of cuts that files may have?
20. Describe the technique you should employ when using a file.

CHAPTER 2

Fasteners

2-1 INTRODUCTION

This chapter covers the most common industrial threaded-type fasteners. Fasteners are among the most important elements of modern industrial production. You will learn how to identify, select, and use standard threaded-type fasteners (see Figure 2.1). Today, parts are assembled using not only threaded fasteners but also riveting, welding, and even adhesive bonding. Covering all these areas could take several books, so this chapter concentrates on just the most common industrial threaded-type fasteners. First, we look at a short history of this type of fastener.

Archimedes used the principle of the screw thread around 250 B.C. He invented a screw-type conveyor to lift water from a river for irrigation (see Figure 2.2). Some countries still use variations of Archimedes' screw for irrigation (see Figure 2.3). The Romans had technology to cut male (external) and female (internal) threads, but each had to be cut and matched by hand. Because these male and female threads were custom-made to match each other, they usually would not work with any other part (i.e., there was no standardization). During the Dark Ages, this technology was mostly abandoned. Instead, handmade nails and wooden pegs were used because they were much easier to fabricate. Finally, during the Renaissance in the 1400s, machinery was invented to make the threaded screws used in the clock manufacturing industry. Leonardo da Vinci (1452–1519) also penned designs for screw thread–cutting machinery in some of his notebooks.

Sometime in the mid-1500s, Jacques Besson, a Frenchman, invented the predecessor of the modern machine lathe, making machine-made threads possible. However, metal screws continued to be made by hand for another century.

The Industrial Revolution brought an increased importance to threaded fasteners. Inventions during this time required gears, shafts, springs, and—equally important—the screws that held everything together. In 1841, Sir Joseph Whitworth presented his paper, "The Uniform System of Screw-threads," to

FIGURE 2.1
A common threaded-type fastener.

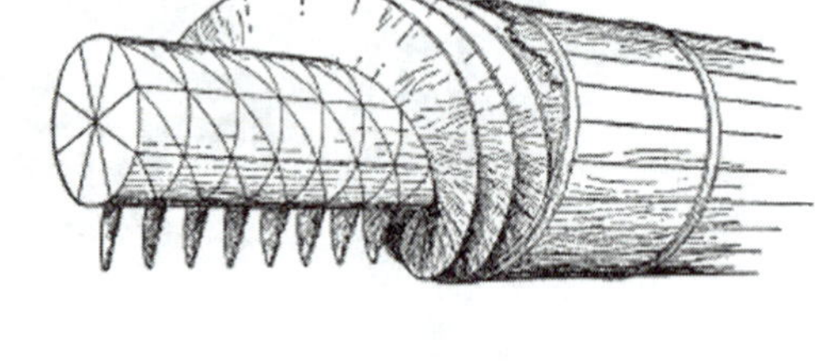

FIGURE 2.2
Archimedes' screw.

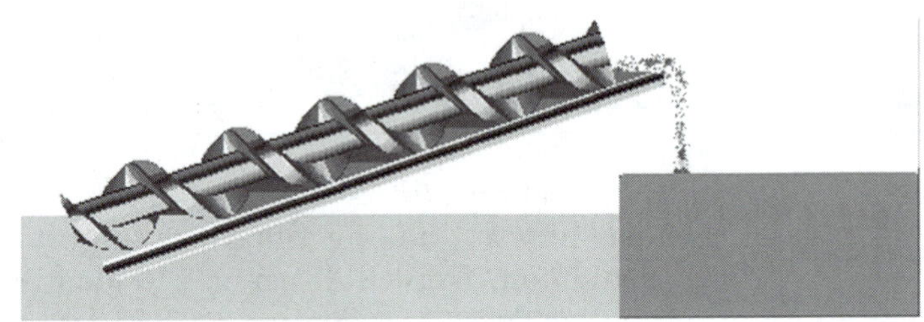

FIGURE 2.3
Archimedes' screw used to pump water.

Great Britain's Institute of Civil Engineers, which led to the first national standard for screw threads. Sir Joseph's standard thread featured a constant thread angle of 55° and became known as the Whitworth thread. About the same time, an American named William Sellers developed a screw thread system based on a standard thread angle of 60°. In 1864 Mr. Seller's system was adopted by the United States as the basis for the American National Standard Thread. As trade increased between the United States and Great Britain, the two different standards became a problem. This problem magnified even more during World Wars I and II. Finally, in 1948, after much effort by representatives of the United States, Canada, and Great Britain, the unified screw thread system was established, combining the best of two systems developed by Whitworth and Sellers.

World trade forced Great Britain to adopt the metric system in 1965. In 1969, the **ISO** (International Standards Organization) established the standard ISO 68, the first international screw thread system. The ISO metric thread is now the most commonly used in the world for general fastening purposes.

2-2 SCREW THREAD NOMENCLATURE

Figure 2.4 illustrates the basic components of a screw thread. The **crest** is the top portion of an individual thread. The opposite of the crest is the **root,** which is the bottom section or the valley of a thread. The **flanks** are the sloped surfaces

FIGURE 2.4
Components of a screw thread.

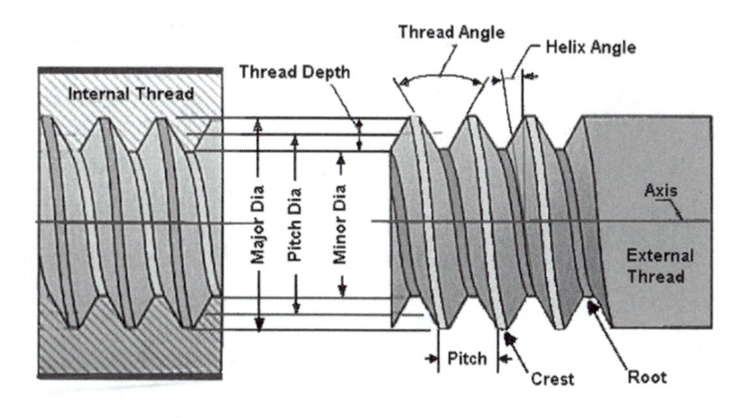

joining the crests and roots. The **major diameter** is the maximum diameter measured from crest to crest of a bolt. The major diameter of a bolt is measured to determine the bolt's nominal size. The **minor diameter** is the minimum diameter measured from the thread valleys. The **pitch diameter** is the imaginary circle between the major and minor diameters, where the force applied by a screw thread is centered.

Screw threads are shaped as a helical curve, or **helix.** A helix is the curving path that a point would follow if it were to travel in an even spiral around a cylinder and in line with the axis of that cylinder (see Figure 2.5). Screw threads may be *internal* or *external.* Threads cut inside drilled stock are internal threads. Tapped holes and nuts are examples of internal threads. Threads on the outside surface of round stock are external threads. Screws, bolts, and threaded rods are examples of external threads. The *helix angle* on a straight thread is the angle made by the helix of the thread and its relation to the thread axis. The *angle of thread* is the angle formed by the sides of the thread. The difference between the angles of thread in the original British standard (55° angle of thread) and the original U.S. standard screw thread (60° angle of thread) made them impossible to interchange.

The **pitch** of a thread is the distance from the center of one crest to the center of the one next to it. Associated with the thread pitch is the number of **threads per inch** (TPI) of a screw. The threads per inch are determined by using a ruler and physically counting the number of threads in 1 in. of bolt. Bolts with more threads per inch are referred to as *fine threads,* and bolts with fewer threads per inch are called *coarse threads.*

FIGURE 2.5
The helix of a screw thread.

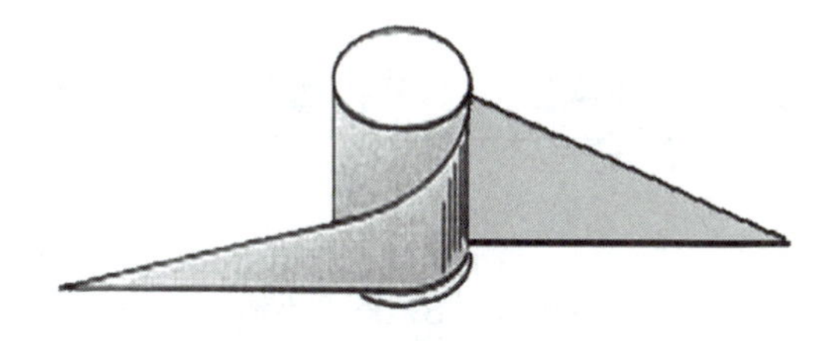

Fine threads offer the following advantages:

- They have more holding power because of the greater surface area of threads in contact with each other.
- They have more shear strength because they have a smaller minor diameter than coarse threads.
- They are easier to tap because less material is being removed.
- They work better for thin-walled materials (thin plate, pipe, or tubing) because they have more threads per inch.
- They work better for fine adjustment because more turns are required for fine threads to move a given distance than for coarse threads.

Coarse threads offer the following advantages:

- They are more economical than fine threads.
- They assemble and disassemble more quickly and easily.
- They tolerate more contamination and abuse.
- They tolerate corrosive attack with less strength loss.
- They have less tendency to *cross-thread* than fine threads. (Cross-threading occurs when mated threaded parts are incorrectly started at an angle so that the threads cross each other, thus becoming ruined.)
- They have better fatigue resistance.

Low-strength materials such as plastics, aluminum, or even cast iron are usually tapped with coarse threads. The deeper thread is stronger for these relatively soft materials.

Fine threads are more common on high-strength steel fasteners. Fine-threaded fasteners are also used where torque must be precise. Gasoline and diesel engines typically use fine threads on components where high strength is necessary.

Screw threads are commonly *right-hand* threads. Right-hand threads screw in, or tighten, when turned clockwise. *Left-hand* threads are used for special applications and screw in when turned counterclockwise. Threads are always considered to be right hand unless marked with the letters *LH,* designating a left-hand thread. Bicycle pedals typically have left-hand threads on the left side and right-hand threads on the right side. This is done to ensure that the force you apply when pedaling will always tend to tighten the pedals, reducing the possibility of accidents.

2-3 FASTENER STANDARDS

In the United States, the American National Standards Institute (ANSI) has played a major part in setting standards for industrial fasteners. In 1982, the ANSI combined with the American Society of Mechanical Engineers (ASME).

Since that time, most American fastener standards have been designated ANSI/ASME. Specialized fasteners may be regulated by the American Society for Testing and Materials (ASTM). A new set of standards (mostly concerning fine threads) was developed specifically for automobiles and other vehicles by the Society of Automotive Engineers (SAE). Many mechanics commonly refer to fine threads as *SAE threads.* Another organization primarily concerned with fasteners used by industry is the Industrial Fasteners Institute (IFI). Most modern manufacturers rely on the ISO to ensure uniform high quality products. Industrial technicians should be familiar with the standards produced by each of these organizations.

2-4 CLASS OF FIT AND BOLT GRADES

Screw threads are divided into *classes* determined by their **tolerances** (the amount of size differences allowed from the exact size) and their **allowances** (how loosely or tightly they fit with their mating parts). For American National Standard screw threads, tolerance is referred to as *class.* For metric screw threads, tolerance is referred to as *fit.* There are three classes for external threads and three classes for internal threads. External threads are designated by a number followed by the letter *A*. Internal threads are designated by a number followed by the letter *B* (see Table 2.1).

Classes 1A and 1B have a large allowance for a loose fit. These classes work well where quick and easy assembly is important. These bolts will tolerate dirt and contamination, nicks, and burrs and still thread easily. Very few fasteners produced in the United States have class 1 threads because of their lower relative strength.

Classes 2A and 2B have a smaller allowance than class 1. These classes are a good compromise between economy of manufacture and fastener performance. These are the most common classes used in the United States.

Classes 3A and 3B have almost no allowances and are manufactured to restrictive tolerances. These classes are used in most socket head cap screws and set screws. They are also used in the aircraft industry and in other applications where safety is a critical issue.

The fit designation for ISO metric screw threads is somewhat more complicated than that for the unified inch classes. The first number–letter

TABLE 2.1
Unified inch thread classes.

	External	Internal
Loose	1A	1B
Medium	2A	2B
Tight	3A	3B

TABLE 2.2
Metric ISO thread fit.

	External	Internal
Loose	8g	7H
Medium	6g	6H
Tight	4h	5H

TABLE 2.3
Metric ISO Allowance.

ISO Allowance	External	Internal
Large allowance	e	E
Small allowance	g	G
No allowance	h	H

combination denotes the pitch diameter tolerance. The second number–letter combination denotes the crest diameter tolerance (see Table 2.2). The number in each set denotes a grade tolerance, and the letter denotes the tolerance position or allowance (see Table 2.3). A capital letter indicates that the threads are internal. A lowercase letter indicates that the threads are external. If the specifications contain only one set of symbols, the pitch diameter tolerance is the same as the crest diameter tolerance. The ISO tolerance grade is designated by the numbers 3 through 9 for external (bolt) threads and 4 through 8 for internal (nut) threads. The larger the number is, the greater the tolerance and the looser the fit.

For the most industrial use, only three fits are common with ISO threads:

Loose fit: 8g/7H
Medium fit: 6g/6H
Tight fit: 4h/5H

The medium-fit ISO fastener 6g/6H is the counterpart to the unified inch class 2A/2B. The 6 in the medium-fit ISO metric fastener signifies a medium-tolerance grade, the lowercase *g* indicates that the pitch diameter has an allowance, and the capital *H* indicates that the crest diameter has no allowance. The strength grades of bolts are shown on the head of the bolt. Unified inch strength grades are indicated by marks (see Figure 2.6).

Unified inch strength grades 1 and 2 have no marks whatsoever (see Figure 2.7). They are made from low-carbon steel with insufficient carbon to permit a strengthening heat treatment. Because of this process, these bolts are

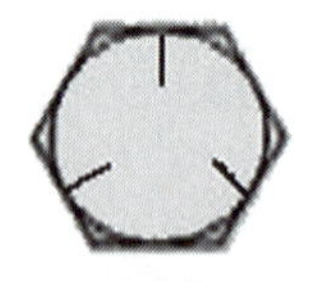

FIGURE 2.6
Strength grades are indicated on the head of the bolt.

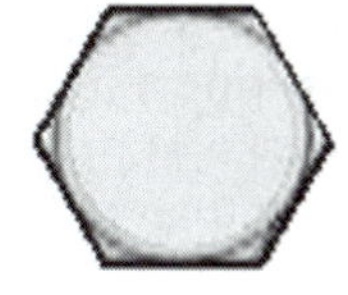

FIGURE 2.7
Unified inch grades 1 and 2 have no marks.

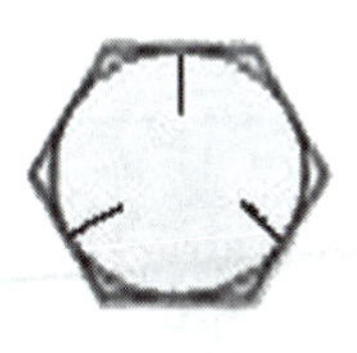

FIGURE 2.8
Grade 5 bolt.

FIGURE 2.9
Grade 8 bolt.

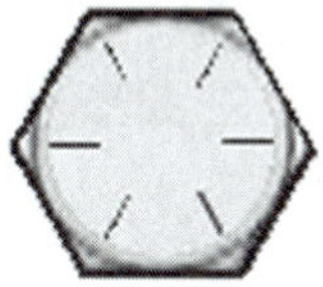

soft and break easily, but they are easy to manufacture. This economical bolt grade is used in applications where bolt strength is not a factor. It is also used with cast-iron products where it is desirable for the bolt to break before excessive force is applied that would break the cast part. Examples are cast-iron pump flanges, valve flanges, and pipe flanges.

Grade 5 bolts have three equally spaced marks (see Figure 2.8). These medium-grade bolts are economical and heat-treatable. When comparing strength to cost, grade 5 bolts offer the best value for most industrial applications. This grade tolerates service abuse and has twice the strength of grades 1 and 2 bolts. It is by far the most popular grade used in industry.

Grade 8 bolt heads have six equally spaced marks for identification (see Figure 2.9). This high-strength bolt is made with alloy steel and is about 1 1/2 times as strong as grade 5 bolts. These high-strength bolts are used in applications where safety and strength are factors. Examples include hoist-and-crane operation, tool-and-die applications, and many aerospace applications.

Grade 8 bolts should not be used in high impact applications. This type of application may cause them to fail without warning.

Metric bolt-strength grades are designated by numbers stamped on the head of the bolt. ISO standards call this number the *property class* of the bolt. The property class of metric bolts consists of two numbers separated by a period (see Figure 2.10). The first number is approximately 1/100 of the specified *minimum* tensile strength in megapascals (MPa). The number following the period is approximately 1/10 of the percentage between the *minimum* yield strength and the *minimum* tensile strength. The yield strength is always

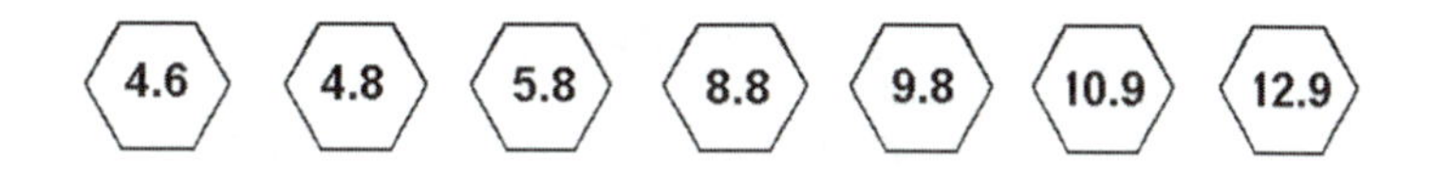

FIGURE 2.10
Metric bolts are designated by numbers.

expressed as a percentage of tensile strength. *Tensile strength* is the maximum tension a bolt can support just before it fractures. *Yield strength* is the maximum tension a bolt can support just before its threads begin to deform. Yield strength is always less than tensile strength. (Threads will always deform before the bolt breaks.)

EXAMPLE

A class 508 steel metric bolt has a specified minimum tensile strength of 500 MPa (5×100) and a specified minimum *yield* strength of 400 MPa (0.8×500).

Note: These numbers designate *approximate* strengths.

The metric property class numbers on the bolt head may not be *exact* tensile and yield values, but they are close approximations.

Metric class bolt strengths do not directly compare to inch grades, but a 5.8 metric class is about the same strength as a grade 2 unified inch bolt (see Figure 2.11). A metric class 8.8 bolt compares to a grade 5 bolt. Class 8.8 is one of the most widely used metric bolts in industrial applications. A class 10.9 metric is approximately equivalent to a grade 8 bolt. Table 2.4 compares some popular metric class bolts to unified inch grades.

You now have all the basic information you need to specify inch or metric bolts. Specifications give exact descriptions for each bolt application. For inch bolts, this description includes bolt size, number of threads per inch, type of thread, tolerance, and grade.

Diameter of bolt
Threads per inch
Type of thread
Tolerance (fit of thread)
Length of bolt

1/4 - 16 U.C. - A × 2

FIGURE 2.11
ISO 5.8 metric bolt.

TABLE 2.4
ISO property class and SE grade comparison.

SE grade	1	2	5	8
Metric class	4.6, 4.8	5.8	8.8	10.9

Screw diameters are given in *nominal* dimensions. This means that the actual diameter may be slightly different. The number 16 in the preceding example tells you that there are 16 threads per inch of bolt. The letters *U.C.* stand for unified national coarse thread. The *A* gives the tolerance of the thread (in this case, it is a loose-fit tolerance). The last number, 2, is the bolt length in inches, not including the head (in this case, 2 in.).

The metric thread designation of bolts is somewhat similar to the unified inch, but there are some exceptions:

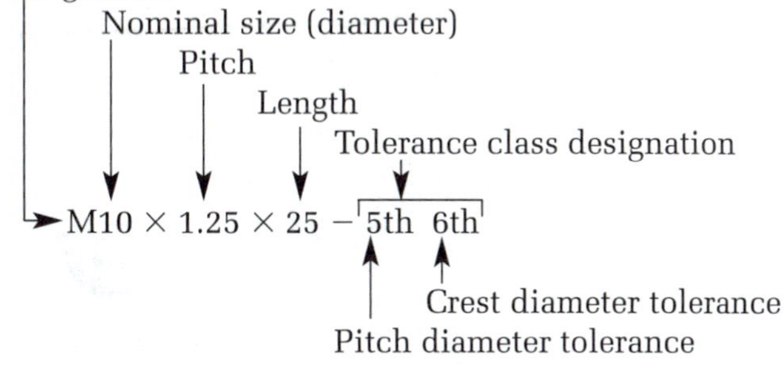

The first letter, *M,* designates a metric thread bolt. The number 10 designates a nominal diameter of 10 mm. The number 1.25 indicates the pitch in millimeters, and the number 25 is the length under the bolt head in millimeters. The next numbers describe the metric bolt tolerance class. The 5th indicates the pitch diameter tolerance, and the 6th indicates the crest diameter tolerance.

2-5 COMMON INDUSTRIAL THREADED FASTENERS

A wide variety of fasteners is available for many different purposes. It would take volumes of books to address them all. This section covers some of the most common types found in industry.

All threaded fasteners can be identified by the following characteristics (see Figure 2.12):

- Driving recess
- Head style
- Shoulder type
- Point style

Threaded fasteners may or may not have a driving recess. Hex-head bolts, for example, use the shape of the head for gripping purposes. The slotted head was the first type of driving recess. It was easy to manufacture and the flat-blade screwdriver used to turn it was also easy to produce. As mechanized production

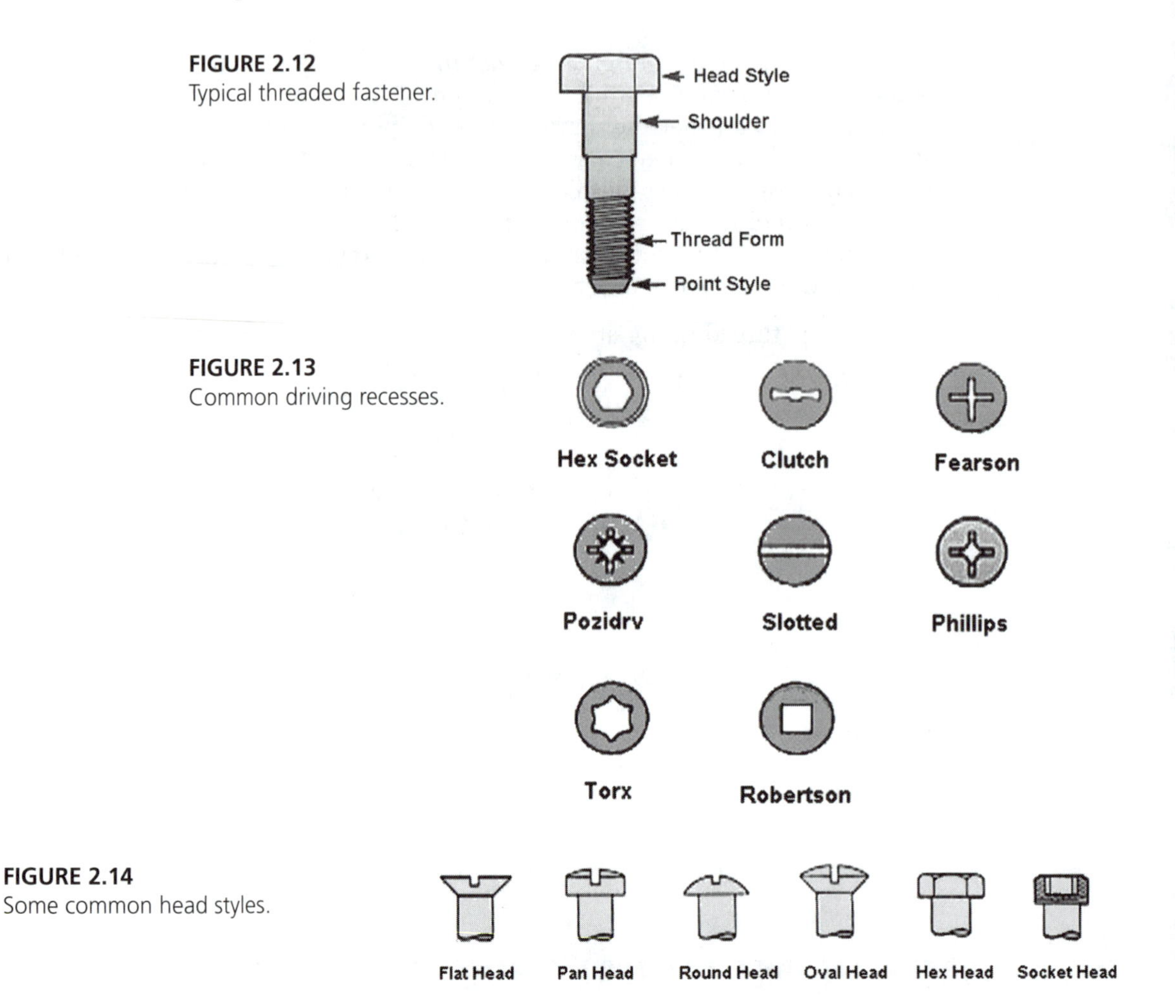

FIGURE 2.12
Typical threaded fastener.

FIGURE 2.13
Common driving recesses.

FIGURE 2.14
Some common head styles.

lines began to be developed, a self-centering driving recess was needed. The Phillips driving recess was invented to meet this need. Today, there are many different types of driving recesses. Figure 2.13 shows examples of the more common ones.

The type, or style, of fastener head refers to the shape of the head. One of the most common fastener head styles is the hex head used for standard bolts. Other common styles include square head, round head, flat countersunk, and pan head. Some of these types have driving recesses, and some do not. Figure 2.14 shows a variety of fastener head styles. For illustration purposes, all driving recesses shown are slotted; other types of driving recesses are available for any fastener.

The shoulder of a fastener is the portion of the fastener located below the head and above the threaded part of the fastener (see Figure 2.12). The fastener may be threaded all the way to the head or threaded for only a small portion of the fastener shank. The shoulder may be a larger diameter than the threaded

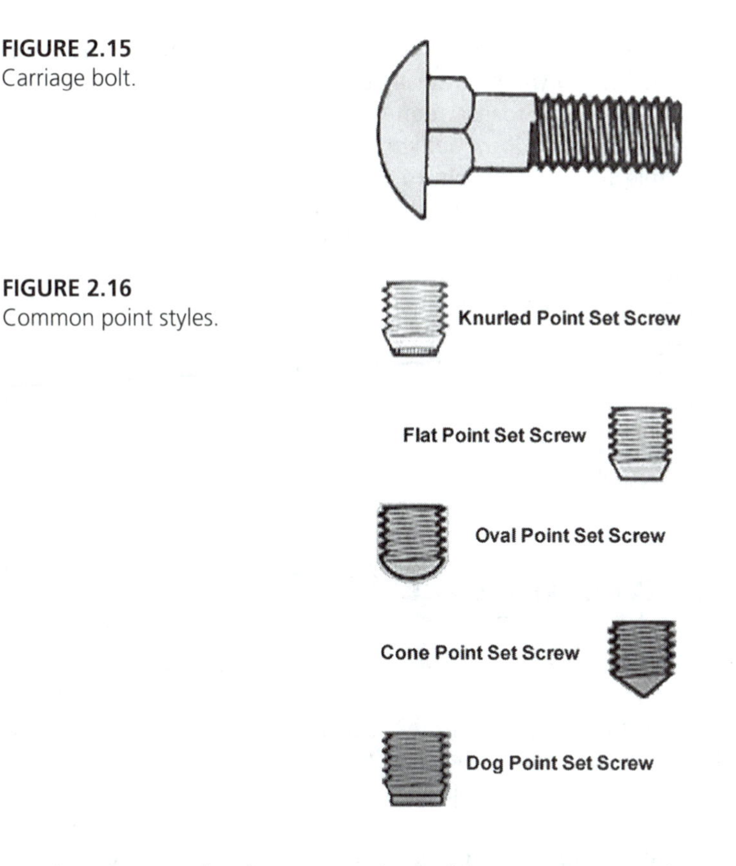

FIGURE 2.15
Carriage bolt.

FIGURE 2.16
Common point styles.

portion. It can also be square for holding purposes. This type of fastener is commonly referred to as a *carriage* bolt (see Figure 2.15).

The threaded shank of the fastener is the part that does the holding. There are many different specifications for threads. Refer to Sections 2-2 and 2-3.

The point style is the last physical part of a fastener we need to discuss. Most fasteners have a *plain* point, because the point is not used as a holding force. The type of point becomes much more critical for *set screws.* A set screw is a headless fastener used to lock an object to a shaft. Typically, sprockets, gears, pulleys, bearings, and set collars require set screws to be attached to a shaft. The holding power of a set screw is determined by the torque applied to the set screw and the point style. Figure 2.16 shows some common point styles.

The *cone point* is used for permanent installations. Its deep penetration gives the highest holding power. It has as much as 15% more holding power than some other types of points, but it also does the most damage to the shaft.

Oval points are the opposite of cone points. They have the least penetration and do the least damage to the shafts. They are used where frequent adjustments are necessary.

The *flat point* holds better than the oval point but not as well as the cone point. It is often used with shafts that have a ground flat place for the flat point to lock onto. Like the oval point, it is used where frequent adjustments are required.

The *dog point* resembles the tip of a dowel pin. This type of point locks in a hole drilled in the shaft or against a ground flat in the shaft. This point style is most often used for permanent mounting.

The *knurled point set screw* has a counter-bored cup point with counter-clockwise locking knurls to resist vibration and prevent the fastener from loosening. This point style leaves a small burr when removed that may have to be removed before reinstalling the part. The knurled cup point is the most commonly used point style for industrial set screws. It does not require any shaft modification and is designed to be a permanent installation.

It is essential to use the proper tool for each different type of fastener. Substitute tools will nearly always cause damage to the fastener. Cheap tools and multi-purpose kits may be used in an emergency but should never be used on a regular basis. Proper fit with the proper tools is essential in preventing physical damage to fasteners.

2-6 REMOVING DAMAGED FASTENERS

Sometimes fasteners cannot be removed with ordinary means. Because of rust, corrosion, contamination, or heat nuts can be very difficult to remove. There are several different methods of dealing with a frozen nut, depending on how difficult it might be to remove.

The first and easiest method you should try is to clean the threads carefully with a wire brush and then apply a good penetrating oil (see Figure 2.17). Allow some time for the penetrating oil to soak into the threads. If you have a good impact wrench you should use it. The hammering action will increase your likelihood of success.

If the penetrating oil doesn't work, you may want to heat the nut. Before heating, check the surrounding area for flammable materials or anything that might be damaged by heat. The heating process can destroy any nonmetal materials, such as plastic or rubber. Seals and bearings can also be damaged by excessive heat. If you heat only the outer edges of the nut, the heat causes the nut to expand and loosen up the fastener. After heating the nut you should

FIGURE 2.17
Try penetrating oil for frozen fasteners.

FIGURE 2.18
Nut splitter.

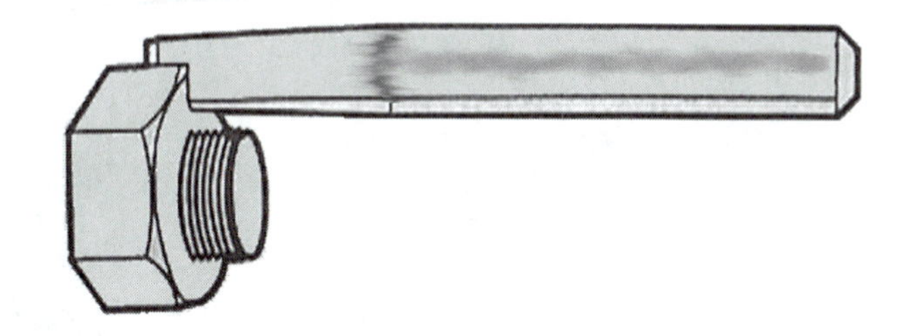

FIGURE 2.19
Cutting a nut using a cold chisel.

quickly remove it before the fastener heats up. If the fastener also becomes heated, both the nut and the fastener will expand; as a result, the nut is not loosened in relation to the fastener.

In some cases heating is not available or desirable. The area may be a fire hazard or have materials close by that will be damaged by the heat. A *nut splitter* is a clamplike device with a chisel point that literally does split the nut (see Figure 2.18). Make sure the nut splitter you are using is large enough for the nut that is to be split. A nut splitter will work on several different sizes of nuts, but it isn't available in larger sizes for extremely large nuts. For these larger sizes, a hydraulic version is available. You should tighten the nut splitter only as much as needed to split the nut; tightening it too much may damage the fastener threads.

When a nut splitter is not available, a chisel or a hacksaw can be used to remove a stubborn nut. Be careful when using either of these tools so you do not damage the thread of the fastener if it is to be reused. Figure 2.19 shows the correct way to cut a nut with a chisel. This method is usually used only on soft-grade nuts. Be sure the chisel is sharp and always use eye protection. Note the chisel cuts *parallel* to the threads on a corner—*not* from the flat side toward the thread, as the nut splitter does. The one-sided pressure of an incorrectly placed chisel can bend or stretch the threads of the fastener.

When using a hacksaw, you should make a straight, clean cut close to the fastener threads. Be careful to not cut into the threads. Soft-grade and hardened nuts may be removed using a hacksaw with the appropriate blade and good safety methods (see Figure 2.20).

Many bolts do not have nuts but are directly threaded into machinery frames. The bolt head may be broken off because of corrosion, rust, vibration,

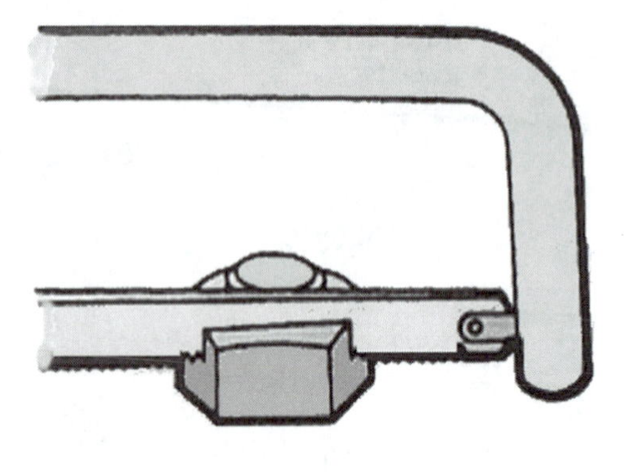

FIGURE 2.20
Cutting a nut using a hacksaw.

FIGURE 2.21
Stud remover.

or even just overtorquing. Whether the bolt is broken or seized, you use the same procedures to remove it.

A bolt or stud that is broken off above the surface can be removed fairly easily. A bolt with as few as three threads showing may be gripped using a special tool called a *stud remover* (see Figure 2.21). This tool uses a fine-toothed eccentric gear to bite into the fastener. The more torque you apply to loosen the fastener, the tighter the teeth dig in and grip it. As with nuts, you should first clean the threads thoroughly with a wire brush and apply a good-quality penetrant. If a stud remover is not available, a good pair of vise grips or a small pipe wrench can sometimes be used successfully.

Sometimes the bolt breaks below the surface of the machine (see Figure 2.22). This type of break is typically more difficult to remove and usually involves drilling into the fastener and using a special tool called an *extractor*. Before getting out the power tools, you might try using a cape (diamond-point) chisel and a pin

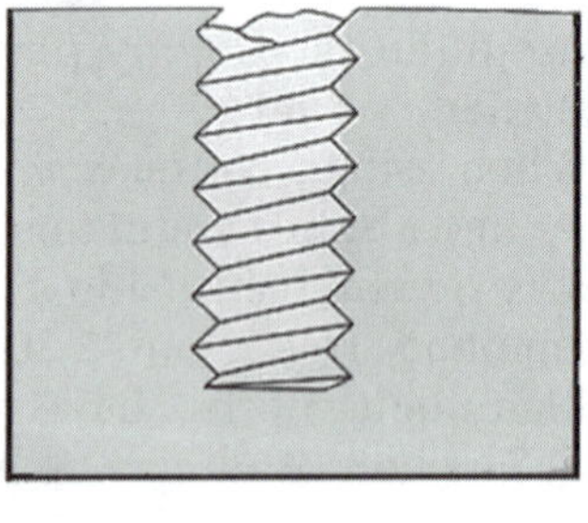

FIGURE 2.22
Sometimes the bolt breaks below the surface.

FIGURE 2.23
Using a cape chisel to remove a broken stud.

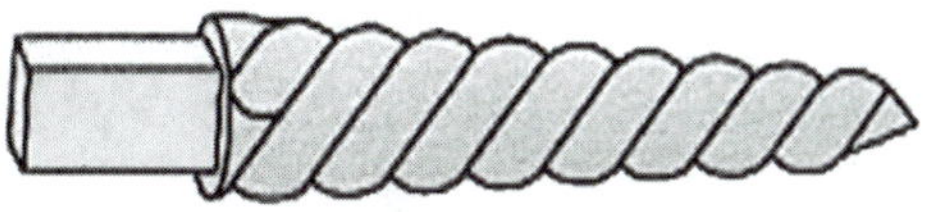

FIGURE 2.24
Screw extractor.

punch to remove the fastener (see Figure 2.23). If the fastener has a nub sticking up, use it to position the chisel or pin punch. Holding the chisel at an angle, tap it with a hammer counterclockwise to remove the fastener. You may need to create your own nub by holding the chisel vertically and tapping it with a hammer.

If you are still unable to remove the fastener, the next step is to drill through the fastener and use an extractor (see Figure 2.24). Very few fasteners break with a smooth, flat surface. Usually, they have a nub or ridge, making it impossible to center a drill. If the fastener is broken above the machine surface, use a file or grinder to create a flat surface.

If the fastener is broken below the surface, use a chisel and a center punch to remove the ridge. Center punch the fastener as close to the center as possible. Next, select a small drill and drill a starter hole. If the starter hole is at the center of the fastener, continue to drill all the way through. If the hole doesn't look centered, use a pin router bit or lathe center bit to move the hole to center.

Select the proper screw extractor for your size of fastener. The manufacturer of the extractor will recommend the correct drill size for the extractor. (Drills will be slightly larger than the small end of the extractor.) Some manufacturers even supply the correct drill to use in a set with extractors. Some maintenance technicians use *left-hand* drills to drill for the extractor. A left-hand drill cuts when rotating counterclockwise. This serves two purposes: (1) The drilling process does not tighten the fastener further, and (2) sometimes the counterclockwise rotation of the drill actually removes the broken fastener without even using the extractor.

After the hole is drilled, you can use the extractor. Insert the extractor and rotate it counterclockwise while applying downward pressure. Be very careful—extractors are very hard and brittle, causing them to break easily. A hardened, broken extractor is *very* difficult to remove. Avoid this situation.

If the extractor is unsuccessful, you must drill out the fastener. Use successively larger drills until you can see threads. A left-hand drill is an excellent choice for this operation. There is still the possibility of it removing the fastener before the threads are damaged. Do not exceed the drill size required by the tap size of the broken fasteners. Clean out the remaining fastener threads, and use the correct tap to rethread. Check the threads for damage caused by removing the fastener. If more than 10% of the threads are destroyed, you may be forced to drill or tap to the next larger size fastener. If the same size fastener must be used, a heli-coil or thread repair insert may be used. A heli-coil is a repair system that enables the user to repair damaged threads with a special insert that replaces stripped or damaged threads.

2-7 CUTTING THREADS WITH TAPS AND DIES

Threads may be cut internally using a tap or externally using a die. Nuts are threaded using a tap, and bolts are threaded using a die. The process of cutting an internal thread is called **tapping.** Using a die to cut external threads is called **threading.**

Taps are made from hardened tool steel. Flutes are cut the length of the thread to form the cutting edge, provide an exit for chips, and allow space for cutting fluid. The square end of the tap is used for holding the tap with a tap wrench. Taps are hard and brittle and can break easily if you do not use them with care. Broken taps can be extremely sharp. Taps that are broken can be extremely difficult to remove. The three basic types of hand taps are the *taper* tap, the *plug* tap, and the *bottom* tap (see Figure 2.25). The only basic difference in the three taps is the number of threads that are ground to a taper on the end

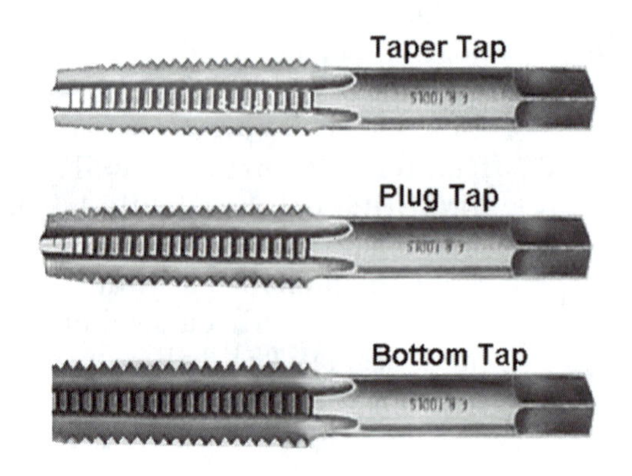

FIGURE 2.25
Three basic types of taps.

of the tap. Taps are ground-tapered in order to start and cut the threads more easily. Most of the cutting of the threads is done by the tapered end of the tap.

The tapered tap has the most taper and the smallest starting end, which makes it the easiest hand tap to start. Eight to ten threads are tapered in the tapered tap. It is used to start the tapping process for another tap. It can also be used by itself to tap open, or *through,* holes. The tapered tap requires less torque to cut threads than either the plug or bottoming taps.

Three to five threads are tapered in the plug tap. It is a good general-purpose tap if you have only one type of tap. It does not start as well as the taper but will usually adequately tap most blind holes.

In the bottoming tap, one to two threads are tapered. It is used after the plug tap for tapping to the bottom of a hole. It should never be used to start a tapped hole. Whenever possible, all three types of hand taps should be used in tapping closed or blind holes. The tapered tap should be used first, followed by the plug tap, and, finally, the bottoming tap. These three taps together make up a *tap set* (see Figure 2.26).

The type of hole you want to tap will determine which hand taps to use. A hole that goes all the way through should be tapped with a tapered tap. A blind, or closed, hole whose threads do *not* go all the way to the bottom should be tapped with the tapered tap and then the plug tap. The blind hole, which requires threads very close to the bottom, requires a tapered tap first, then a plug tap, and a bottom tap to finish.

Dies are used to cut external threads on bolts or studs. The die is turned onto the bolt, cutting threads as it is rotated. Dies may be solid or the adjustable round split type (see Figure 2.27). The solid die is most often used for changing

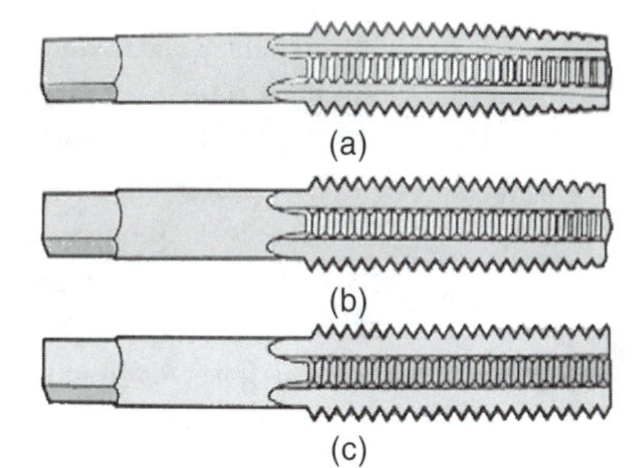

FIGURE 2.26
A tap set: (a) taper tap; (b) plug tap; (c) bottom tap.

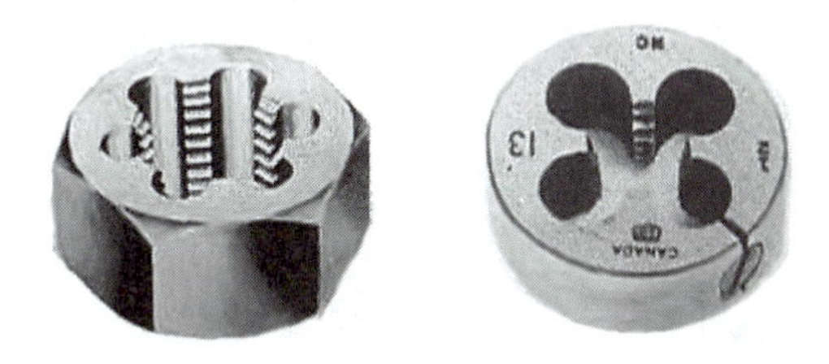

FIGURE 2.27
External thread dies.

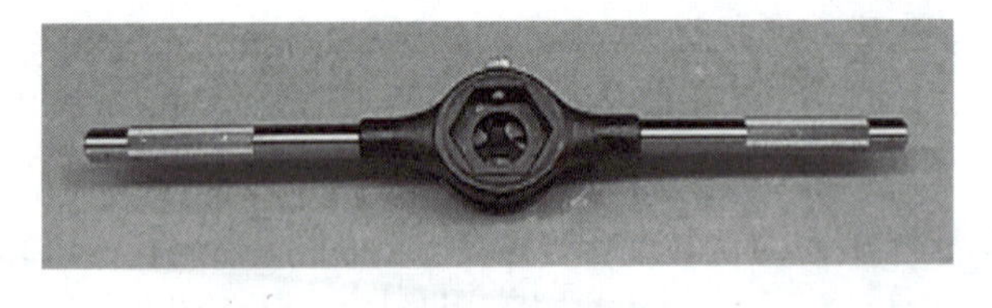

FIGURE 2.28
Die stock.

or recutting damaged threads. A standard hand wrench is used to turn it. This nonadjustable die should not be used for heavy cuts or new threads. Use plenty of cutting fluid to keep the solid die well lubricated. Adjustable split dies allow the maintenance technician to cut shallow or deep threads. The adjustable die requires a special holding tool called a *die stock* (see Figure 2.28). When threading, it is recommended that you **chamfer,** or bevel, the rod by 30° to 60°. This allows the die to start easily and eliminates a burr on the rod end. Be sure to place the tapered side of the die square on the beveled end. Rotate the die clockwise while applying even downward pressure. You should reverse the die occasionally to break the metal chips. Keep cutting fluid on the cutting threads at all times. Remove the die, and try the correct nut size on the threaded rod. If necessary, adjust the die and rethread.

QUESTIONS

1. Who is credited for the first thread standard?
2. What is the pitch of a thread?
3. List the advantages of fasteners with fine threads.
4. List the advantages of fasteners with coarse threads.
5. To tighten right-hand threads you should turn them ______________.
6. What is the most recently formed organization concerned with fastener standards?
7. What is the most common class of fit used by the unified system?
8. How will the head of a grade 5 unified bolt be marked?
9. What is the property class of a metric bolt?
10. List the common types of driving recesses used by threaded fasteners.
11. What is a set screw?
12. Which fastener point gives the greatest holding power?
13. What is the main disadvantage of the cone point?
14. What is the correct way to cut off a hex nut with a hammer and chisel?
15. What are the three basic types of hand taps?
16. What is the most common use for solid dies?

Chapter 3

Basic Principles of Mechanical Systems

3-1 INTRODUCTION

We use energy every day as we work and play and don't even realize it. Whenever you see motion, heat, or light, you are experiencing energy at work. Technically, energy is the ability to do work. There are six forms of energy: mechanical, electrical, heat, light, chemical, and nuclear. Any of these six forms of energy can be changed into any other form. Some conversions are easy. Applying your car brakes changes mechanical energy (motion) into heat energy. Electrical energy applied to a light bulb becomes light and heat energy. A typical battery converts chemical energy into electrical energy. Other conversions are more difficult. Nuclear energy can be converted into heat, which can then be converted into mechanical energy, which is finally converted into electricity generated by a power plant. In this book, we are concerned only with mechanical energy.

3-2 MECHANICAL ENERGY

Mechanical energy is probably one of the most common forms of energy that we deal with every day. Almost all motion is a form of mechanical energy. Mechanical energy at work can be a very useful thing. A car moving down the

FIGURE 3.1
Energy can be kinetic or potential.

FIGURE 3.2
Natural energy sources.

road, a belt driving a sheave, a chain driving a sprocket, and a sled sliding down a snow-covered hill are all using mechanical energy. All these examples of mechanical energy in motion are called **kinetic energy** (see Figure 3.1). Kinetic energy is energy in motion. But what about *stored* energy? Stored energy is called **potential energy.** A good example of potential energy is the water stored in a lake. As the water is released into turbines, it becomes kinetic energy. Potential energy can be defined as energy available for use (see Figure 3.2). Whenever we want to use potential energy, we change it into kinetic energy. Sources of potential energy can sometimes be surprising. A wrench lying on top of a ladder would not seem to be a source of potential energy, but if it falls off the ladder and hits you on the head, you quickly realize its potential energy. There are many sources of potential energy that may not be immediately obvious.

3-3 MECHANICAL FORCE

Force in a linear mechanical system consists of a *push* or *pull* being applied to an object. The SI, or metric, system measures mechanical force in *newtons* (N). The English system of measurement, which is still used in the United States, measures mechanical force in *pounds*. One pound is equal to 4.45 N and you can do a conversion for any application. Force and weight are interchangeable. The mass of any body interacts with the gravitational attraction of Earth to cause the two bodies to attract each other. This attraction is called *weight*. Applying force can be as simple as stepping on a bathroom scale in the morning to check your weight (see Figure 3.3). The scale tells you how much force you are applying, in pounds. Once the scales stop moving, the force you apply to the scales and the force the scales apply to hold you up become *balanced*. When all the forces acting on a body are balanced, they cancel each other out, and there is no movement. This special state of balanced force is called **equilibrium.**

An object can be in equilibrium and be moving. If an object is moving in a straight line at a constant speed when balanced forces are applied, it continues to move without speeding up, slowing down, or changing its direction. An example of this type of equilibrium is an asteroid moving in space. As long as no other force is applied, it moves in a straight line at a constant speed. All forces are balanced, and the asteroid is in equilibrium. Another example is a person holding a rope tied to a boat in a fast-flowing river (see Figure 3.4). The force of the river

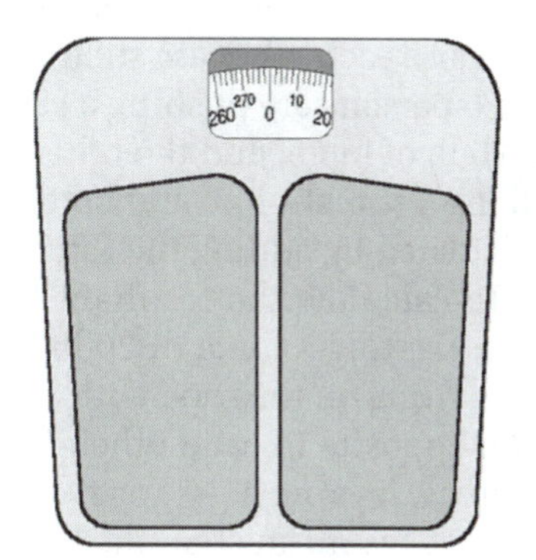

FIGURE 3.3
Weight measured by bathroom scales.

FIGURE 3.4
Balanced forces equals no movement.

tries to move the boat downstream, but the force applied by the person holding the rope keeps the boat from moving. The forces are equal and cancel each other out; they are balanced, and the boat is in equilibrium. If the person starts pulling the boat upstream, then the forces become *unbalanced.* The effect of an unbalanced force results in a *change* of an object's speed, direction, or shape.

Mechanical force has two distinct qualities: the force's strength, or *magnitude,* and the force's *direction.* Using these two physical qualities of mechanical force, we can analyze all the forces acting on an object and compute a single net, or *resultant,* force. A force with a magnitude and direction is called a **vector.** Other vectors may include momentum, velocity, and acceleration. Forces that are described without direction are called **scalars.** Mass, temperature, and pressure are some examples of scalars. An air tank can contain a force of 20 $lb/in.^2$ (pounds per square inch) of compressed air; describing this as "the pressure in this tank is 20 $lb/in.^2$ *south*" makes no sense. In this example, pressure as a scalar is a force with no direction. An airplane flying 200 mi/h (miles per hour) describes a scalar (no direction is given). The same airplane flying *southwest* at 200 mi/h describes a vector (it has direction and magnitude).

Computing Resultant Force

It is easy to calculate the resultant force if all force is along a straight line. Forces applied in the same direction are simply added to get the resultant force. Forces that oppose each other are subtracted to get the resultant force. Consider this example: two persons are pushing a car to the nearest gas station. One person is pushing with 20 lb of force, and the other person is pushing with 15 lb of force. The forces are along a straight line and both are in the same direction. The resultant force can be calculated by adding the forces, which gives 35 lb of force applied to the car.

To calculate the resultant force in our boat example, let's suppose that the river is exerting a force of 20 lb trying to force the boat downstream and the person is pulling on the rope with a force of 25 lb. Both forces are along a straight line but are opposite to each other. We can find the resultant force by subtracting the two forces, leaving a net, or resultant, force of 5 lb in favor of the person pulling the rope. This means that the boat will move closer to the person (see Figure 3.5).

Forces along a straight line (180°) are easy enough, but what about forces at right angles (90°) (see Figure 3.5)? The resultant of two forces at right angles can be computed by using the Pythagorean theorem: for any right triangle (one with a 90° angle), where a and b are the sides adjacent to the 90° angle (the shorter sides) and c is the side opposite to the 90° angle (the longest side),

$$c^2 = a^2 + b^2$$

Forces that act at *any* angle can be computed by using a graphical method. For example, suppose we have two people trying to pull a car out of the mud.

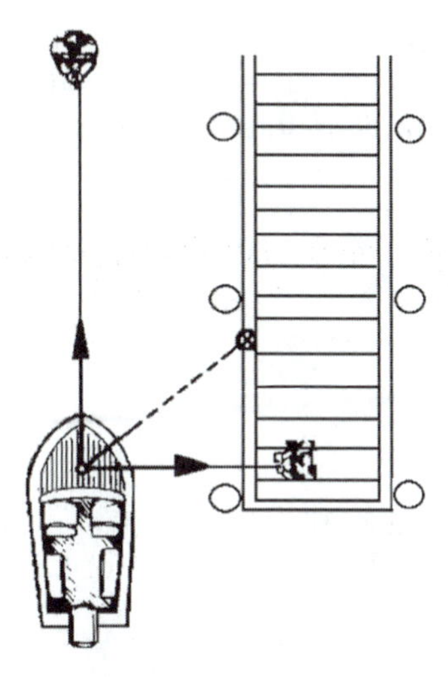

FIGURE 3.5
The boat will move at an angle.

One is pulling north with a force of 30 lb and the other is pulling east with a force of 40 lb. To find the resultant force on the car, first draw a horizontal line on a piece of graph paper, which is the reference line. Determine a reasonable scale for the values represented. For this example 1/2 in. equals 10 lb works well. Label the units in pounds, starting with zero on the left. Now, draw the first vector of 40 lb with an arrow pointing to the right. Label this vector **A.** On the left side of **A,** draw a vector to represent the person pulling north. Because north is 90° from east, the vector should be drawn at an angle of 90°, with the arrow pointing straight up. Its value is 30 lb, and you can label it **B.** Complete the triangle with a line labeled **C.** Vector **C** is 37° from vector **A,** between the two vectors **A** and **B.** Determine its force by measuring the length of line **C** and converting it into pounds using the scale of 1/2 in. equals 10 lb. The resultant force should be 50 lb at a 37° angle (see Figure 3.6).

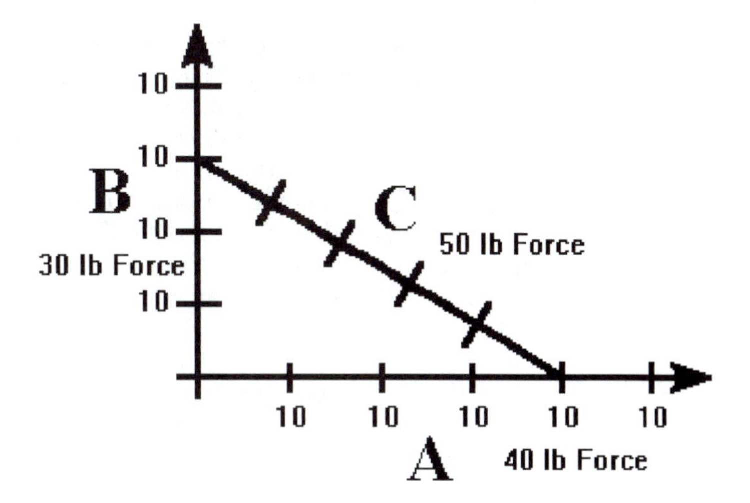

FIGURE 3.6
Use vectors to find the resultant force.

Because our example created a triangle with a 90° angle, we can also use the formula $c^2 = a^2 + b^2$. The values for the vectors adjacent to the 90° angle are 40 and 30. Therefore,

$$c^2 = 40^2 + 30^2$$
$$c^2 = 1{,}600 + 900$$
$$c^2 = 2{,}500$$
$$c = \sqrt{2{,}500}$$
$$c = 50$$

Remember, this formula works only for triangles with right angles (90°), but the graphical method works for all cases.

3-4 TORQUE

Force in a rotating mechanical system is called **torque.** The rotation of motors, gears, sprockets, nuts, and bolts is caused by torque applied to them. Torque is found by multiplying the force applied by the length of the lever arm:

$$T = F \times L$$

where T = torque in lb-ft or N·m
F = force in pounds or N
L = length of lever in ft or m

In the English system, the units used are ounce-inches (oz-in.) or pound-feet (lb-ft); in the metric system it is measured in newton-meters (N·m).

An example of a common torque application is using a wrench to apply torque to a bolt. The center of the bolt is the axis of rotation. From the axis of rotation, we measure the distance at which the force is applied. If that distance is 2 ft and the amount of force is 10 lb, the torque applied is 20 lb-ft. Torque is the force multiplied by the length of the lever arm; therefore, torque is doubled if we double the length of the lever arm without changing the force. This principle is illustrated by an amateur mechanic using a piece of water pipe on a wrench to loosen a stubborn bolt. Torque is increased, and the bolt turns (barring any unforeseen accident).

You can determine the torque in a rotating wheel in much the same way. The lever arm is the radius of the wheel, and the force is the applied force at the edge of the wheel. An example of this is a pull starter on a small engine (see Figures 3.7 and 3.8). You supply the force by pulling on the rope that is

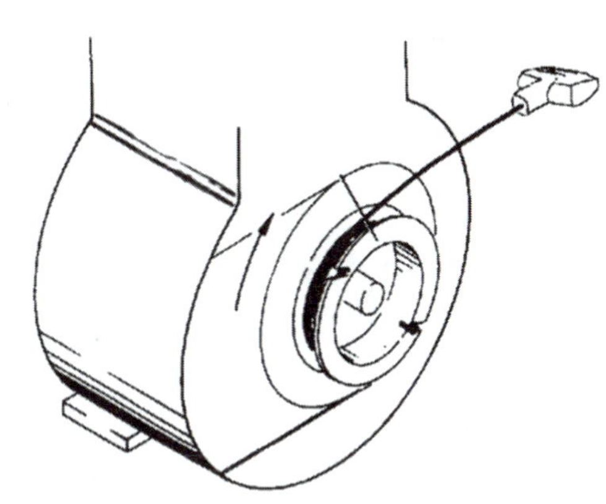

FIGURE 3.7
Pull starter on a small engine.

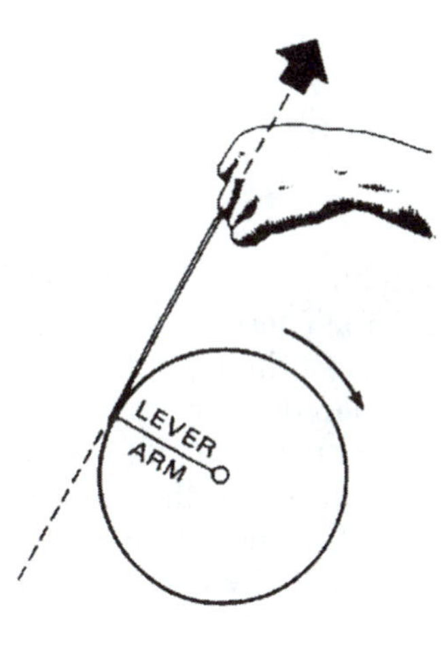

FIGURE 3.8
The lever arm is equal to the radius.

wrapped around the wheel, causing it to rotate. The force applied to the edge of the wheel causes the torque at the motor shaft. If you apply a 30-lb force by pulling the rope and the radius of the wheel is 6 in. (0.5 ft), the torque applied to the motor shaft is 30 lb multiplied by 0.5 ft, or 15 lb-ft. Sprockets, pulleys, and gears all work in the same manner.

3-5 MECHANICAL WORK

Mechanical work results when a force causes an object to move. The technical definition of mechanical work has two parts:

1. An applied force must act on an object. Just having an applied force is not enough,
2. The object must *move* while the force is applied.

FIGURE 3.9
No movement equals no work.

FIGURE 3.10
Movement equals work.

A 100-lb bucket resting on the floor is obviously *not* an example of work (see Figure 3.9). There is, however a 100-lb force applied by gravity pushing the bucket against the floor. If, on the other hand, you were handed the same 100-lb bucket and told you must hold it 3 ft above the floor, you might be inclined to call it work (see Figure 3.10). Just holding the bucket, though, cannot technically be called work because there is no movement involved. However, the act of moving the 100-lb bucket 3 ft above the floor is a good example of mechanical work. There is a force involved (100 lb), and there is movement (3 ft).

$$W = F \times D$$

where W = mechanical work in ft-lb or N·m
F = force applied to an object in lb or N
D = distance moved in ft or m

When you are just *holding* the bucket, mechanical work is determined as follows:

$$W = F \times D$$
$$W = 100 \text{ lb} \times 0 \text{ ft} \quad \text{(no movement)}$$
$$W = 0 \quad \text{(no work is done)}$$

In the preceding example, when you *move* the bucket, mechanical work is determined as follows:

$$W = F \times D$$
$$W = 100 \text{ lb} \times 3 \text{ ft}$$
$$W = 300 \text{ ft-lb} \quad \text{(work is accomplished)}$$

Let's review what has happened. When the bucket is at rest on the floor, the forces on the bucket are balanced. The bucket is said to be in *equilibrium.* This is true because the force of gravity pulls the bucket down with a force of 100 lb while the floor holds it up with an equal and opposite force of 100 lb. No movement is involved; therefore, no work is done. When you grasp the bucket handle and begin to lift it, your upward pull becomes a little stronger than the downward pull of gravity. The bucket moves upward, and work is now accomplished on the bucket. After you have lifted the bucket as far as you want, the bucket stops moving. You must still exert a force of 100 lb to hold it in this position, but this holding force does no work on the bucket because there is no movement. You may become tired and feel like you are working. However, by definition, there is technically no work being done.

Now that we have discussed work as force multiplied by distance, how do you apply that to an electric motor? One force, a rotational system that we have already studied, is called *torque.* Distance in a rotational system is the angle of rotation measured in units called **radians** (rad). Angles are often measured in degrees, where one-quarter of a circle is 90°, one-half of a circle is 180°, and a full circle is 360°. The measure of an angle can also be given in radians. Radians work with the constant *pi,* pronounced "pie." The symbol for *pi* is the Greek letter "π." *Pi* is a constant that is associated with circles. Its value never changes. Although it is a never-ending number, *pi* is usually rounded to 3.1416, which is accurate enough for our purposes (see Figure 3.11).

The circumference of a circle is equal to 2π times the radius of the circle. There are 360° circle and 2π rad in a complete revolution, as illustrated in Figure 3.11. The formulas we use call for radian measure instead of degrees. It is easy to convert from degrees to radians. Because 2π rad equals 360°, 1π rad equals 180°, and π/2 rad equals 90°. We thus see that 1 rad is about 57.3°. The radian measure of an angle is the ratio of the length of an arc to the length of its radius (see Figure 3.12) and is not really a true unit like a pound or a foot. In mathematics, you must be careful to solve for the correct units as well as the correct numerical value. You should never just randomly drop or change units.

FIGURE 3.11
Radians compared to degrees.

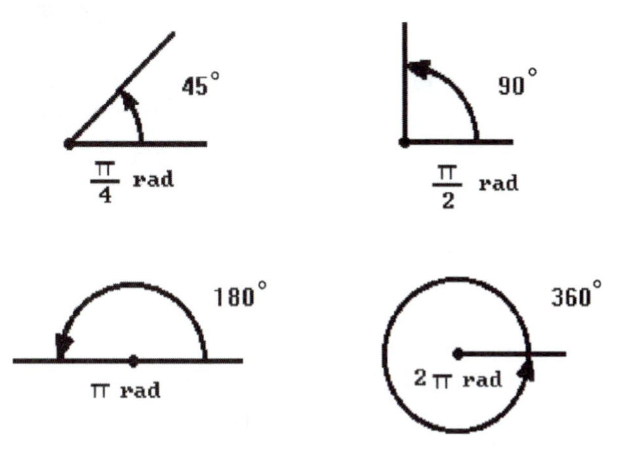

FIGURE 3.12
One radian equals 57.3°.

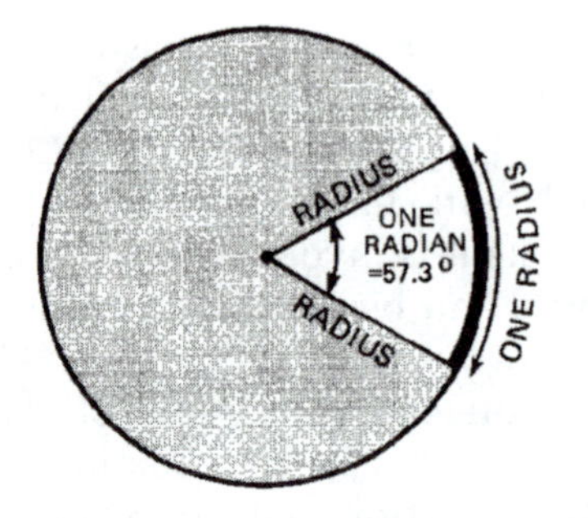

The radian is the exception to this rule. Because it is not a true unit, it can be dropped without changing the result.

Let's review. A radian (rad) is a unit of angle measure. It can be directly converted to degrees and is equal to 57.3°. The radian can be defined in terms of a piece of a circle. An angle of one radian cuts an arc on the circle equal to the radius of the circle.

$$\text{One complete circle} = 360°$$
$$\text{One complete circle} = 2\pi \text{ rad } (6.28 \text{ rad})$$
$$1 \text{ rad} = 57.3°$$

Now that we've discussed the basics of radians, let's look at work done by torque.

$$W = T \times \theta$$

where W = work done by a torque in ft-lb or N·m
T = force applied by the torque in lb-ft or N·m
θ = angle of rotation, measured in rad, through which the torque is applied ("θ" is the symbol for the Greek letter theta.)

EXAMPLE

Given: A drill press has a table that is raised and lowered by a hand crank attached to a rack-and-pinion gear. A hand crank has a handle 1 ft long. A force of 10 lb is required to turn the crank.

Find: The amount of work done if the crank is turned five revolutions.

SOLUTIONS

First find the torque.

$$T = F \times L$$
$$T = 10 \text{ lb} \times 1 \text{ ft}$$
$$T = 10 \text{ lb-ft}$$

Now find the total angle of revolution in radians:

$$\theta = (5\ \cancel{\text{rev}})\frac{(2\pi\ \text{rad})}{1\ (\cancel{\text{rev}})}\ \text{(revs cancel)}$$

$$\theta = 10\pi\ \text{rad (or 31.4 rad)}$$

Next find the work done:

$$W = T \times \theta$$

$$W = 10\ \text{lb-ft} \times 10\pi$$

$$W = 10 \times 10 \times 3.14\ \text{ft-lb}$$

$$W = 314\ \text{ft-lb}$$

Note: It is customary for units of work to be written as ft-lb. Units for torque are written as lb-ft.

Let's review what we did with the units in the preceding example. First, we found the torque (T) in pound-feet. Next, we found the angle (θ) in radians (one revolution equals 2π rad; therefore, five revolutions equals $5 \times 2\pi$ rad, or 10π rad). Then, we substituted these values in the work equation, dropped *rad* because it is not a true unit, and came up with an answer in lb-ft. Because we solved for work instead of torque, lb-ft was changed to ft-lb as is customary. (Mathematically $A \times B = B \times A$, so really nothing changes.)

3-6 BASIC MECHANICAL MACHINES

Most machines, no matter how complicated, are a combination of six or fewer simple machines. These simple machines are the *lever,* the *wheel and axle,* the *pulley,* the *inclined plane,* the *wedge,* and the *screw.*

Simple machines have one purpose: to change the mechanical force and distance produced by a person or machine. Typically, a small force is changed to a larger force by applying the small force over a greater distance. Simple machines are used to produce a *mechanical advantage,* or MA. Mechanical advantage is the name given to the ratio of the input to the output force converted by a simple machine. By using mechanical advantage, a person is able to move a 1,000-lb rock, as shown in Figure 3.13. In this example,

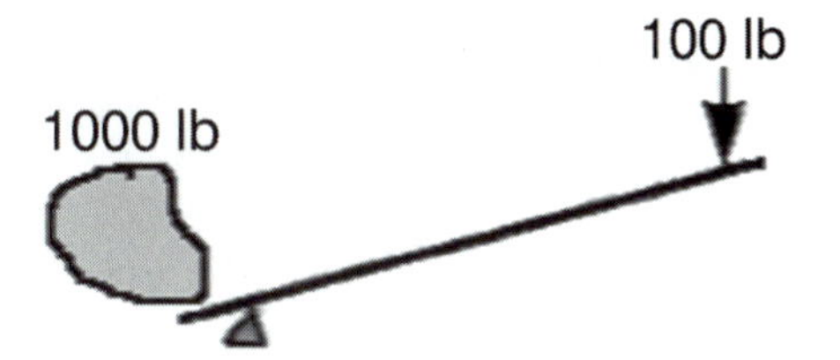

FIGURE 3.13
Mechanical advantage 10:1.

a lever is used to provide the mechanical advantage. When pushing down on the lever, the person exerts a force of 100 lb for a distance of 60 in. If you divide 1,000 lb by 100 lb, you get a mechanical advantage of 10:1. The mechanical advantage can also be found by dividing 60 in. by 6 in. For every pound of force applied, the person lifts 10 lb of rock; however, he or she must apply that force 10 times as far as the rock moves. In using a simple machine, there is a change in mechanical advantage, but there is no change in the amount of work done.

$$\text{Input work} = \text{output work}$$
$$\text{(Input) force} \times \text{distance} = \text{force} \times \text{distance (output)}$$
$$100\text{ lb} \times 60\text{ in.} = 1{,}000\text{ lb} \times 6\text{ in.}$$
$$6{,}000\text{ in.-lb} = 6{,}000\text{ in.-lb}$$

The preceding formula does not consider friction, which we discuss later. Note that simple machines do not increase the amount of work done. Mechanical advantage is gained at a price. In this case, the price is distance moved.

Lever

A lever in its simplest form is a rigid bar moving about a fixed point called the fulcrum (see Figure 3.14). Some examples of the lever are the pry bar, shovel, wrench, pliers, wheelbarrow, and many other everyday machines. The principle of the lever is also used in two other simple machines: the wheel and axle and the pulley.

A lever can serve two purposes: (1) It can provide mechanical advantage, and (2) it can change the direction of the applied force. Levers can be catalogued in three different classifications: *first class, second class,* and *third class.* In first-class levers, the fulcrum is located between the input force (also called *effort* force) and the output force (also called *load*). First-class levers provide mechanical advantage and change in direction. A child's seesaw is an example of a first-class lever. Figure 3.15 shows a pair of pliers another example of a first-class lever.

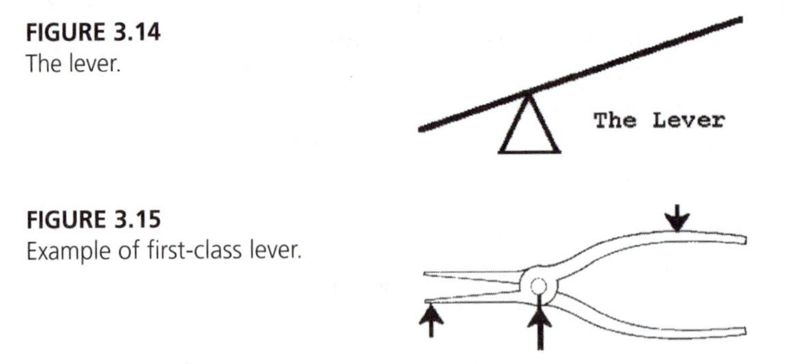

FIGURE 3.14
The lever.

FIGURE 3.15
Example of first-class lever.

FIGURE 3.16
Example of second-class lever.

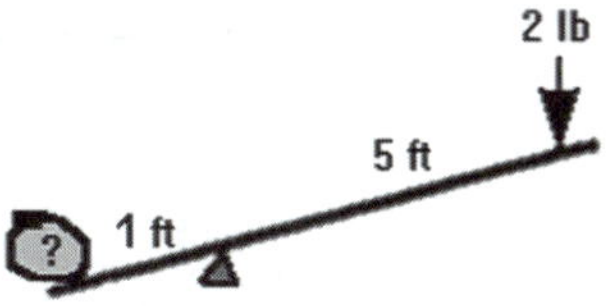

FIGURE 3.17
How much weight can be lifted?

With second-class levers, the load is applied between the fulcrum and the effort force. Second-class levers provide mechanical advantage with no change of direction. A wheelbarrow is a good example of a second-class lever (see Figure 3.16).

In third-class levers, the effort force is closer to the fulcrum, between the fulcrum and the load. The effort force is greater than the load, but the distance moved is less than the load distance moved. Third-class levers, like second-class levers, provide mechanical advantage with no change in direction. When you throw a ball, your arm is a good example of a third-class lever. A small amount of muscle movement causes a large amount of arm movement. The effort is applied close to the fulcrum, and the ball is thrown very fast with a short, strong muscle movement.

The mechanical advantage of a lever is easy to determine. It is the ratio of the lengths of the lever arms or the ratio of the forces of the load and the effort (see Figure 3.17).

EXAMPLE

Given: A first-class lever with an applied effort force of 2 lb. The lever arm on the effort force side is 5 ft, and the lever arm on the load side is 1 ft.

Find: The mechanical advantage and the load that this effort force can lift.

(continued on next page)

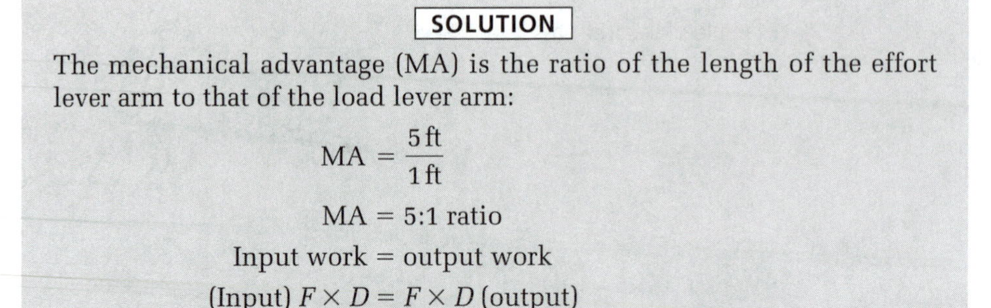

SOLUTION

The mechanical advantage (MA) is the ratio of the length of the effort lever arm to that of the load lever arm:

$$\text{MA} = \frac{5\,\text{ft}}{1\,\text{ft}}$$

$$\text{MA} = 5\text{:}1 \text{ ratio}$$

$$\text{Input work} = \text{output work}$$

$$(\text{Input})\ F \times D = F \times D\ (\text{output})$$

$$2\ \text{lb} \times 5\ \text{ft} = F \times 1\ \text{ft}$$

$$\frac{(2\,\text{lbs} \times 5\,\text{ft})}{1\,\text{ft}} = F\ (\text{Divide both sides by } 1\,\text{ft.})$$

$$10\ \text{lb} = \text{force out, or load}$$

Wheel and Axle

The wheel and axle is one of people's oldest machines. It consists of a wheel attached to an axle and can be thought of as a round lever (see Figure 3.18). It can provide mechanical advantage in the same way as a second-class lever does. A steering wheel on a truck is a good example. The lever arm is the radius of the wheel, and the effort force is applied as you turn the wheel.

The wheel and axle can also provide an increase in speed, using the principles of the third-class lever. The automobile wheel and axle illustrates this principle. By applying force to the axle, the outer edge of the wheel moves at a greater speed and for a greater distance than the axle does.

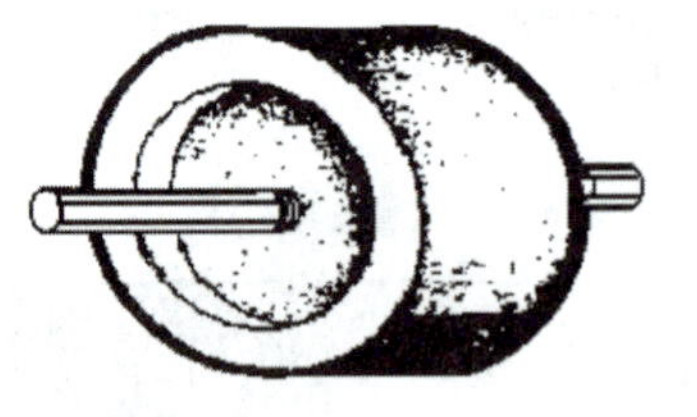

FIGURE 3.18
Wheel and axle.

EXAMPLE

Given: A 20-in.-diameter wheel with an effort force of 10 lb applied at its edge turns a 2-in.-diameter axle.

Find: The mechanical advantage of the wheel and axle.

SOLUTION

First, find the radius of both the wheel and axle in inches. The radius is always one-half the diameter:

$$20 \times 1/2 = 10\text{-in. wheel radius}$$
$$2 \text{ in.} \times 1/2 = 1\text{-in. axle radius}$$

Now:

$$MA = 10 \text{ in.} \div 1 \text{ in.}$$
$$MA = 10{:}1 \text{ ratio}$$

Note: If you use diameter instead of radius, you still get the correct ratio.

If you want to know the torque applied, you must use the lever arm, which is always the radius.

EXAMPLE

Given: The same information as in the preceding example.

Find: The torque of the axle.

SOLUTION

Torque equals force times lever arm.

$$T = F \times L$$
$$T = 10 \text{ lb} \times 10 \text{ in.}$$ (Lever arm of a wheel is always the radius.)
$$T = 100 \text{ lb-in.}$$

EXAMPLE

Given: A wheel and axle on an automobile; the axle has an applied torque of 200 lb-ft, and the diameter of the wheel is 24 in.

Find: The force applied at the edge of the wheel.

SOLUTION

Notice that the torque is in lb-ft and the diameter of the wheel is in inches. One of these units must be converted to the other so they can be combined. Convert 24 in. to 2 ft. You must also convert the diameter (d) to the radius (r) of the wheel.

(continued on next page)

$$r = 1/2 \times d$$
$$r = 1/2 \times 2 \text{ ft}$$
$$r = 1 \text{ ft}$$

Now we are ready to find the force applied:

$$\text{Torque} = \text{force} \times \text{lever arm}$$
$$T = F \times L$$

Rearranging to solve for F:

$$F = T \div L$$
$$F = 200 \text{ lb-ft} \div 1 \text{ ft}$$
$$F = 200 \text{ lb}$$

Pulleys

Pulleys can be found at work for us in many different places. They are also common in most industrial plants. In industry a grooved pulley is called a *sheave.* No matter what name it's given, this simple machine is vital to our modern way of life. A pulley consists of a wheel mounted in a frame or block so it can turn freely (see Figure 3.19). Pulleys are often combined in pairs to increase mechanical advantage. This type of pulley arrangement is known as a *block and tackle.* Pulleys not only provide mechanical advantage but are often used to change the direction of an applied force as well.

Figure 3.20 illustrates the manner in which pulleys gain mechanical advantage. In the first illustration, a downward pull on the rope produces an equal upward movement of the load. This simple pulley changes the direction of motion, but it provides no mechanical advantage. Mechanical advantage is usually written as a ratio. When there is no mechanical advantage the ratio is 1:1.

In the second drawing, the pulley now moves with the load. There is no directional change, but note that the load is supported by two ropes. As a result, the load is equally divided between the two supporting ropes, each holding 60 lb of the 120-lb load. When a force of 60 lb is applied to the rope, the load will move upward, and the rope will move twice the distance that the load moves. This pulley system provides a mechanical advantage of 2:1.

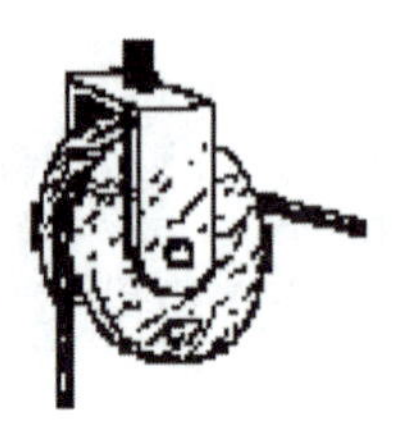

FIGURE 3.19
Pulley.

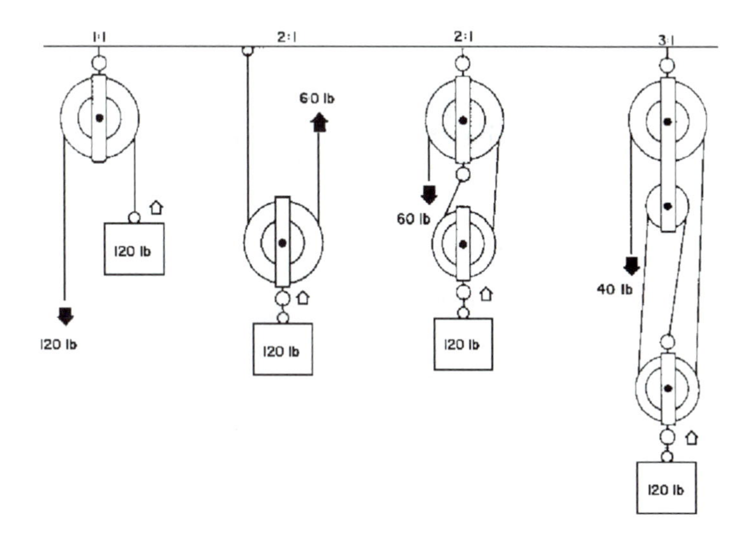

FIGURE 3.20
Block and tackle used to increase mechanical advantage.

The third illustration adds a fixed pulley to the second illustration. This provides a change of direction but does not change the 2:1 mechanical advantage.

The last illustration adds a third pulley. Because three ropes now support the load, each rope supports 40 lb. The mechanical advantage is 3:1, and a force of 40 lb must be applied to a distance of 3 ft to move the load up 1 ft.

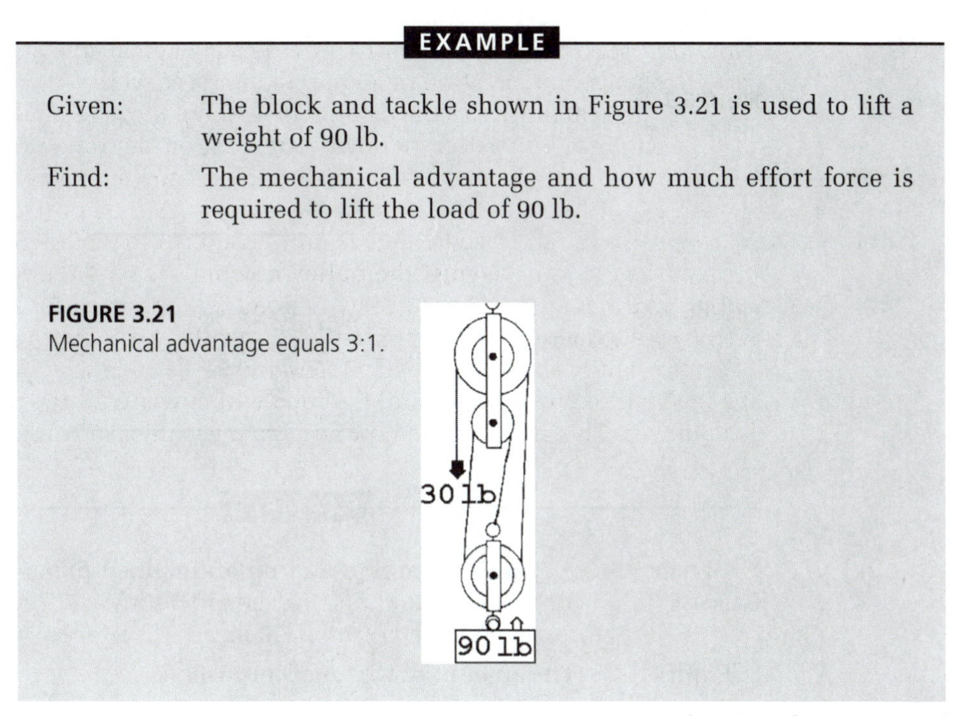

EXAMPLE

Given: The block and tackle shown in Figure 3.21 is used to lift a weight of 90 lb.

Find: The mechanical advantage and how much effort force is required to lift the load of 90 lb.

FIGURE 3.21
Mechanical advantage equals 3:1.

(continued on next page)

SOLUTION

Because there are three ropes equally supporting the weight of the load, the ratio is 3:1. Because the ratio is 3:1, input force will equal output force divided by the ratio.

$$\text{Input force} = 90 \text{ lb} \div 3$$

$$\text{Input force} = 30 \text{ lb}$$

Inclined Plane

Inclined planes were used extensively in ancient Egypt. They gave the Egyptians the mechanical advantage needed to move the heavy stones required to build the pyramids (see Figure 3.22). The last two simple machines—the wedge and the screw—are adaptations of the inclined plane. The mechanical advantage of an inclined plane is determined by comparing the length of the inclined plane to the vertical height the load has been moved. Lifting a 250-lb drum and putting it on a 2-ft-high platform is a difficult task for the average person. By using an inclined plane to achieve a mechanical advantage of 5:1, it requires an effort force of only 50 lb to move the drum to the platform (Figure 3.23).

Remember, the amount of work done is the same, whether the drum is picked up and set on the platform or you use an inclined plane, exerting a force of 50 lb over a longer distance.

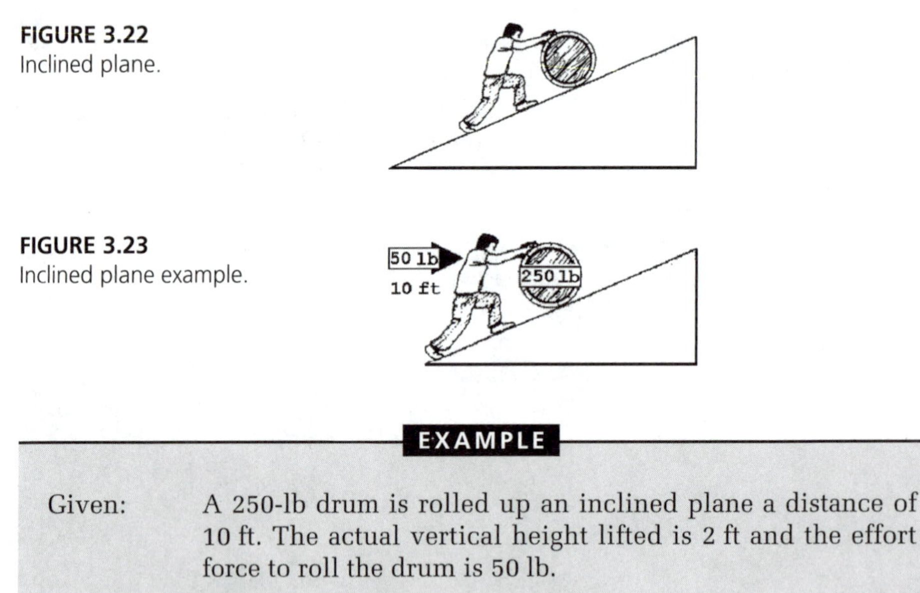

FIGURE 3.22
Inclined plane.

FIGURE 3.23
Inclined plane example.

EXAMPLE

Given: A 250-lb drum is rolled up an inclined plane a distance of 10 ft. The actual vertical height lifted is 2 ft and the effort force to roll the drum is 50 lb.

Find: The amount of work accomplished.

SOLUTION

Work is equal to force times distance.

$$W = F \times D$$
$$W = 50 \text{ lb} \times 10 \text{ ft}$$
$$W = 500 \text{ ft-lb}$$

Note: We found the value for work by using *input* force times *input* distance. We could have used *output* force times *output* distance to get the same value because we haven't taken into account any losses due to friction, which is discussed later.

Wedge

The wedge is so much like the inclined plane that it is often forgotten as a simple machine. For all practical purposes it can be considered as two inclined planes put together. The wedge is just a small inclined plane used as a tool (see Figure 3.24). The splitting wedge is a tool that uses the force imparted by a hammer to split firewood. In industry, tapered bushings are used with pulleys, sheaves, sprockets, and couplings to ensure a secure, positive grip on a shaft.

FIGURE 3.24
The wedge.

Screw

The principle of the screw is used wherever objects have threads. A screw can be thought of as an inclined plane wrapped around a shaft (see Figure 3.25). Extremely high mechanical advantages can be obtained by using a large number of threads per inch on a bolt. As with the inclined plane, the ratio can be calculated by taking the total distance of the incline and dividing it by the vertical distance traveled. Another common example is a propeller. Whether on a boat or an airplane, a propeller is governed by the same rules as a simple screw.

FIGURE 3.25
The screw.

3-7 MECHANICAL EFFICIENCY

So far, all the examples and problems we have discussed have used *theoretical mechanical advantage*, or TMA. This ratio has been expressed in several different ways:

- The relationship between the length of the force arm and the length of the load arm.
- The relationship between the distance the effort force moves and the distance the load moves.
- The relationship between the length of the incline and the rise of the incline.
- The relationship between the radius of the wheel and the radius of the axle.

We have also discussed the methods to find other unknowns about a machine by using the formula

$$\text{Input work} = \text{output work}$$

In the real world, the input work is *always* greater than the output work because no machine is 100% efficient. Actually, the formula is

$$\text{Input work} = \text{output work} + \text{friction loss}$$

In practice, the *actual mechanical advantage,* or AMA, takes into account these losses. The ratio of actual output work and input work is called *efficiency*, and it can be found by the formula

$$\%\ \text{efficiency} = \frac{\text{actual output work}}{\text{actual input work}} \times 100\%$$

Efficiency can also be found using TMA and AMA:

$$\%\ \text{efficiency} = \frac{\text{AMA}}{\text{TMA}} \times 100\%$$

EXAMPLE

Given: The block and tackle shown in Figure 3.26 is used to lift a load of 600 lb 1 ft high. The machine has to pull with an effort force of 104 lb.

Find: Efficiency.

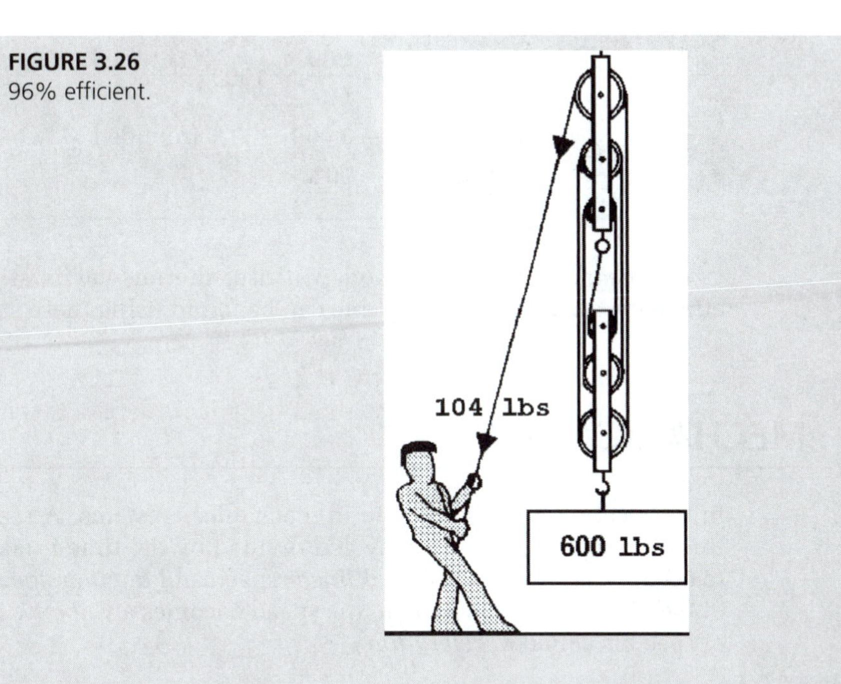

FIGURE 3.26
96% efficient.

SOLUTION

Because the block and tackle has six ropes supporting the load, the TMA is 6:1. The AMA can be calculated by dividing the actual output force by the actual input force.

$$\text{AMA} = \frac{600}{104} = 5.77 \text{ (rounded to two decimal places)}$$

Therefore,

$$\%\ \text{efficiency} = \frac{\text{AMA}}{\text{TMA}} \times 100\%$$

$$\%\ \text{efficiency} = \frac{5.77}{6.0} \times 100\%$$

$$\%\ \text{efficiency} = 0.96 \times 100\% \text{ (rounded to two decimal places)}$$

$$\%\ \text{efficiency} = 96\%$$

Alternatively,

$$\%\ \text{efficiency} = \frac{\text{actual work output}}{\text{actual work input}} \times 100\%$$

$$\%\ \text{efficiency} = \frac{600\,\text{lb} \times 1\,\text{ft}}{104\,\text{lb} \times 6\,\text{ft}} \times 100\%$$

(continued on next page)

$$\% \text{ efficiency} = \frac{600}{624} \times 100\%$$

$$\% \text{ efficiency} = 0.96 \times 100\% \text{ (rounded to two decimal places)}$$

$$\% \text{ efficiency} = 96\%$$

No matter which method you use, the answer is still the same. The efficiency of any simple machine can be found using these same methods.

3-8 MECHANICAL RATE

In this section we discuss rate in mechanical systems. A mechanical rate is a linear or angular distance traveled divided by the time it takes the movement to occur. These rates are called *linear speed* and *angular speed.* If the direction of the displacement is known, the speed becomes a *velocity.* The rate of change of speed is called *acceleration.*

Speed and Velocity

We have already discussed the definition for scalar and vector. Let's see how these terms apply to mechanical rate. Speed in mechanical systems is important to the industrial mechanic. Speed is used to define how fast something is moving. It doesn't describe the direction in which an object moves. One unit often used for speed is miles per hour (mi/h). A speed such as 60 mi/h tells us nothing about the direction in which the speed occurs; by definition that makes speed a scalar quantity. Speed is calculated using the distance an object moves and the time it takes to accomplish that movement. Stated in terms of an equation,

$$\text{Speed} = \frac{\text{distance} \quad \text{(how far)}}{\text{total time} \quad \text{(how long)}}$$

Velocity is sometimes used interchangeably with speed. Technically, speed and velocity are not the same. Remember, speed has no direction and is a scalar. Velocity is a speed with direction, which makes velocity a vector quantity. We have already learned that speed is equal to the distance divided by the total time of travel. The formula for speed can be written as follows:

$$S = \frac{d}{t}$$

where S = speed (units are ft/s, mi/h, m/s, or km/h)
d = total distance traveled (units are ft, mi, m, km, and so on)
t = total time traveled (units are s, min, or h)

EXAMPLE

Given: A train travels 90 mi west in 2 h

Find:
a. The speed of the train.
b. The velocity of the train.

SOLUTION

a. the equation for speed is

$$S = \frac{d}{t}$$

$$S = \frac{90\,\text{mi}}{2\,\text{h}}$$

$$S = 45\ \text{mi/h}$$

b. Velocity is speed with a direction (a vector). The velocity is 45 mi/h west.

Sometimes we need to find the average of two or more different speeds. This can be accomplished by using the following formula:

$$S_{ave} = \frac{S_1 + S_2 + S_3 + \cdots}{n}$$

where S_{ave} = average speed
S_1, S_2, S_3 = speeds to be averaged
n = number of speeds averaged

Because we are still finding speed when we find average speed, the units for average speed are those used for speed.

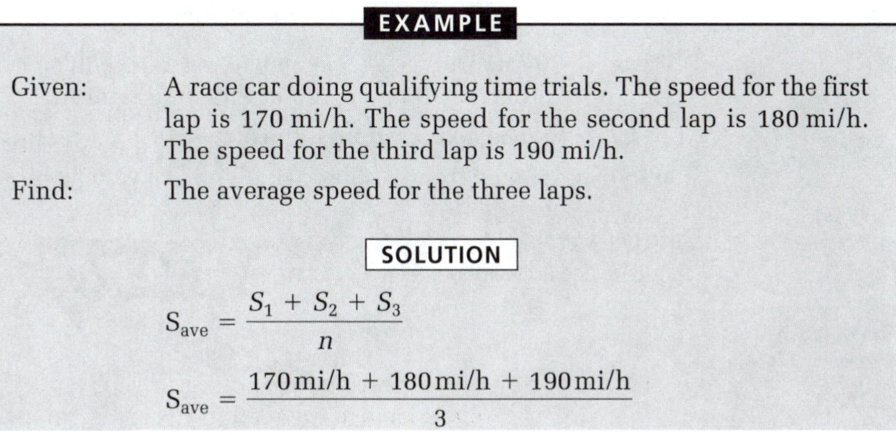

EXAMPLE

Given: A race car doing qualifying time trials. The speed for the first lap is 170 mi/h. The speed for the second lap is 180 mi/h. The speed for the third lap is 190 mi/h.

Find: The average speed for the three laps.

SOLUTION

$$S_{ave} = \frac{S_1 + S_2 + S_3}{n}$$

$$S_{ave} = \frac{170\,\text{mi/h} + 180\,\text{mi/h} + 190\,\text{mi/h}}{3}$$

(continued on next page)

$$S_{ave} = \frac{540\,\text{mi/h}}{3}$$

$$S_{ave} = 180\ \text{mi/h}$$

Acceleration is another quantity we often need to know in mechanical systems. When a speed is increasing, acceleration tells us how fast the speed is changing. *Deceleration* describes a decreasing speed. Both acceleration and deceleration are calculated using the same formula and tell us how fast the speed is changing. If the answer is a positive number, then acceleration is indicated. A negative number indicates deceleration.

$$\text{Acceleration} = \frac{\text{final speed} - \text{initial speed}}{\text{time elapsed}}$$

$$a = \frac{S_f - S_i}{t}$$

where a = acceleration (or deceleration if negative)
S_f = final speed
S_i = initial speed
t = time elapsed

Units for acceleration are units of speed divided by units of time. The acceleration of a falling object due to gravity is 32 ft/s^2. This means that the speed increases by 32 ft/s in 1 s (this is why seconds are squared).

In the SI system the common unit for acceleration is meters/second squared, or m/s^2.

EXAMPLE

Given: An airplane accelerates to reach takeoff speed from a rolling start of 10 ft/s to 120 ft/s in 40 s (Figure 3.27).

Find: The acceleration of the airplane.

FIGURE 3.27
Find airplane acceleration.

SOLUTION

$$a = \frac{S_f - S_i}{t}$$

$$a = \frac{120\,\text{ft/s} - 10\,\text{ft/s}}{40\,\text{s}}$$

$$a = \frac{(120 - 10)\,(\text{ft})}{40\,(\text{s})\,(\text{s})}$$

$$a = \frac{110}{40}\,\text{ft/s}^2$$

$$a = 2.75\ \text{ft/s}^2$$

The airplane accelerates at a rate of 2.75 ft/s during each second of the 40-s elapsed time.

Rotational Speed

So far, we have discussed speed only in a straight line, or linear speed. In mechanical systems rotational, or angular, speed is also very important. Unlike linear speed or rates, angular speed involves rotating an object around its own axis (see Figure 3.28). The equation for angular speed is very similar to that for linear speed:

$$\text{Angular speed} = \frac{\text{angular distance}}{\text{total time}}$$

Common units for angular speed are revolutions per minute (rev/min) and radians per second (rad/s). Angular distance can be the number of revolutions the object turns or the number of radians of an object's rotation. Total time

FIGURE 3.28
Angular speed is rotation.

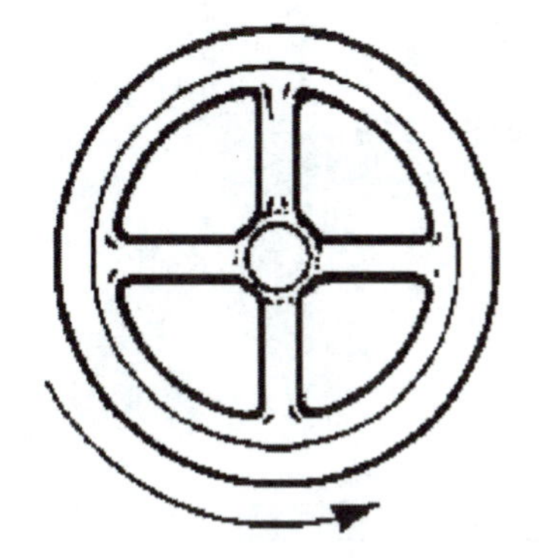

is usually measured in seconds, minutes, or hours. The formula for angular speed is also similar to linear speed:

$$\omega = \frac{\theta}{t}$$

where ω = angular speed
θ = revolutions, or angle moved through
t = time elapsed

Angular speed (ω, the Greek letter omega) is a scalar quantity. Because velocity is a vector quantity, it must have a direction. There are only two choices for direction in angular velocity, clockwise and counterclockwise.

EXAMPLE

Given: A gearbox turns a sprocket clockwise 150 rev in 5 min (see Figure 3.29).

Find: a. Angular speed

b. Angular velocity

SOLUTION

a. First, find angular speed.

$$\omega = \frac{\theta}{t}$$

$$\omega = \frac{150\,\text{rev}}{5\,\text{min}}$$

$$\omega = 30 \text{ rev/min} \qquad \text{(angular speed)}$$

b. To find angular velocity, use the speed and add the direction. The angular velocity in this example is 30 rev/min (speed), in a clockwise direction.

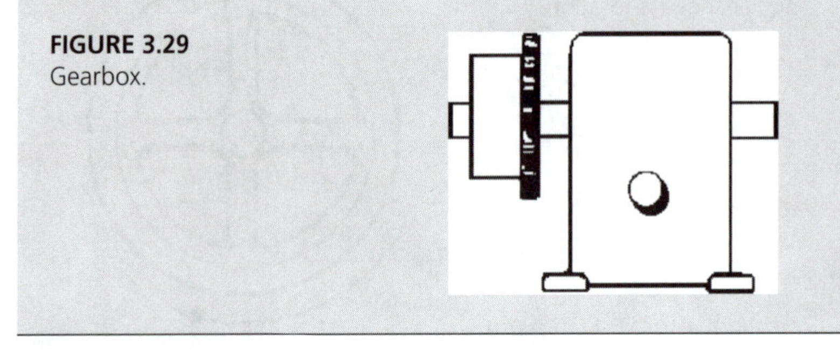

FIGURE 3.29
Gearbox.

Sometimes we need to convert revolutions per minute to radians per second. This is easily accomplished using the following facts:

$$1 \text{ rev} = 360° = 2\pi \text{ rad}$$
$$1 \text{ min} = 60 \text{ s}$$

Because 2π rad = 1 rev, we have the following:

$$1\,\text{rev/min} = \frac{(1\,\text{rev})\,(2\pi\text{rad})}{(1\,\text{min})\,(1\,\text{rev})}$$

Because 1 min = 60 s, this can be added without changing the equation:

$$1\,\text{rev/min} = \frac{(1\,\text{rev})\,(2\pi\text{rad})\,(1\,\text{min})}{(1\,\text{min})\,(1\,\text{rev})\,(60\,\text{s})}$$

Cancel 1 min and 1 rev:

$$1\,\text{rev/min} = \frac{(\cancel{1\,\text{rev}})\,(2\pi\,\text{rad})\,(\cancel{1\,\text{min}})}{(\cancel{1\,\text{min}})\,(\cancel{1\,\text{rev}})\,(60\,\text{s})}$$

$$1\,\text{rev/min} = \frac{2\pi\ \text{rad}}{60\,\text{s}}$$

$$1\,\text{rev/min} = \frac{(2)\,(3.14)\,\text{rad}}{60\,\text{s}}$$

$$1 \text{ rev/min} = 0.105 \text{ rad/s}$$

Any given units can be changed to other units using this method.

Angular acceleration (α, the Greek letter alpha) is calculated in a manner similar to linear acceleration.

$$\text{Angular acceleration} = \frac{\text{final angular speed} - \text{initial angular speed}}{\text{elapsed time}}$$

$$\alpha = \frac{\omega_f - \omega_i}{t}$$

where α = angular acceleration (rev/min^2 or rad/s^2)
ω_f = final angular speed (rev/min or rad/s)
ω_i = initial angular speed (rev/min or rad/s)
t = elapsed time

Let's look at an example.

EXAMPLE

Given: An electric power-plant turbine accelerates from an idle of 50 rad/s to full speed of 500 rad/s in 30 s.

Find: The angular acceleration produced.

SOLUTION

$$\alpha = \frac{\omega_f - \omega_i}{t}$$

$$\alpha = \frac{500\,\text{rad/s} - 50\,\text{rad/s}}{30\,\text{s}}$$

$$\alpha = \frac{450\,\text{rad}}{30\,(\text{s})\,(\text{s})}$$

$$\alpha = 15\ \text{rad/s}^2$$

If the answer in such an example is a negative number, it is called *negative acceleration,* or *deceleration.*

3-9 MECHANICAL RESISTANCE

In mechanical systems, resistance occurs in all types of machinery. We have already studied force; mechanical resistance can be thought of as opposition to mechanical force. In this section, we discuss a mechanical resistance commonly known as *friction.* Friction can be defined as the resistance opposing any effort to slide or roll one object on another. Friction affects all six simple machines we have studied. Because of friction, machines never operate at 100% efficiency; therefore, input is always greater than output.

The results of friction are as follows:

- Energy loss
- Generation of waste heat
- Wear on parts

Friction is commonly felt to be a detriment in all mechanical systems, but friction also provides some very important functions. Without friction, brakes would not stop an automobile, clutches would not work, belts would not grip pulleys, and tires would just spin because they would never get the traction that friction provides. In fact, you could not walk without friction because your feet would slip as you push your body forward, just as if you were on super slippery ice.

There are three main causes of friction in mechanical systems:

1. Surface finish.
2. Cohesion and adhesion of molecular attraction between surface materials.
3. Weight or force holding surfaces together.

A fourth possible cause may be electrostatic attraction, but we discuss only the first three.

Surface Finish

The smoothness, or finish, of a surface greatly affects how much friction it produces. The intermeshing of surface irregularities causes grinding action similar to sandpaper on a smaller scale. This results in the surface being worn away and heat being generated.

Cohesion and Adhesion of Molecules

When all the air is removed from between two surfaces of the same material, they tend to stick together and become very hard to separate. Thus cohesion occurs on a molecular level. Adhesion occurs between materials that are unlike each other. A popular example is adhesive tape.

Weight or Force Holding Surfaces Together

This cause of friction is self-explanatory. The more an object weighs, the harder it is to push because of the increased friction created by its weight. Sanding a piece of wood is another example. The harder you press against the sandpaper, the more difficult it becomes to move it. The two types of friction we need to deal with are static friction and kinetic friction. If you try to slide a heavy box across the floor, you will discover that it takes more force to start it sliding than to keep it moving (see Figure 3.30). The initial resistance to

FIGURE 3.30
Friction resists movement.

beginning movement between two contacting surfaces is called static friction. When an object is moving, the resistance opposing continued movement is called kinetic friction. Kinetic friction is always less than static friction.

3-10 MECHANICAL POWER

During the Industrial Revolution, an engineer named James Watt made substantial improvements to the steam engine so that it became widely accepted as a replacement for horses to do work (see Figure 3.31). He needed a standard unit for mechanical power to compare the output of different steam engines. Because the horse was the accepted "engine" of that time, Watt did tests to see what the average rate of work a horse could produce. He found, on the average, a horse could move 550 lb for a distance of 1 ft in 1 s, or, in other words, it can work at a rate of 550 ft-lb/s. He called this rate of doing work *one horsepower* (abbreviated 1 hp).

$$1 \text{ hp} = 550 \text{ ft-lb/s}$$

In the SI system, since work is measured in newton-meters (N·m), the unit for power is newton-meters per second (N·m/s). Because a joule (J) equals a newton-meter, power in SI is also expressed in joules/second (J/s).

The unit of power for electricity was later named the *watt* in honor of James Watt, for his work with mechanical power. Watts can be converted to horsepower by the following equation:

$$1 \text{ hp} = 746 \text{ W} = 550 \text{ ft-lb/s}$$

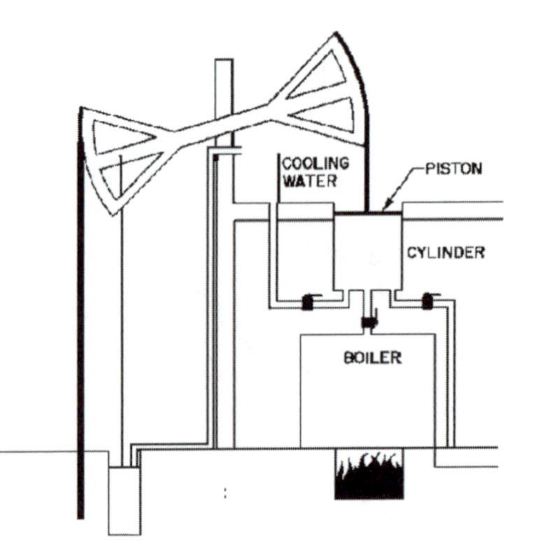

FIGURE 3.31
Early steam engine.

Watts can also be converted to SI units using an easy conversion:

$$1\ \text{W} = 1\ \text{N·m/s}$$

or

$$1\ \text{W} = 1\ \text{J/s}$$

Recall that force is a push or pull on an object and work is equal to force times the distance traveled:

$$W = F \times D$$

Power is the *rate* of accomplishing work and can be found with the equation

$$P = W/t$$

where P = power (ft-lb/s, N·m/s, or J/s)
W = work done (ft-lb, or N·m, or J)
t = time needed to accomplish work (usually s)

Because work equals force times distance, W in the formula above can be replaced with $F \times D$ to get another formula for power:

$$P = \frac{F \times D}{t}$$

EXAMPLE

Given: A winch lifts a load of 500 lb to a height of 5 ft in 5 s (Figure 3.32).

Find: Power, in horsepower.

SOLUTION

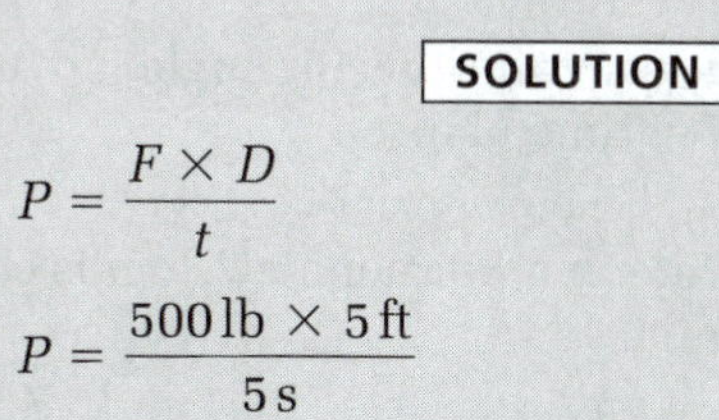

$$P = \frac{F \times D}{t}$$

$$P = \frac{500\,\text{lb} \times 5\,\text{ft}}{5\,\text{s}}$$

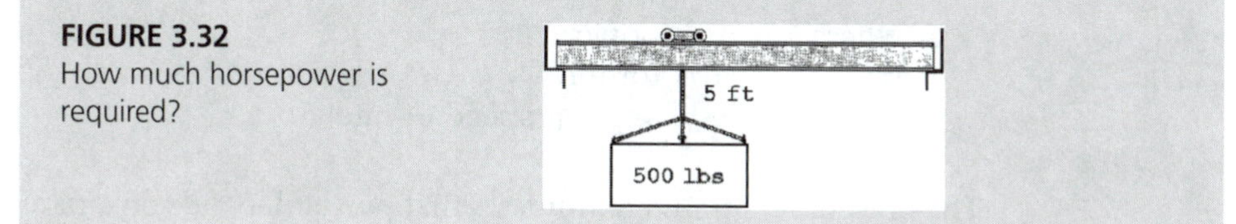

FIGURE 3.32
How much horsepower is required?

(continued on next page)

$$P = \frac{(500)\,(5)}{5}\text{ ft-lb/s}$$

$$P = 500\text{ ft-lb/s}$$

Because 1 hp = 550 ft-lb/s

$$P = \frac{500\text{ ft-lb/s}}{1} \times \frac{1\text{ hp}}{550\text{ ft-lb/s}}$$

$$P = \frac{500\,\cancel{\text{ft-lb/s}}}{1} \times \frac{1\text{ hp}}{550\,\cancel{\text{ft-lb/s}}}$$

$$P = \frac{500}{550} \times 1\text{ hp}$$

$P = 0.91$ hp (rounded off to two decimal places)

The winch produces 0.91 hp.

Remember the formula for work in rotational mechanical systems is

$$\text{Work} = \text{torque} \times \text{revolutions (or angle moved through)}$$

$$W = T \times \theta$$

Because $P = W/t$ and $W = T \times \theta$, W can be replaced to give the basic formula for power in rotational systems:

$$P = \frac{T \times \theta}{t}$$

where
P = power
T = torque
θ = revolutions (or angle moved through)
t = time elapsed

Because the rate in a rotational system is equal to θ/t, it can be substituted in the power formula:

$$P = T \times \omega$$

where
P = power
T = torque
ω = angular speed or rate

The preceding formula reminds us that power depends on a rate, because angular speed is a rate.

EXAMPLE

Given: An electric motor has a shaft torque of 1.2 lb-ft as the shaft rotates at 1,800 rev/min (see Figure 3.33).

Find: The horsepower developed at the motor shaft.

SOLUTION

Using the formula

$$P = T \times \omega$$

first find the angular speed (ω) in rad/s:

$$\omega = 1{,}800 \frac{\text{rev}}{\text{min}} \times \frac{1\,\text{min}}{60\,\text{s}} \times 6.28 \frac{\text{rad}}{\text{rev}}$$

Rearranging:

$$\omega = \frac{1{,}800 \times 6.28}{60} \frac{\text{rev}}{\text{min}} \times \frac{\text{min}}{\text{s}} \times \frac{\text{rad}}{\text{rev}}$$

Canceling units:

$$\omega = \frac{1{,}800 \times 6.28}{60} \frac{\cancel{\text{rev}}}{\cancel{\text{min}}} \times \frac{\cancel{\text{min}}}{\text{s}} \times \frac{\text{rad}}{\cancel{\text{rev}}}$$

$$\omega = 188.4 \text{ rad/s}$$

Next, find power:

$$P = T \times \omega$$

$$P = 1.2 \text{ lb-ft} \times 188.4 \text{ rad/s}$$

$$P = 226 \text{ ft-lb/s (dropping rad)}$$

Because 1 hp = 550 ft-lb/s

$$P = \frac{226 \text{ ft-lb/s}}{1} \times \frac{1\,\text{hp}}{550 \text{ ft-lb/s}}$$

$$P = \frac{226 \cancel{\text{ft-lb/s}}}{1} \times \frac{1\,\text{hp}}{550 \cancel{\text{ft-lb/s}}}$$

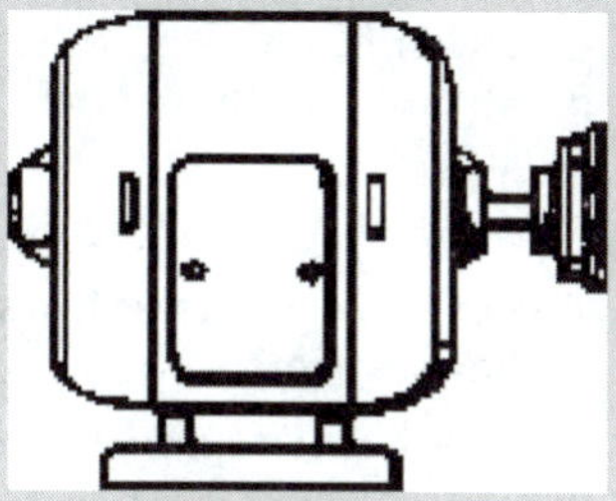

FIGURE 3.33
Electric motor.

(continued on next page)

$$P = \frac{226}{550}\text{ hp}$$

$$P = 0.41\text{ hp}$$

QUESTIONS

1. What are the six forms of energy?
2. What is kinetic energy?
3. What is potential energy?
4. What units are used to measure mechanical force in the SI and English systems?
5. What does the term *equilibrium* mean?
6. What is the difference between a scalar and a vector?
7. Define the term *torque.*
8. What is mechanical work?
9. Convert 180° to radians.
10. How many degrees are in 1 rad?
11. What are the six simple machines?
12. What is the main purpose of a simple machine?
13. A man uses a 10-ft lever to move a 500-lb stone. The man has to apply 100 lb of force to get the stone to move. What is the mechanical advantage?
14. Give an example of a first-, second-, and third-class lever.
15. A women applies a 5-lb force to a 24-in.-diameter wheel mounted on a 1-in.-diameter shaft. What is the mechanical advantage of this wheel and axle?
16. What is the torque applied to the axle in Question 15?
17. A block and tackle is used to lift 100 lb. The load is equally supported by four ropes. Assuming no friction, how much effort will it take to raise the load?
18. A 500-lb pushcart is rolled up an inclined plane that is 20 ft long. The vertical distance the object is raised is 4 ft. Assuming no friction, how much effort will it take to push the cart up the incline?
19. Label each of the following values as either a scalar or a vector quantity:
    a. 75° F
    b. 100 kg
    c. 80 km/h, south
    d. 50 mi/h, north
    e. 10 ft/min
20. A car travels 130 mi in 3 h. What is the car's average speed?
21. A conveyor accelerates to a speed of 60 ft/min in 3 s. What is the start-up acceleration of the conveyor?
22. Find the angular speed and velocity of a sprocket that turns 200 rev in 3 min clockwise.
23. An elevator lifts 100 lb to a height of 20 ft in 10 s. Find the horsepower required to operate the elevator (neglecting friction).

CHAPTER 4

Lubrication

4-1 INTRODUCTION

Lubrication is used to reduce the effects of friction, to prevent wear and corrosion, and to protect against contamination. The lubricant forms a barrier between moving metal surfaces, decreasing friction and wear. In the past, vegetable oils and animal fats were the primary sources of lubricants. Since the Industrial Revolution, more than 90% of all lubricants have come from petroleum and shale oil. Liquid lubricants are most often applied by mechanical means for better control, usually by rotating rings or chains, immersion devices, splash devices, or pumps. Grease-type lubricants are applied by packing, pressing, or pumping. The method of application can be as important as the proper selection of the lubricant for long machinery life.

4-2 LUBRICATION PRINCIPLES

The primary function of any lubricant is to prevent metal-to-metal contact of moving surfaces, whether they are sliding or rolling (see Figures 4.1 and 4.2). The technician must be familiar with several terms and principles to understand and select the proper lubricant.

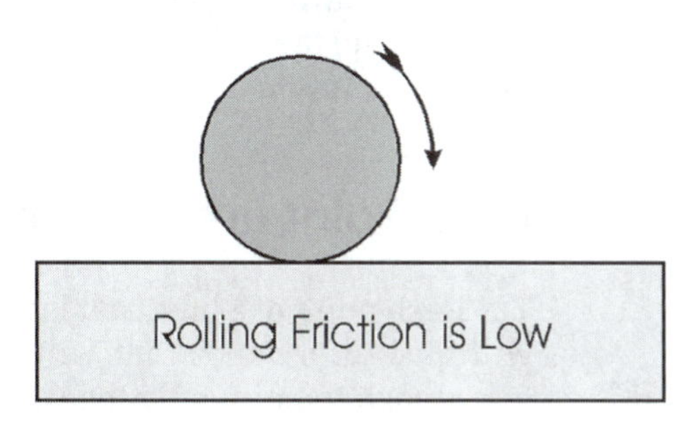

FIGURE 4.1
Rolling friction is low.

FIGURE 4.2
Sliding friction is high.

Viscosity

Viscosity is the property of a fluid, semifluid, or semisolid substance that causes it to resist flow. Technically, viscosity is defined as the shear stress on a fluid element divided by the rate of shear. Lubricants in general become thinner, or less viscous, as the temperature increases, and become thicker, or more viscous, as the temperature decreases. A lubricant's viscosity must have the ability to separate moving parts at the operating temperature of the machine. It would seem that the ideal lubricant should be very thick, or viscous, for maximum separation of metal parts. This is not true. More viscous lubricants require more power to overcome the friction of the oil itself. If the lubricant is extremely viscous, the heat generated by lubricant friction can destroy the lubricant and, ultimately, the machine it serves to lubricate. The ideal lubricant is the one with the lowest viscosity that can maintain an oil film between metal surfaces at the operating temperature of the machine. Lower temperature, lighter loads, and higher speeds generally require thinner, lighter, lower-viscosity lubricants. Higher temperatures, heavier loads, and lower speeds generally require thicker, heavier, higher-viscosity lubricants.

Viscosity Index

The **viscosity index** is a measure of the rate at which the viscosity of a lubricant changes as the temperature changes. The higher the viscosity index, the more resistant the lubricant is to thinning out with increased temperature. Higher-viscosity-index lubricants are more desirable if the operating temperature of the machinery varies. When operating temperatures of the machinery remain constant, the viscosity index is not an important consideration in selecting a lubricant.

Flash Point and Fire Point

The **flash point** of a lubricant is the minimum temperature at which the lubricant will give off a vapor that will "flash" into flame. The lubricant itself is not hot enough to burn, only the vapor or flames. The **fire point** of a lubricant is the minimum temperature at which the lubricant will burn continuously. For this to happen, the lubricant must continuously vaporize to maintain combustion. In any case, for either the flash point or the fire point, combustion is possible only when there is enough oxygen available. Because the fire point temperature is higher than the flash point, this information is sometimes not included on the lubrication label. In general, the lubricant fire point is usually about 86°F (30°C) above the flash point. Fortunately, storage and operating temperatures for most lubricants are low enough to eliminate any possibility of fire, but flash point and fire point are important considerations in higher-temperature environments.

Pour Point

The **pour point** of a lubricant is the point at which the lubricant becomes so thick that it no longer flows. If a lubricant must flow in low-temperature environments, it should have a lower pour point than the coldest expected temperature. Oils containing hydrocarbons tend to form wax crystals at lower temperatures. The wax crystals prevent the oil from flowing. This causes machinery damage if start-up temperatures are lower than the pour point of the oil. A pour point depressant additive can be added to oils operating in extreme lower temperatures. This type of additive modifies the wax crystal structure, resulting in better low-temperature performance and a lower pour point.

Oxidation Inhibitors

Petroleum-based lubricants are made up of a compound of hydrogen and oxygen called *hydrocarbons*. When hydrocarbons are exposed to air and heat, they absorb oxygen from the air. This process of oxidation causes chemical changes in the lubricant, resulting in thickening lubricant, increasing acidity, and metal corrosion. **Oxidation inhibitors** are additives that slow the rate of a lubricant's natural tendency to oxidize.

Rust and Corrosion Inhibitors

Rust inhibitors are additives that improve a lubricant's ability to stick to metal surfaces. This coating action protects the metal surfaces from oxidation and rust formation by preventing moisture penetrating into the protective film. Rust inhibitors can be found in almost all lubricants.

Detergent and Dispersant Additives

A **detergent additive** holds tiny particles, too small for filtration, in suspension to help keep metal surfaces clean by preventing deposit formation. This is extremely important for internal combustion engines, where high temperatures and combustion wastes produce soot, sludge, and other small particles detrimental to machinery. The purpose of detergent additives is to keep an engine or machine parts clean rather than to remove deposits already in the engine or machine. Therefore, it is important to follow manufacturers' recommendations for oil-change intervals because the eventual depletion of detergent additives will allow contaminants to be deposited throughout the engine or machine. A companion additive called a **dispersant additive** assists the detergent by keeping these contaminants suspended in the oil until the next oil change. The dispersed material is rendered virtually nonabrasive and does not damage any bearing surfaces if they are properly maintained.

Extreme-Pressure (EP) Additives

Extreme-pressure additives are commonly found in almost all gear lubricants. This chemical compound increases the load-carrying ability of the lubricant to cushion the shock and rubbing action associated with gearboxes. Extreme-pressure additives are most effective in reducing friction and wear at high temperatures. When gear teeth mesh, extremely high temperatures can cause galling, scoring, seizure, and accelerated wear. The extreme-pressure additive forms a protective, easily sheared film between the gear teeth. Sulfur, phosphorus, and chlorine are common EP additives. Sometimes combinations of compounds, such as sulfur-phosphorus, are used to improve the performance of the additive.

Antifoam Additives

Detergent additives tend to trap air and can cause increased foaming in lubricants. Entrapped air is detrimental to lubricating qualities of oil. It also reduces the efficiency of hydraulic systems. **Antifoam additives** break up the air bubbles, preventing the oil from foaming.

4-3 LUBRICANT FILM

The lubricant film separates the surfaces of two rubbing or sliding parts. How well this film separates these parts determines how much friction and wear will occur. The following are the four main types of lubricant film conditions:

1. No lubricant film (dry friction condition)
2. Mixed lubricant film (partial hydrodynamic condition)
3. Boundary lubricant film
4. Full lubricant film (full hydrodynamic or hydrostatic lubrication)

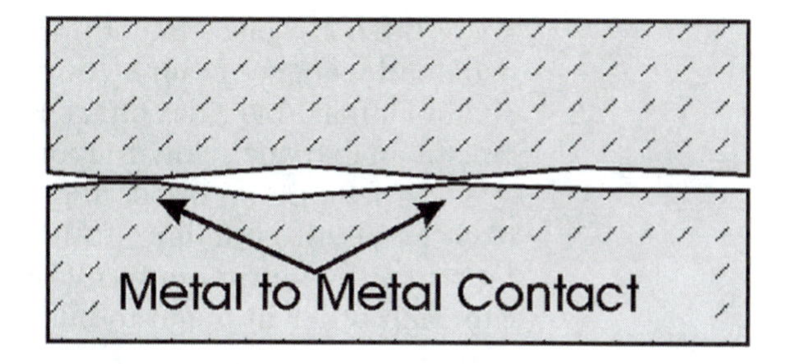

FIGURE 4.3
Metal-to-metal contact is not good.

No lubricant film describes the condition in which there is absolutely no lubricant between two surfaces. In reality, this condition is more theoretical than practical, because most materials have some kind of film that reduces friction, even if it is only dust.

Mixed lubricant film describes the condition in which there is a thin lubricating film. Because of the film's thinness or the surface's roughness (or a combination of the two), a heavy load or shock causes metal-to-metal contact of the high points of the surfaces (see Figure 4.3). This also occurs with ball bearings. As the ball runs along the raceway, a film of oil separates the metal surfaces. However, when a shock load occurs, the lubricant is momentarily forced away from the point of high-pressure contact, allowing metal-to-metal contact of high points. This condition is also called *partial hydrodynamic lubrication.*

Boundary film lubrication is very similar to mixed lubricant film except the film in boundary lubrication is somewhat thicker, but it is not thick enough to afford full protection from metal-to-metal contact. In boundary lubrication, surface high points are separated, but only by the thickness of one lubricant molecule. Sometimes an additive is included in the lubricant that acts to coat the surfaces so that metal-to-metal contact is reduced.

Full lubricant film describes the lubrication condition in which the two bearing surfaces are completely separated by a film of lubricant at all times. Both bearing surfaces are in contact with the lubricant only, even the high points of the surfaces. There are two types of full-film lubrication, **hydrodynamic** and **hydrostatic.** With *hydrodynamic lubrication,* the coating action of the lubricant and the rotation of the shaft keep it in place, separating the metal surfaces. In Figure 4.3, note that metal-to-metal contact exists when the shaft is not moving; however, shaft rotation causes the hydrodynamic action, and the shaft becomes fully supported by a full film of oil (see Figure 4.4). *Hydrostatic lubrication* also prevents metal-to-metal contact with a full oil film, but it is kept in place by an outside pressure source. An example is an oil pump in an engine. The pump

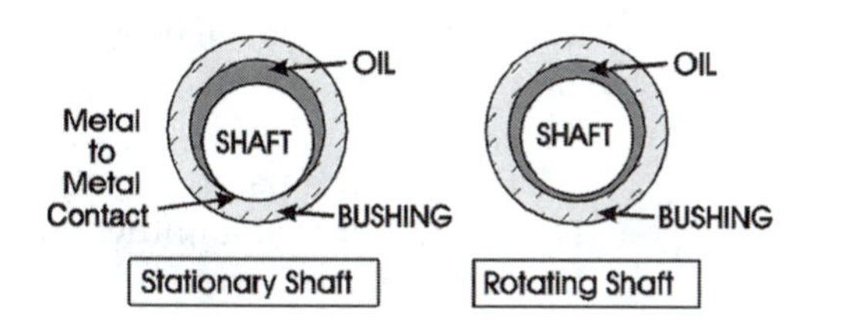

FIGURE 4.4
Hydrodynamic lubrication.

keeps the oil in place under pressure, thus avoiding metal-to-metal contact and, eventually, engine failure. An industrial application is air bearings, where the shaft of a machine rides on a cushion of air pumped into the bearing area. This creates an extremely low-friction application.

Surface finish of bearing surfaces affects any type of lubrication condition. For hydrodynamic, full-film lubrication, metal high points should be less than the minimum film thickness. If not, the high points of surfaces will rub each other in metal-to-metal contact. The result is wear, higher friction, and increased temperature. Generally, better surface finishes are required for full-film lubrication, whether hydrodynamic or hydrostatic, to avoid metal-to-metal contact. Boundary film lubrication surface finishes don't have to be as good as full film because the "wear-in" smoothes out the high points. Better surfaces are usually needed for higher rotational speeds, higher loads, and harder materials.

4-4 INDUSTRIAL OIL LUBRICATION

Industrial lubricating oils can be classified into three main types:

1. Animal and vegetable oils
2. Mineral or petroleum-based oils
3. Synthetic oils

The Egyptians used animal fat to lubricate their chariot axles 3,500 years ago. Vegetable oils come from several different plants and seeds; some examples include olive oil, cottonseed oil, linseed oil, castor oil, and soybean oil. Animal and vegetable oils do not work well with today's demanding machinery. They tend to break down chemically and are typically unstable.

Mineral oils are refined from crude petroleum oil. Since prehistoric times, people have used petroleum products for different purposes. They were first obtained from surface pools of crude oil and tar. In 1859, the first oil well was drilled. Since then, petroleum-based lubricants have been the most widely used lubricants in the world. Today, refined crude oil produces many types of lubricating oils. Crude oil from different locations can vary greatly. Different types of crude oil can vary from light grades, excellent for gasoline, to block crudes, which can be almost solid, like asphalt. It is estimated that crude oil contains more than a million different hydrocarbon compounds. Only about a thousand of these have been cataloged. There are three basic types of common crude oil stock:

1. Paraffinic: a paraffin-base crude oil
2. Naphthenic: a naphthene-base crude oil
3. Asphaltic: an asphalt-base crude oil

Paraffinic oils have a relatively high viscosity index, which means their viscosity changes less with temperature fluctuations than other oils do. Paraffin-base oils typically have a higher pour point, making them less suitable for lower-temperature environments. They also tend to contain waxes if not completely refined. This type of crude oil can typically be found in Pennsylvania oil wells.

Naphthenic oils, unlike paraffinic types, have a lower viscosity index. Their viscosity varies as the temperature changes. Again unlike paraffinic oils, naphthene-base oils have a low pour point. They contain almost no wax and flow at low temperatures, making them a good choice for refrigeration oils. A very important property of these oils when compared to paraffin-base oils is that they are less likely to form hard-carbon deposits in high temperature environments. Naphthenic crudes are typically found in oil wells around the Gulf of Mexico.

Asphaltic oils come from black crudes and contain heavy tarlike materials. They produce a heavy, inexpensive, lubricant that typically has been used to lubricate slow-speed open gears. Asphalt-base oils are often used in roofing materials, highway-paving materials, and even as fuel in some power plants.

Some experts would add a fourth type of crude called *mixed-base oils*. Most crude oil contains a mixture of paraffinic and naphthenic compounds that can range from highly paraffinic with little naphthenic to just the opposite. These mixed-base oils may also contain asphaltenes, resins, and aromatics. Modern refineries can separate any of these types of oils. The degree of refining depends on the type of crude oil, the demand for better quality, and the cost. Oils are usually blended to produce specific oils for different applications. For example, a paraffin-base oil may be blended with a naphthene-base oil to increase the viscosity index to be used in an environment with changing temperatures.

Synthetic oils became important during World War II, when petroleum products became difficult to obtain. Some of the problems with synthetics were high cost, incompatibility with seal and gasket materials, incompatibility with petroleum-base oils, and an inability to readily accept additives. Most of these problems have been solved by modern technology. Today's synthetics provide superior performance at extremely low and high temperatures. They are much more efficient and fire resistant. They provide greater protection from oil breakdown and deterioration and have a longer service life than petroleum counterparts. Synthetic oils can greatly reduce friction, because the molecules of synthetic oil are all the same size (see Figure 4.5). Petroleum oils have a complex mixture of different-size molecules containing impurities that are difficult and expensive to completely remove when refined. The uniform molecules in synthetic oil make them slicker, reducing friction and increasing efficiency.

FIGURE 4.5
Synthetic oil molecules are more uniform than petroleum oil molecules.

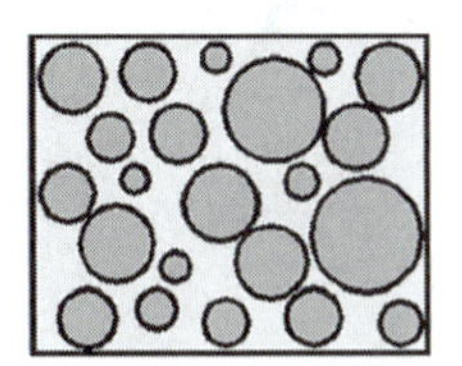

Petroleum Oil Molecules

Synthetic Oil Molecules

4-5 OIL VISCOSITY CLASSIFICATIONS

Oil viscosity is the first thing to consider when selecting the correct oil for your application. It is important that you understand oil viscosity classifications and the difference between them.

In the United States, industrial lubricants' viscosity is measured in Saybolt universal seconds (SUS) or **Saybolt seconds universal (SSU).** Both names are used and mean the same thing. The American Society for Testing and Materials (ASTM) is the governing body for this standard. SSU viscosities are determined by a Saybolt universal **viscometer** (also called *viscosimeter*). A standard container called a *Saybolt flask* is filled from another container through an orifice (see Figure 4.6). The amount of time needed to fill the Saybolt flask to a mark etched on its neck determines the SSU viscosity of the oil. An oil that takes 60 s to fill a Saybolt flask is rated at a viscosity of 60 Saybolt seconds universal, or 60 SSU. In addition, to maintain accuracy, a standard temperature of 100°F (38°C) was established. If the oil takes more than 1,000 s to fill the flask, the test temperature of the oil is increased to 210°F (99°C).

In recent years, a new international standard—ISO 3448—established a series of lubricant viscosity grades (VG) based on kinematic viscosities rated at temperature standards of 40°C (104°F) and 100°C (212°F). This international standard uses *centistokes* (cSt) and Celsius metric measures for specifications, classifications, and various petroleum tests (see Table 4.1). Kinematic viscosities are commonly established using glass capillary tubes and timing the flow of oil at temperature standards. ISO standard 3448 has been widely accepted by the petroleum industry throughout the world.

Another oil viscosity classification has been established by the Society of Automotive Engineers (SAE) for use with engine oils. Table 4.2 gives

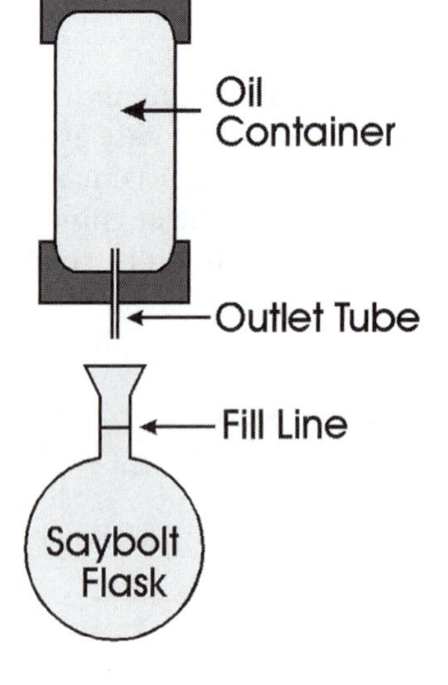

FIGURE 4.6
Saybolt universal viscometer.

TABLE 4.1
Comparison of different viscosity grade systems. (Approximate midpoint viscosity at 40°C/104°F)

ISO Standard Viscosity Number	Centistokes	SUS
ISO VG 2	2	32
ISO VG 3	3	36
ISO VG 5	5	40
ISO VG 7	7	50
ISO VG 10	10	60
ISO VG 15	15	75
ISO VG 22	22	105
ISO VG 32	32	150
ISO VG 46	46	215
ISO VG 68	68	315
ISO VG 100	100	465
ISO VG 150	150	700
ISO VG 220	220	1,000
ISO VG 320	320	1,500
ISO VG 460	460	2,150
ISO VG 680	680	3,150
ISO VG 1000	1,000	4,650
ISO VG 1500	1,500	7,000

TABLE 4.2
Comparison of SAE viscosity to centistokes. (Approximate viscosity at 100°C/212°F)

SAE Viscosity Grade	Centistokes
5W	≥ 3.8
10W	≥ 4.1
20W	≥ 5.6
20	≥ 5.6; < 9.3
30	≥ 9.3; < 12.5
40	≥ 12.5; < 16.3
50	≥ 16.3; < 21.9
60	≥ 21.9; < 26.1

approximate comparisons of SAE viscosity grades to ISO centistoke viscosity. Most engine oils today are not single-weight oils but are compounded to react as lightweight oils in cold temperatures and heavyweight oils at high temperatures. These *multigrade* oils can replace as many as four or five single-grade oils. They are designed to lubricate, protect, and run efficiently under a wide range of temperatures, thanks to today's technology. Multigrade oils use a base oil with a lower-viscosity classification to which viscosity-index improvers are added. These additives do not change low-temperature performance but increase viscosity as temperature increases. Typical multigrade oils include 5W-30, 10W-30, 10W-40, 15W-40, 15W-50, and 20W-50. An SAE 5W-30 oil acts like an SAE 5—weight oil with cold temperatures—and thickens to an SAE 30—weight oil under high temperatures.

Remember, viscosity is not a measure of the quality of an oil. All the viscosity classifications discussed measure only viscosity, not quality.

The American Petroleum Institute (API) has established performance ratings and standards for SAE engine oils. The API engine oil service classification system is actually a joint venture with the American Society for Testing and Materials and the Society of Automotive Engineers organizations. The purpose of these API service classifications is to set quality and performance standards for the petroleum industry, the engine manufacturer, and the consumer. There are two categories for the classifications: *S* classifications are for spark ignition or gasoline engines and *C* classifications are for compression ignition or diesel engines. Table 4.3 gives a description of common API service ratings.

Note that newer classifications exceed the performance ratings of older classifications and usually replace them. A later classification can be used in an older engine requiring an older classification, but an older classification should never be used in a newer engine that requires a newer classification.

All automobiles and many industrial machines rely on gears to transfer and transform energy. Gears require special considerations when you select a good lubricant; meshing teeth tend to squeeze and roll lubricant away. Gear teeth must have an oil film heavy enough to dampen shock and prevent metal-to-metal contact. However, if the lubricant viscosity is too high, excessive friction and heat will result in damage to gears and bearings.

TABLE 4.3
American Petroleum Industry service ratings for engine oils.

API S Classifications	API C Classifications
SA: Moderate service for older engines. Obsolete classification.	**CA:** Moderate service using higher-quality fuels in diesel engines introduced in 1940. Obsolete classification.
SB: Light duty and minimum protection. Obsolete classification.	**CB:** Moderate service using lower-quality fuels in diesel engines introduced in 1949. Obsolete classification.
SC: Mild detergent oil recommended for 1964–1967 engines. Obsolete classification.	**CC:** Moderate service in diesel engines with a high ash level that can cause problems in newer engines introduced in 1961. Obsolete classification.
SD: Detergent with some additives introduced in 1968. Obsolete classification.	**CD:** Severe service for turbocharged or supercharged diesel engines using higher-quality fuels introduced in 1955. May soon be obsolete.
SE: Detergent with rust, wear, and oxidation additives introduced in 1971. Obsolete classification.	**CE:** Severe service for turbocharged or supercharged diesel engines that may be used to replace all older classifications. Introduced in 1983.

Continued

TABLE 4.3
continued.

API S Classifications	API C Classifications
SF: Improved SE that may be used where SC, SD, and SE oils are required. Introduced in 1980.	**CF-4:** Severe service for high speed turbocharged or supercharged diesel engines. Improved CE may be used to replace all older classifications. Introduced in 1990.
SG: Provides protection against oil thickening, wear, sludge buildup, varnish, and rust. SG replaces SE, SF, and CC. Introduced in 1989.	**CG-4:** Severe duty for high-speed, four-cycle, diesel engines using fuel with less than 0.5% weight sulfur that can be used in place of CD, CE, and CF-4 oils. Introduced in 1995.
SJ: Improved SG that provides protection against oil thickening, wear, sludge buildup, varnish, and rust. Can be used to replace all older classifications. Introduced in 1996.	**CH-4:** For high-speed, four-cycle, diesel engines designed to meet 1998 emission standards using fuel with up to 0.5% weight sulfur that can be used to replace all older classifications. Introduced in 1998.
SL: For all automotive engines presently in use. SL oils are designed to provide better high-temperature deposit control and lower oil consumption. Some of these oils may also meet the latest ILSAC specification and/or qualify as energy conserving. Introduced in July 1, 2001.	

Because there are many different sizes and configurations of gears, the correct lubricant can vary greatly. Gear oils may be lightweight with no EP (extreme pressure) additives or could be higher viscosity with a high percentage of EP additives. The application governs which gear oil to use; in any case manufacturers' recommendations should always be followed. The API has designated classifications for gear oil types. Table 4.4 gives a general description of these classifications.

TABLE 4.4
American Petroleum Industry service ratings for gear lubricants.

API Designation	General Service Type
GL-1	For spiral-bevel and worm gears under mild service. Straight mineral oil.
GL-2	More severe service than GL-1, with antiwear additive.
GL-3	Moderately severe service with mild EP additive.
GL-4	For hypoid gears in normal service without severe shock loading. Meets MIL-L-2105.
GL-5	For hypoid gears in severe service, including shock loading. Meets MIL-L-2105C, D.

4-6 GREASE LUBRICATION

Grease lubricants account for about 10% of the total lubricants in the world. Oil is considered to be a better lubricant than grease because it produces a more uniform film, generates less friction, and is a better coolant than grease. Grease is generally used in applications where oil would leak out or not reach the point where lubrication is needed. The primary advantages of grease are as follows:

- It allows bearings to be sealed, keeping lubricant in and contaminants out.
- It stays in place where oil would run out.
- It generally withstands higher temperatures.
- It decreases the frequency of lubrication maintenance.

Lubricating grease is typically made up of a petroleum-base oil and a thickening agent called a *soap*. This kind of soap is not the same soap you use to wash your hands. The function of the grease thickener, or soap (also sometimes called the *base* of the grease), is to hold the oil and release it slowly to provide the lubricating action. Grease usually consists of 80 to 90% oil; the second main ingredient is soap, and the third is usually an additive to enhance the grease's lubricating properties. There are many different types of grease. The following are some of the more common ones used in industry:

- Calcium soap grease
- Sodium soap grease
- Lithium soap grease
- Complex soap grease
- Synthetic grease

Calcium Soap Grease

Calcium soap grease was the first grease to be produced in large quantities in the United States. It is still a popular low-quality, low-cost grease. Calcium soap grease is also sometimes called *lime soap grease.* This type of grease contains 1 to 3% water to help combine the soap with the oil. Because of the water content, it cannot be used at temperatures above 180°F (82°C). At higher temperatures, the water evaporates and causes the soap to separate from the oil. This grease works better in lower-temperature, slow-speed applications. It is also water-resistant and works well in wet environments.

Sodium Soap Grease

Sodium soap grease, like calcium soap, is a low-cost, general-purpose grease. It can handle temperatures up to 250°F (121°C). It has a fibrous texture and is good for rust protection. It works well lubricating sliding parts; however, like calcium soap grease, it should be used only for slow-speed applications. Sodium soap grease is water soluble and absorbs water. It should not be used in wet environments.

Lithium Soap Grease

Lithium soap grease was originally developed during World War II for use on aircraft parts. Lithium soap grease is probably the best general-purpose grease and is often referred to as a *multipurpose grease.* It is good for higher temperatures, up to 300°F (149°C). With additives, it is also an excellent cold-environment grease. Lithium soap grease has good water-resistant properties and a smooth butterlike texture. It is widely used in industry in many different applications.

Complex Soap Grease

Complex soap grease uses metal-base thickeners to replace conventional soap thickeners. It is primarily a special-purpose grease and is used for high-temperature applications. It does not work as well as other greases with additives and is usually much more expensive.

Synthetic Grease

Synthetic grease is typically a superior, more expensive grease. In today's applications it is used in general purpose as well as special applications. Synthetic grease is a very heat-resistant, low-oxidation, highly efficient grease. The following are the four main types of synthetic grease:

1. *Organic esters:* Good for wide temperature environments. Found in some aircraft applications.
2. *Polyglycols:* Excellent for high temperatures but has poor water resistance. Found in industrial oven conveyor systems.
3. *Silicones:* Good for high temperature and also for wide temperature range. Has good water resistance. Superior grease but is usually found only in special applications (such as sealed-for-life bearings) because of high cost.
4. *Synthetic hydrocarbons:* Not quite as good as other synthetics but economical in cost. Better than petroleum-base greases in all respects.

Common Grease Extreme-Pressure Additives

Extreme-pressure additives are the most common grease additives. Most grease applications involve shock loads and require the high film strength of EP additives. Chlorine, phosphorus, and sulfur are typical EP additives. Molybdenum disulfide and graphite are also added to grease as EP additives but are actually lubricants themselves. They coat and fill small bearing-surface irregularities, making the surfaces smoother and slicker. When the grease thins, this film remains, preventing metal-to-metal contact. Copper and zinc are EP additives that, as nonferrous powdered metals, prevent ferrous (steel) metals from *seizing* (binding or welding together under extreme pressure loads).

Bearing Lubrication with Grease

The greases available today are much improved compared to the greases of the past. Because of this and of the increased technology of precision-bearing manufacturing, a very small amount of grease can lubricate a bearing for a lifetime. Many modern bearings are *sealed for life* with a high-quality grease (typically synthetic) and never need any maintenance. When bearings are not sealed, they should be relubricated periodically. You should never mix different types of grease. Always follow manufacturers' recommendations and always relubricate with the same type of grease. If the grease type must be changed, the bearing must be flushed out with solvent and cleaned thoroughly before applying new grease. If you are relubricating with the same type of grease, you should open the drain plug, force out *all* the old grease, and refill with the new grease.

Contamination is one of the main causes of bearing failure. Flushing out the old grease also flushes out most contaminants. You should clean grease fittings, clean grease-applying equipment, and keep new grease clean to help avoid contamination in bearings. Do *not* overfill bearings with grease. It is better to underfill than overfill any bearing. Excessive grease causes bearings to overheat; high temperatures cause grease failure and, eventually, bearing failure. After the old grease is forced out by the new grease, rotate the bearing slowly for a short time with the vent plug removed. This will allow the excess grease to be forced out. The following are five basic rules for good grease lubrication:

1. Use the correct type of grease.
2. Use the proper intervals of lubrication.
3. Use the correct type of grease-applying equipment.
4. Use the proper methods of grease lubrication.
5. Use the correct amount of grease.

QUESTIONS

1. What is the primary function of any lubricant?
2. ______________ is the property of a fluid, semifluid, or semisolid substance that causes it to resist flow.
3. Higher temperatures, heavier loads, and lower speeds generally require ______________ viscosity lubricants.
4. ______________ ______________ is a measure of the rate at which the viscosity of a lubricant changes as the temperature changes.
5. What is the flash point of a lubricant?
6. What is an oxidation inhibitor?
7. What does a detergent do for lubricants?
8. What do extreme-pressure additives do for lubricants?
9. Name the four main types of lubricant film conditions.
10. What are the three main types of industrial lubricating oils?
11. In the United States, the viscosity of industrial lubricants is measured in ______________ ______________ ______________ .
12. Hypoid gears in severe service, including shock loading, require an API designation of ______________.
13. Name two advantages of grease over oil.
14. Lubricating grease is typically made up of a petroleum-base oil and a thickening agent called a ______________.
15. What are the five rules for good grease lubrication?

CHAPTER 5

Bearings

5-1 INTRODUCTION

A **bearing** is the part of any machine that is used to reduce the friction of a rotating shaft or the friction between two moving surfaces. When the wheel was invented thousands of years ago, it was mounted on an axle, and a bearing was used to connect the wheel and axle. Bearings became even more important during the Industrial Revolution, when round steel shafts were first used with wooden, bronze, and cast-iron bearings.

There are two main types of bearings: *plain bearings* and *rolling-element bearings*. A **plain, or journal, bearing** is a cylindrical sleeve that supports a rotating or sliding shaft (see Figure 5.1). The inner lining, called the *bushing,* is usually made of a metal softer than that of the shaft so that any wear occurs in the replaceable bushing and not in the shaft.

A **rolling-element bearing,** shown in Figure 5.2, is also called an antifriction bearing because the friction created by this bearing is rolling friction rather than the sliding friction created by the plain bearings. The rolling-element bearing is a cylinder containing a moving inner ring of steel balls or rollers.

Bearings may be used for radial loads, axial loads, or a combination of both. A radial load is a load applied perpendicular to the axis of the shaft (see Figure 5.3). **Axial,** or **thrust, loads** are loads applied parallel to the axis of the shaft (see Figure 5.4).

FIGURE 5.1
Journal bearing.

FIGURE 5.2
Ball bearing.

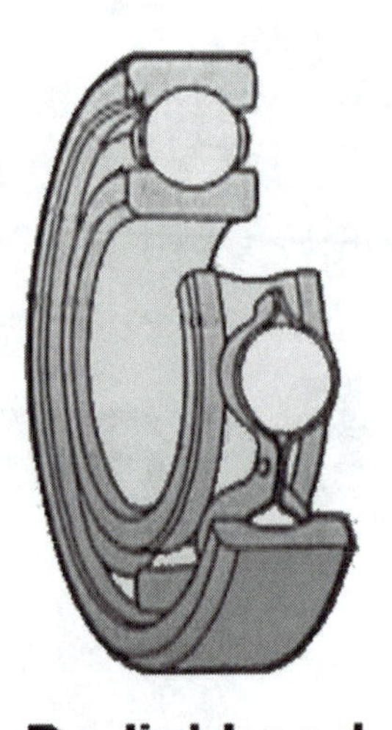

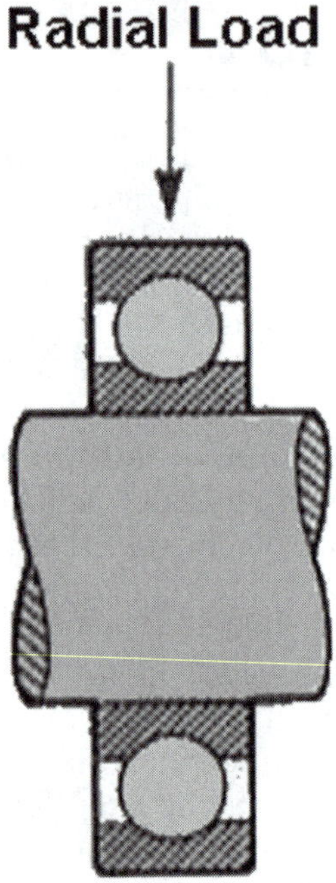

FIGURE 5.3
Radial load.

Radial Load

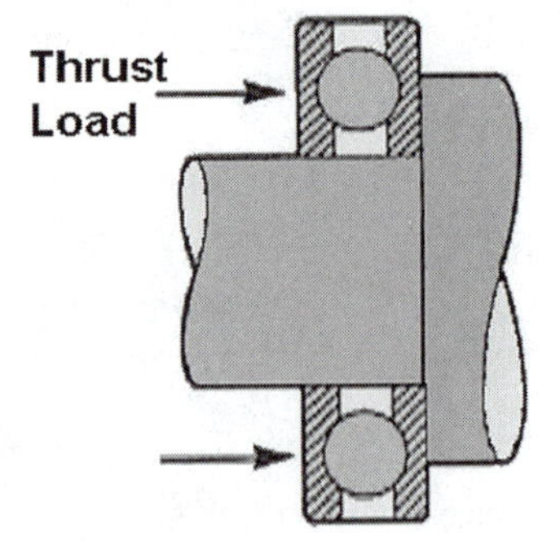

FIGURE 5.4
Thrust load.

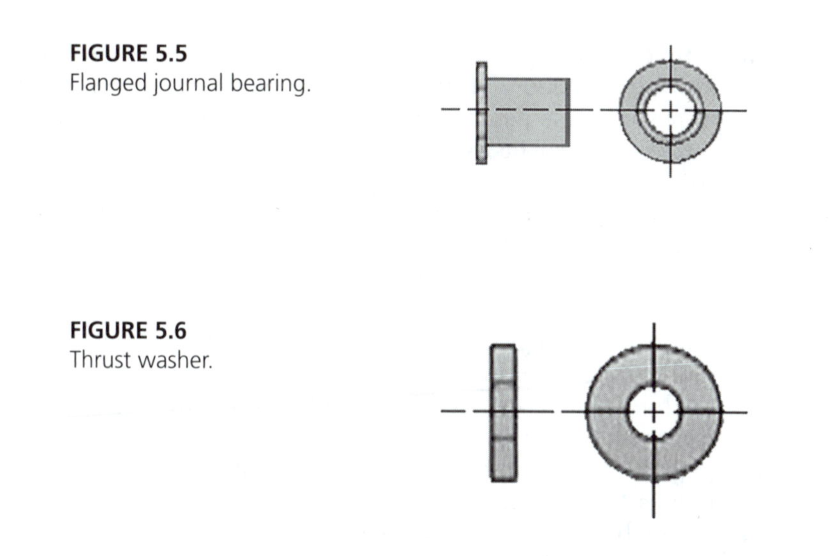

FIGURE 5.5
Flanged journal bearing.

FIGURE 5.6
Thrust washer.

5-2 PLAIN BEARINGS

A *plain bearing* is any bearing using a sliding action rather than a rolling action. It may or may not be lubricated. Plain bearings are sometimes referred to as *journal,* or *sleeve,* bearings. Figure 5.5 shows a typical flanged journal bearing.

Plain bearings are typically cylindrical-shaped bearings designed to carry radial loads. The terms “journal” and “sleeve” are often used interchangeably; sleeve refers to the general configuration, and journal refers to the part of the shaft in contact with the bearing. Plain bearings may also be *thrust bearings,* or *thrust washers,* because they look like thin, disclike washers (see Figure 5.6).

Plain bearings are categorized into three classes: class I, class II, and class III. Class I bearing systems are lubricated from an outside source. An example of a class I bearing system is a plain bearing requiring a liquid lubricant.

Class II bearing systems have internal lubrication. This class includes porous powder metal impregnated with oil that requires no outside lubrication. Class III bearing systems have graphite, PTFE [Teflon® (a registered trademark of E. I. duPont de Nemours & Company)], or plastic bearings that require no lubrication. The following are some of the more common plain bearing materials:

- **Babbitt** is a widely used class I bearing. Babbitt is a lead- or tin-based alloy with copper and antimony. These bearings provide dependable service even in moist or chemical environments. They are relatively soft and are good for low-temperature [below 200°F (93°C)] and low- speed applications.

- **Solid bronze** is harder than babbitt and is used where higher temperatures [up to 300°F (149°C)] are required. Because it is harder, solid bronze can also damage shafts if not properly lubricated.
- **Sintered bronze bearings** are common class II bearings. They are made from a powdered bronze composite and impregnated with oil. This oil is dispersed throughout the bearing and is drawn to the area of sliding contact by a capillary action produced by the heat and pressure of the shaft on the bearing surface. Sintered bronze bearings can be relubricated and operate at temperatures up to 200°F (93°C).
- **Cast-iron** bearings are plain bearings used primarily for slow, light applications. Their major advantages are their low cost, long life, and low (or no) maintenance requirement. With slow speeds, they can be used with no outside lubrication, using the self-lubricating quality of graphite present in the cast iron. Application temperatures may be as high as 1,000°F (538°C).
- **Carbon** bearings may be lubricated or be used as self-lubricating bearings. They work well with temperatures up to 700°F (371°C) and have good chemical and moisture resistance. Their main disadvantage is their low tolerance for dirty environments and contaminants and their inability to handle shock loads because of their brittleness.
- **Plastic** bearings require no lubrication, have good wear resistance, and have long life. The most commonly used bearing plastics are polypropylene, polyethylene, phenolic, and fluorocarbons. The large variety of plastic-bearing materials offers many different combinations of heat and chemical resistance.

5-3 PLAIN-BEARING LUBRICATION

Plain bearings may use hydrodynamic lubrication or hydrostatic lubrication (see Figure 5.7). *Hydrodynamic lubrication* uses the coating action of the lubricant and the rotation of the shaft to keep metal surfaces separated.

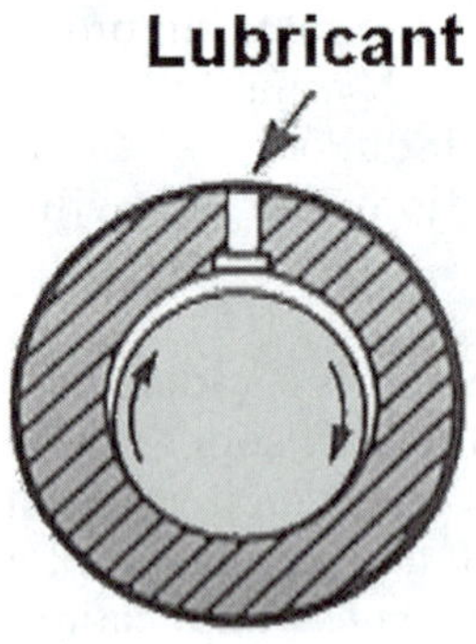

FIGURE 5.7
Plain-bearing lubrication.

Lubricant

Hydrostatic lubrication uses an outside pressure source (an oil pump is an example) to keep a continuous stream of lubricant to separate moving metal surfaces.

The following are the three modes of hydrodynamic lubrication in plain bearings:

1. In *boundary lubrication,* only an extremely thin film of lubricant is present to separate bearing surfaces.
2. In *mixed-film lubrication,* part of the bearing surface is supported by a boundary film and part is supported by a full film of lubrication.
3. In *full-film lubrication,* a continuous thick film of lubrication separates moving metal parts.

When a plain bearing starts up, it goes through all three modes of lubrication. Before start-up, metal surfaces may contact each other. When rotation begins, boundary lubrication starts. As the shaft increases speed, it passes through the mixed-film mode of lubrication. When operational speed is obtained, hydrodynamic action produces full-film lubrication in a well-designed application (see Figure 5.8).

By using an outside pressure source for lubrication, hydrostatically lubricated bearings will have full-film lubrication at any shaft speed (see Figure 5.9).

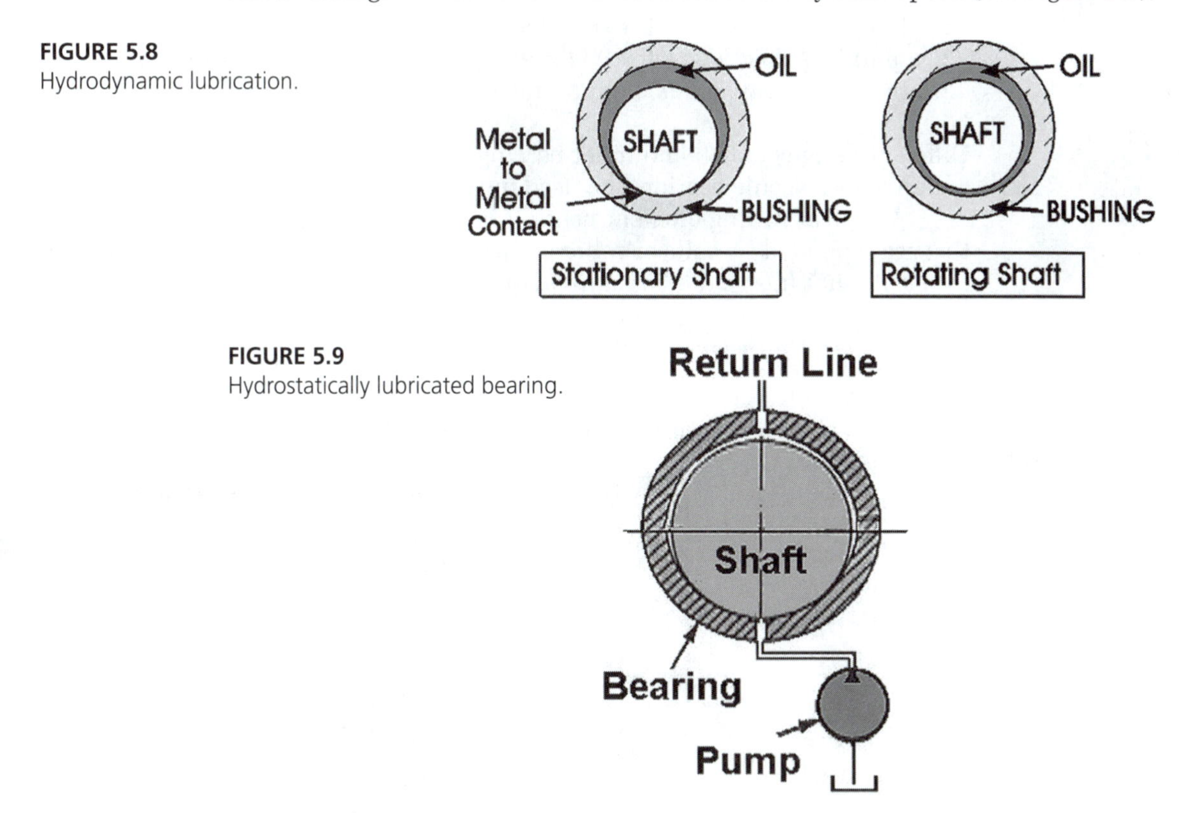

FIGURE 5.8
Hydrodynamic lubrication.

FIGURE 5.9
Hydrostatically lubricated bearing.

Hydrostatic lubrication is commonly used on slow rev/min and heavily loaded machinery. The advantages of hydrostatic lubrication are that it is more efficient and has a longer life. The main disadvantages of hydrostatic lubrication are the expense and problems associated with having an external pumping system.

5-4 ROLLING-ELEMENT BEARINGS

Plain bearings rely solely on lubrication to reduce friction. Rolling-element bearings have balls or rollers for increased efficiency. Rolling friction is always less than sliding friction. The following are the three basic types of rolling-element bearings (see Figure 5.10):

1. Ball bearings
2. Roller bearings
3. Needle bearings

There are three basic types of loads that bearings are required to handle (see Figure 5.11):

1. In *radial loads,* force is exerted at 90° to the shaft axis.
2. In *axial thrust loads,* force is exerted parallel to the shaft axis.
3. In *combination loads,* both radial and axial loads are present.

Different designs of ball and roller bearings can handle radial, axial, and combination loads. Needle bearings are used only for radial or axial loads.

A typical rolling-element bearing has inner and outer rings called *races* that are separated by balls or rollers. These rolling elements are equally spaced by a *separator* (also called a *retainer,* or *cage*).

FIGURE 5.10
Three basic types of rolling-element bearings.

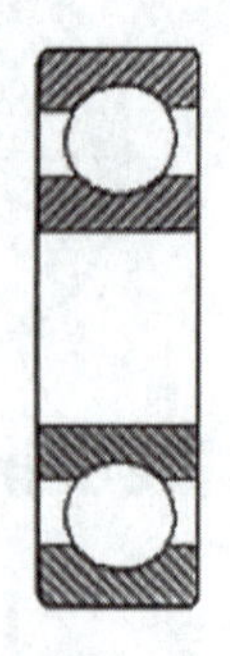

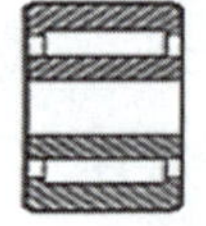

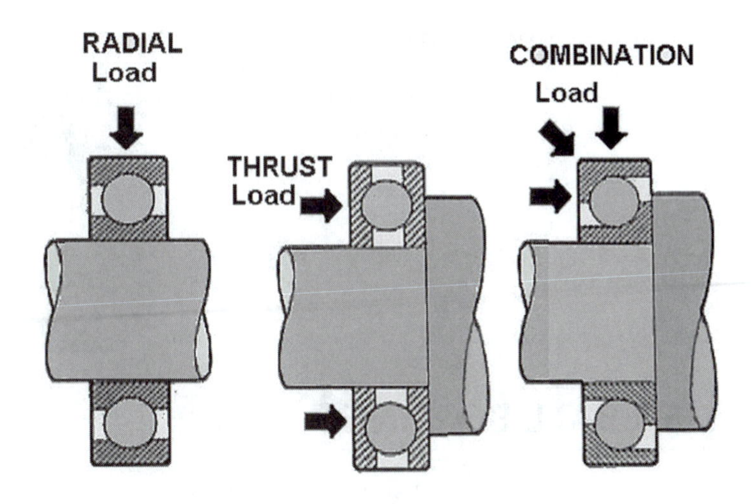

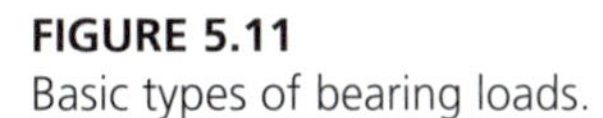
FIGURE 5.11
Basic types of bearing loads.

A long, useful service life is important when selecting any bearing. Selecting the correct bearing for any application starts with determining its load-carrying capacity. Bearing manufacturers use two basic types of load ratings:

1. **Dynamic load rating:** This rating factor gives the life expectancy of a bearing rotating under load. It expresses the bearing radial load that gives a basic rating life of 1 million revolutions. The actual life depends on the magnitude of the imposed load, lubrication, and operating conditions.
2. **Static load rating:** This rating factor gives the life expectancy of a bearing with stationary or slow-moving loads. This rating factor must also be considered when heavy shock loads occur on a rotating bearing.

The life of a rolling-element bearing is determined by the number of revolutions or the number of hours at a standard speed that a bearing operates before the first sign of fatigue failure occurs. Even identical bearings show some variance under the same conditions; therefore, statistics are used to predict bearing life.

The *rating life,* or *minimum life,* of a bearing is the length of operating time that 90% of a group of bearings is expected to exceed before failing. Manufacturers often refer to rating life as *L10 life.*

The life that 50% of a group of bearings will exceed is called the *median life,* or *L50 life.* The L50 life of a bearing is about five times the L10 life.

The actual life of any bearing is called its *service life.* Unlike L10 life or L50 life, service life it is not determined by fatigue failure of the bearing. It is determined by the load and the conditions affecting the bearing, such as lubrication, alignment, corrosion, contamination, temperature, and other considerations.

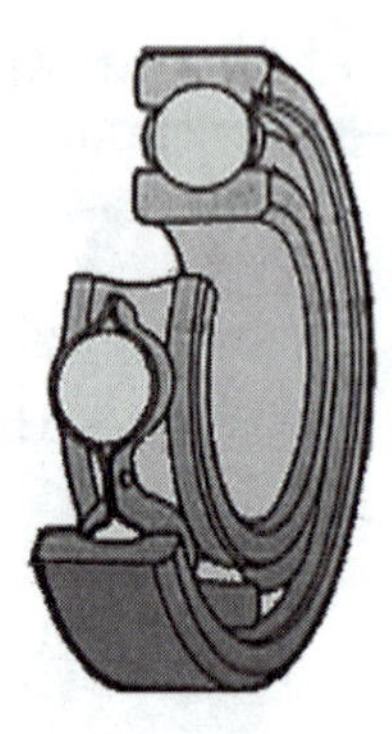

FIGURE 5.12
Ball bearing.

5-5 BALL BEARINGS

A typical ball bearing is shown in Figure 5.12. Ball bearings are manufactured in three basic configurations:

1. Single-row radial
2. Single-row angular contact
3. Thrust ball bearings

Single-Row Radial Ball Bearings

The most common form of this type of ball bearing is the *nonfilling slot.* It is also called the *Conrad,* or *deep-groove,* ball bearing. The raceways of this bearing have deep grooves that can support high radial loads as well as axial loads. It is not, however, intended for applications with axial loads only. This bearing is not self-aligning, and accurate alignment between the shaft and bearing mounting is required. Figure 5.13 shows how this bearing is assembled. The inner raceway is located to give space to insert the bearing balls. The balls are then equally spaced, and the bearing cage is installed to keep them in place.

The single-row radial ball bearing is also manufactured with a *filling slot* (see Figure 5.14). This type of bearing is also called a max-type ball bearing because it allows the manufacturer to load the maximum number of balls for a given bearing. Because this type of bearing has more elements than the Conrad nonfilling ball bearing, it is capable of carrying heavier loads. The max-type ball bearing is not as good as the Conrad type for axial or thrust loads. If an

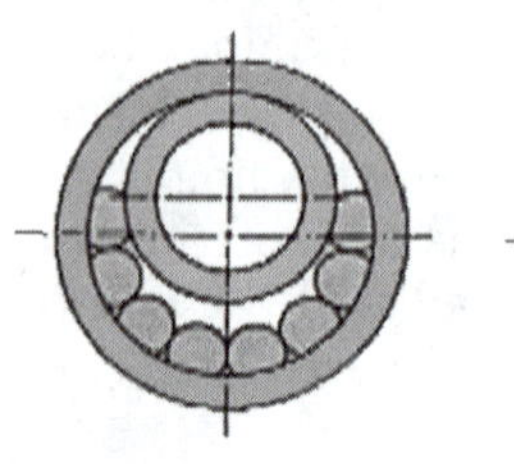
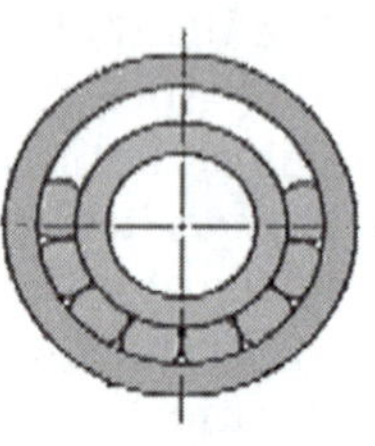
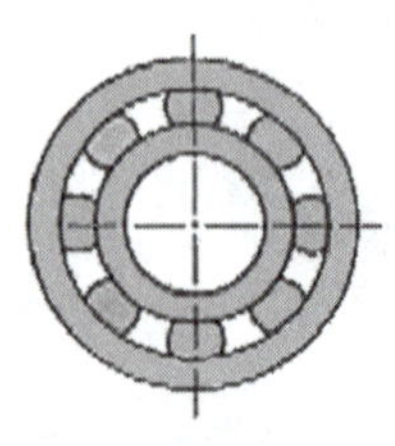

FIGURE 5.13
Conrad ball bearing.

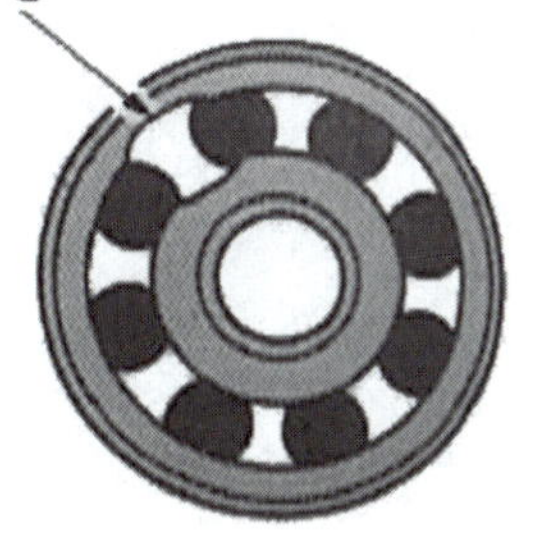

FIGURE 5.14
Max (filling slot) ball bearing.

axial load forces the balls to the side of the raceway, they will run across the loading slot, causing bearing damage.

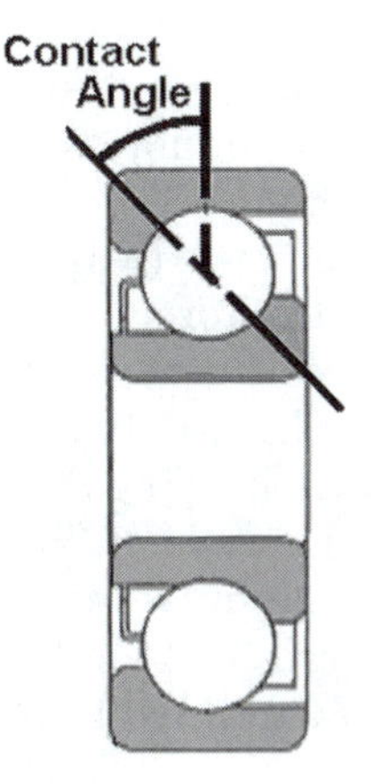

FIGURE 5.15
Angular contact bearing.

Single-Row Angular-Contact Ball Bearings

The single-row angular-contact ball bearing is designed to accommodate up to 300% more axial (thrust) load than Conrad bearings can (see Figure 5.15). However, because of its construction, this ball bearing can take axial loads in *one* direction only. Mounting this bearing the wrong way results in a short bearing life. Angular-contact ball bearings are marked for correct installation (see Figure 5.16). The steep contact angle of the raceway on the inner race and on the opposite side on the outer race ensures a high-thrust capacity along with good radial-loading capacities.

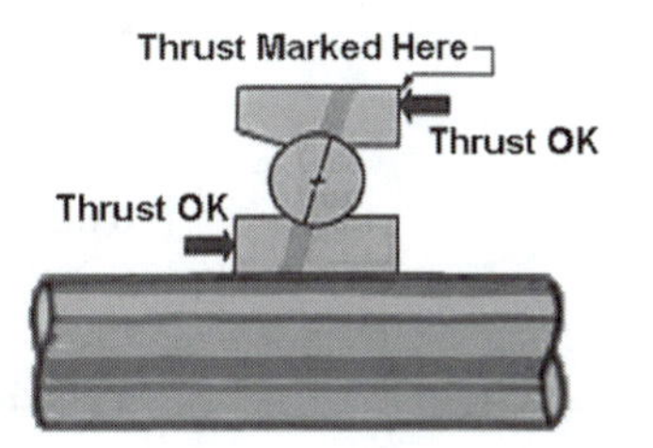

FIGURE 5.16
Axial thrust on one side only.

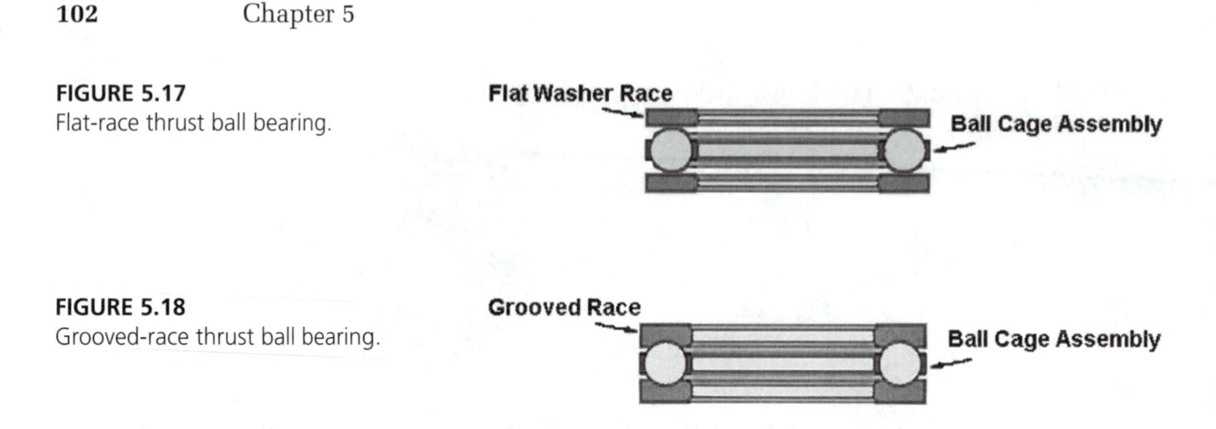

FIGURE 5.17
Flat-race thrust ball bearing.

FIGURE 5.18
Grooved-race thrust ball bearing.

Thrust Ball Bearings

Thrust ball bearings are designed for applications where only axial or thrust loads occur. They cannot handle radial load. A *flat-race*-type thrust ball bearing is shown in Figure 5.17. This flat-seat bearing is made up of two flat washer races, balls, and a cage (ball retainer). Contact stresses are high because of the small contact area. Torque resistance is low, and the shaft can flex or wander because there are no grooves to constrict movement. This type of thrust bearing is best suited for light-duty applications.

Thrust ball bearings are more commonly made with *grooved-race* construction. This type of thrust ball bearing may be *single-acting* or *double-acting,* depending on whether the bearing can accept a load in one or two directions. Figure 5.18 shows single-acting, grooved-race, thrust ball bearings. Grooved-race thrust bearings have about twice the load capacity of flat-race bearings.

5-6 ROLLER BEARINGS

The primary difference between roller bearings and ball bearings is that ball bearings are designed for lighter duty and higher speeds compared to roller bearings. The point of contact in ball bearings and the race is small compared to roller bearings. Figure 5.19 shows the contact point of both ball bearings and

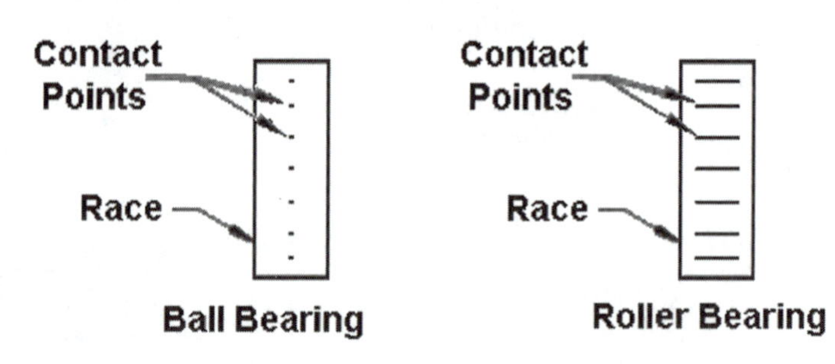

FIGURE 5.19
A roller bearing has a larger contact area.

roller bearings. The roller bearing's longer point of contact makes it capable of higher loads but usually at lower speeds. Rolling resistance is also greater because of the higher contact area.

Roller bearings are manufactured in three basic configurations:

1. Cylindrical roller bearings
2. Tapered roller bearings
3. Spherical roller bearings

Cylindrical Roller Bearings

The cylindrical roller bearing has symmetric, straight-side rollers. This roller bearing is usually a single-row-type bearing (see Figure 5.20). It is designed to carry only radial loads. The raceways may be manufactured with flanges to provide some restraint to any minor axial thrust. Any axial load will push the roller into the flange of the raceway and will be supported by the flange. The sliding action of the roller against the flange will increase resistance to rotation and will generate heat.

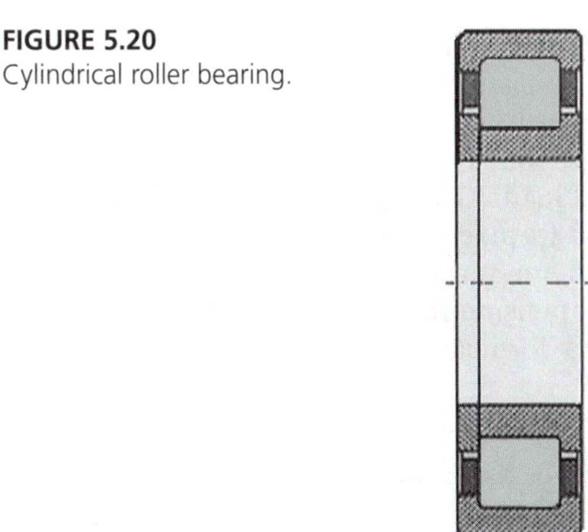

FIGURE 5.20
Cylindrical roller bearing.

Tapered Roller Bearings

Tapered roller bearings have straight-side, tapered rollers (see Figure 5.21). Designed for slower speeds, these bearings can handle high radial and thrust loads because of their unique construction. However, they cannot withstand any appreciable amount of misalignment. Note that the guide flange and cage is on the inner raceway only, allowing for easy disassembly of these bearings.

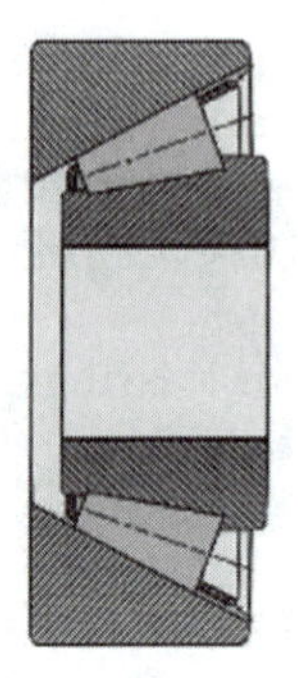

FIGURE 5.21
Tapered roller bearing.

The cage holds the rollers in place as the inner raceway is removed. Tapered roller bearings are commonly used as automobile wheel bearings. The running clearance of tapered roller bearings is adjustable, but it must be *preloaded* for each specific application. Preloading is accomplished by squeezing the rollers between the inner and outer raceway. Adjusting nuts or shims are used to tighten the bearings to a specific torque, compressing the rollers and thus preloading the bearing.

Spherical Roller Bearings

Spherical roller bearings use barrel-shaped rollers (see Figure 5.22). These bearings may be single- or double-row bearings. They are designed to handle heavy radial and axial loads. They also function well with some misalignment. The rollers are locked in place on the inner raceway with flanges and a cage. The outer raceway is a one-piece spherical design that allows the inner bearing unit to swivel to compensate for misalignment. This type of bearing is used extensively in industry for heavy-duty applications.

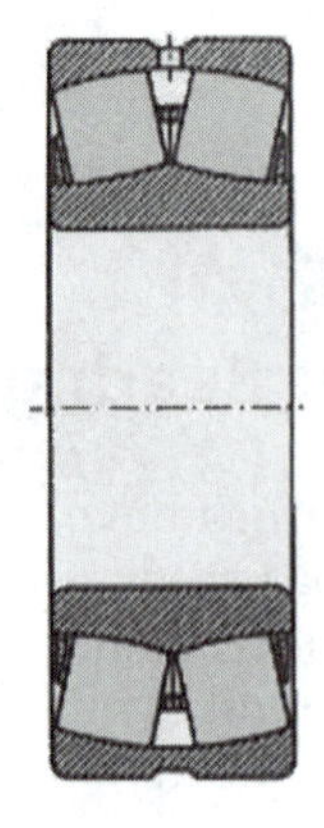

FIGURE 5.22
Spherical roller bearing.

FIGURE 5.23
Needle bearing.

5-7 NEEDLE BEARINGS

Needle bearings are a special type of cylindrical roller bearing (see Figure 5.23). The rollers are typically smaller than 1/4 in. in diameter, and the length of the roller is 6 to 10 times the diameter of the roller. Needle bearings can be used to handle higher loads when only a small space is available to install a bearing. Needle bearings have a large load capacity-to-size ratio.

Needle bearings are typically used in radial or axial loading (see Figure 5.24); radial needle bearings cannot handle axial loads, and axial needle bearings cannot tolerate radial loads. Most needle bearings do not have an inner raceway; rollers make direct contact with the shaft. The shaft must be hardened to a recommended value of Rockwell C58; a softer shaft would reduce the bearing-load capacity.

The original *cageless* needle bearing had no cage to retain and separate the rollers. Because there are no spaces between adjacent rollers, no cage is needed. The extra rollers of the needle bearing give it more load capacity, but friction is increased because the rollers rub each other. This also causes this type of needle bearing to have a lower maximum speed limit.

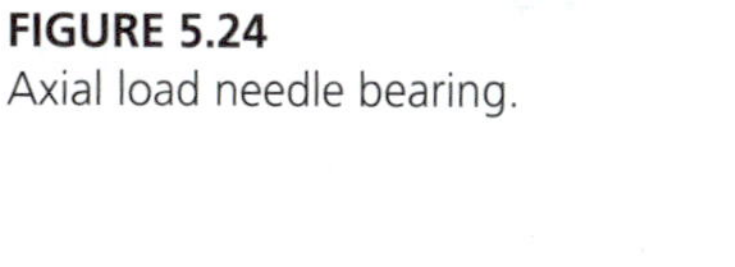

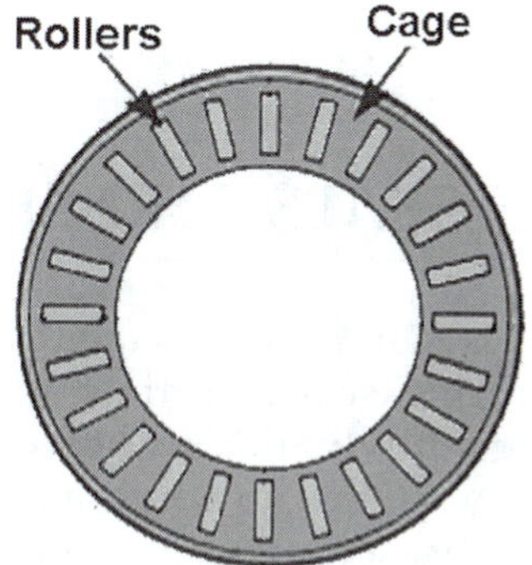

FIGURE 5.24
Axial load needle bearing.

FIGURE 5.25
Cageless needle bearing.

Cageless needle bearings use a drawn cup to retain and house the rollers. Figure 5.25 shows rounded-end rollers that are retained in the drawn cup by the adhesive qualities of the grease that is required. After the bearing runs for a while, the grease loses its stickiness; if the shaft is removed, the rollers may fall out. Some cageless needle bearings are mechanically retained by the cup. This bearing has one end that is hardened and marked (the end is also flat). This is the side that must be used when pressing the bearing into place. The other side is rounded and is *not* hardened. If pressure is applied to this side, it becomes flattened, and the rollers become pinched. Needle bearings should always be carefully inspected before installation.

Caged needle bearings have less than half of the rollers that cageless needle bearings do. This means that loads must be lighter, but the maximum speed may be higher because of less friction. Caged needle bearings may not always have an outer raceway as well as an inner raceway. In some cases, the machine's housing acts as the outer raceway, and the shaft becomes the inner raceway (see Figure 5.26).

FIGURE 5.26
Caged needle bearing.

5-8 BEARING SEALS AND SHIELDS

Bearings may be completely open with a closed lubrication system requiring no bearing seals, or they may be completely sealed (see Figure 5.27). Permanently lubricated bearings seal all contaminants out and lubrication in. They require no additional lubrication or maintenance for the life of the bearing. Most bearings, however, fall between these two extremes and usually have some kind of seal.

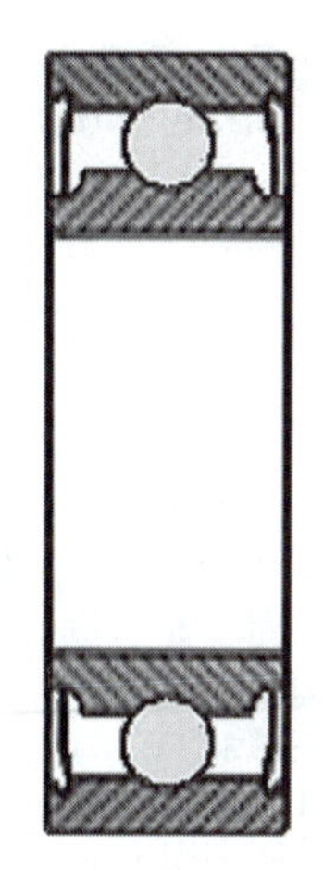

FIGURE 5.27
Sealed bearing.

Seals have two functions:

1. Keep contaminants out of the bearing.
2. Keep lubrication in the bearing.

Some seals are required only to keep out dirt, moisture, and other contaminants. They allow lubrication to escape if overlubrication occurs. Overlubrication can cause excessive friction, heat, and—ultimately—bearing failure. As a general rule, it is better to underlubricate than to overlubricate.

The following are the two basic types of bearing seals:

1. *Contacting seals:* The stationary part of the seal is pressed into the bearing or housing, and a soft, usually synthetic, material contacts the rotating shaft for a positive seal. The effectiveness of this type of seal depends on a smooth contact surface and lubrication. If excessive friction occurs, damage to the seal and possibly the shaft occurs.
2. *Noncontacting seals:* These seals have no rubbing parts contacting the shaft. The bearing shield is the simplest of these. The *bearing shield* is a single metal seal with a small clearance between the shaft and the seal. A more sophisticated version of this is the **labyrinth seal.** The labyrinth seal does not touch the shaft but has a series of teeth that have a small clearance between the teeth and the shaft. A straight labyrinth seal is shown in Figure 5.28.

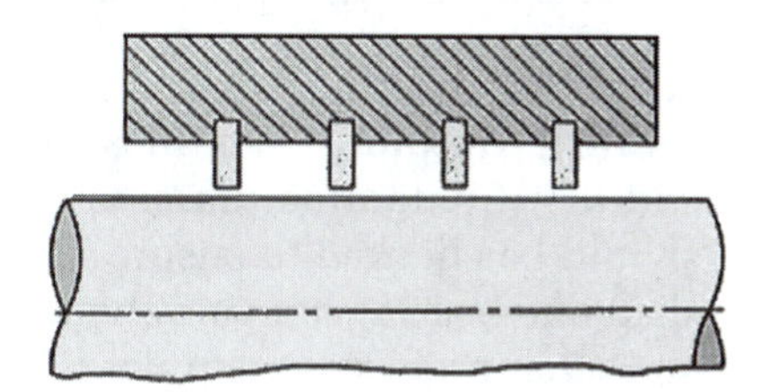

FIGURE 5.28
Labyrinth seal.

There are many types and material for seals. To learn more about them, see Chapter 6 for information about seals, gaskets, and packing.

5-9 WORKING WITH BEARINGS

Bearings have a long, useful life if properly handled, installed, and maintained. These general precautions should be followed when handling or installing a bearing:

- Cleanliness is very important. Even the smallest amount of dirt or contaminant can damage a bearing and shorten its life. Work with clean tools. Handle bearings with clean, dry hands.
- Keep bearings in their original container or wrapping until used. Do not wash or wipe new bearings before installing them. The oil shipped on bearings combines with the bearing lubricant and provides lubrication during the initial start-up.
- Check shafts, housings, and all other parts associated with the bearing for damage and dimensional tolerances. All parts should be thoroughly cleaned and inspected before reinstallation.
- *Never* spin a bearing with compressed air because it damages bearings. The brittle outer raceway may explode because of the centrifugal force. If you must use compressed air, hold both raceways so the bearing cannot spin. Always use clean, dry air.
- Never force, press, strike, or otherwise damage the seal or shield on factory-sealed bearing.
- Make sure that the bearing is properly aligned before applying force to press it on a shaft or housing.
- Press *only* on the ring or raceway being installed. Do *not* press on the inner raceway if the bearing is being pressed into a housing. Do *not* press on the outer raceway if the bearing is being pressed onto a shaft.
- Always follow manufacturers' instructions for handling and assembling bearings.

5-10 INSTALLING BEARINGS

Installing a bearing involves more than just hammering it onto a shaft and lubricating it. Absolute cleanliness is critical. The work area and tools must be clean and dry. Contamination shortens bearing life considerably. The new bearing should be checked to ensure it is a proper replacement for the old bearing. The shaft and bearing housing should be cleaned and checked for burrs, deformities, and other problems that might interfere with mounting or bearing operation.

New bearings are shipped with a protective, anticorrosive lubricant. You should never wash it off. It is compatible with any lubricant and protects the bearing during the initial start-up.

Take time to review the procedure for mounting the bearing. Carefully study the machine-drawing details of the correct assembly sequence.

5-11 INTERFERENCE FIT BEARINGS

A common method of bearing mounting is a *press fit,* or *interference fit,* mounting. The interference produced by tight fits expands the inner ring or contracts the outer ring. This change holds the bearing securely in place. Care must be taken to ensure that the fit is not too loose or too tight. Wear or incorrect machining may cause the fit to be too loose, which can allow the ring to rotate, thus creating problems. Fits that are too tight cause a reduction of internal radial clearance in the bearing (rings press the balls or rollers too tightly). This excessive tightness may cause the bearing to run hot and fail prematurely. Refer to the tables provided by bearing manufacturers for the maximum and minimum shaft and housing diameters for proper interference fits.

An *out-of-round* condition or high spot can cause the bearing raceway to be deformed in an interference fit. When checking shaft and housing diameters, you should also check for out-of-round conditions and flat spots. Correct any problems before attempting to mount the bearing.

5-12 MOUNTING INTERFERENCE FIT BEARINGS

Interference fit bearings use two methods for mounting: *mechanical* methods and *thermal* methods. Whichever method you use, you should always be careful to make sure the bearing is installed facing the right direction. Some bearings accept thrust in one direction only, and some needle bearings must be installed with the flat side (usually the side with numbers) facing out.

5-13 MECHANICAL METHODS OF MOUNTING BEARINGS

Common methods for mechanically mounting bearings are as follows:

- Hammer and sleeve
- Arbor or hydraulic press
- Integral threaded mounting

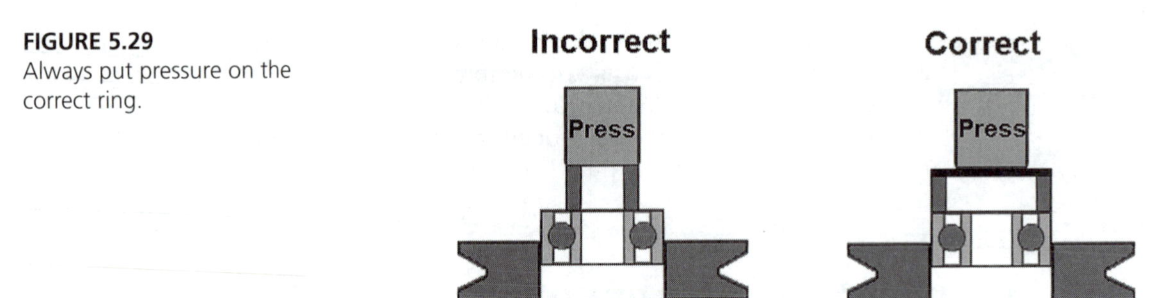

FIGURE 5.29
Always put pressure on the correct ring.

No matter which method is used, it is very important to remember to apply the force to the ring with the interference fit. Do *not* press on the outer ring if the inner ring is the tight fit (see Figure 5.29). Pressing on the wrong ring causes unnecessary stress on the bearing rollers and probable damage.

The bearing should always be accurately aligned with the shaft and housing. Make sure you use the right sleeve or arbor to prevent damage to the bearing.

Smaller bearings and needle bearings are commonly mounted using a hammer and sleeve or block. The bearing should be accurately lined up and the correct-size sleeve should be selected to apply force to the ring with the tighter fit. A flat block made of a softer material can sometimes be used to mount the bearing; however, wood or any type of material that could leave chips or otherwise contaminate the bearing should never be used. Brass is typically used because it is soft enough not to damage bearings but hard enough to force the bearing into place accurately. A soft steel is usually used for sleeves. Never use a hammer directly on the bearing or shaft.

Arbor and hydraulic presses are great for mounting small or medium-sized bearings. Presses provide a steady push to force bearings into a housing or onto a shaft. Sometimes, the same sleeve or block used with a hammer is used with a press. Proper sleeves and good alignment are still required. A coat of light oil applied to the bearing makes mounting easier.

Some bearing mountings have an integral threaded mounting. These bearings use a threaded lockout on the shaft to force the bearing in place. Other bearings have an adapter ring with bolts that are tightened to seat the bearing. Again, a coating of light oil makes mounting easier.

In summary, when mechanically mounting bearings, keep the following things in mind:

- Maintain cleanliness at all times.
- Use proper tools and methods.
- Ensure good bearing alignment before forcing.
- Always apply force squarely and evenly.
- Apply force *only* to the ring or raceway with the interference fit.
- Use a light coat of oil to make mounting easier.

5-14 THERMAL METHODS OF MOUNTING BEARINGS

Because steel expands when hot and contracts when cool, an interference fit bearing can be mounted more easily when heated or cooled, depending on the application.

A bearing may be cooled by putting it in a freezer, but this is not the best choice because of moisture condensation and because most freezers are not cold enough to work effectively. Dry ice (solid carbon dioxide) is a much better choice for cooling bearings to make mounting easier. Always follow the correct safety procedures. Personal protective gear is necessary, and you should never use this method without proper training. You should also wrap the bearing in a protective plastic bag to prevent moisture condensation. When working with any thermal method, you must work quickly to install bearings before the temperatures equalize.

Mounting bearings by heating them is more common than by cooling them. However, you should never use a torch to heat a bearing. The temperature of the flame far exceeds the maximum temperature to which a bearing should be heated. The bearing may not look damaged, but the temper of the bearing steel may be destroyed and result in a reduction of hardness and wear life. The following are some of the more common methods used to heat bearings:

1. Oil-bath bearing heaters are a common and reliable method for heating bearings (see Figure 5.30). For even heating, the bearing should not touch the sides or bottom of the oil bath. A wire basket is usually used to keep the bearing from touching the sides or bottom and also protect the bearing from contaminants that tend to settle to the bottom. Accurate temperature control is essential, and most oil baths usually maintain temperatures well. Bearings should not be heated above 250°F (120°C). The oil used should have a flash point above 480°F (250°C) for safety reasons.
2. Heating plates, hot-air cabinets, and electric ovens are also used to heat bearings. With hot plates, the bearing must be turned over several times to ensure even heating. Hot-air cabinets or convection ovens are clean and easy to use. Their disadvantage is that heat-up times are relatively long. Electric ovens must be kept clean to prevent bearing contamination. The bearings should be suspended for more even heating.

FIGURE 5.30
Oil-bath bearing heater.

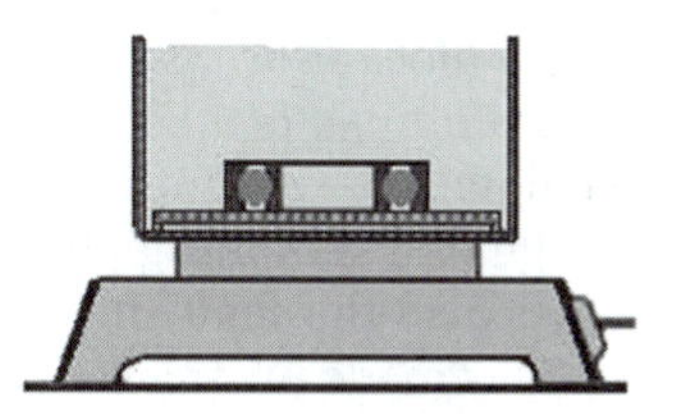

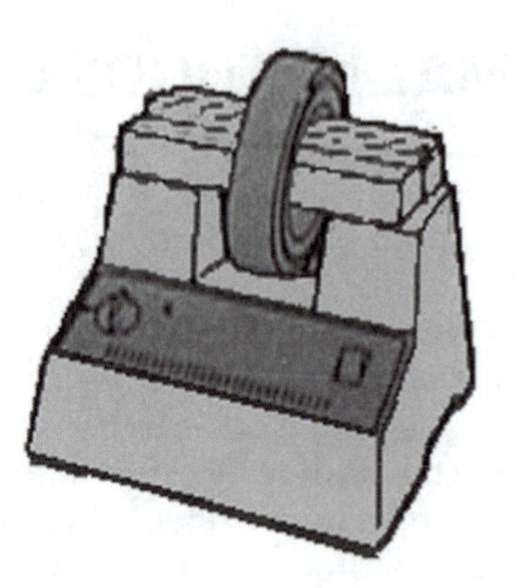

FIGURE 5.31
Induction bearing heater.

3. Induction heaters are fast and easy to keep clean (see Figure 5.31). Electric current is passed through a coil with an iron core that induces a high current and low voltage into the bearing, causing it to heat quickly. A temperature sensor controls the temperature. This process causes the bearing to become magnetized, but a demagnetizing cycle is built into the induction heater to eliminate this problem.

FIGURE 5.32
Pillow block bearing.

5-15 PREMOUNTED BEARINGS

Premounted bearings consist of a bearing element and a housing that can be bolted to a machine frame. This self-contained unit provides protection and lubrication for the bearing. The bearing element has the same characteristics and limitations as unmounted types. They are made in a variety of designs to fit different shaft sizes and mounting requirements. The following are the three most common types:

1. *Pillow block bearings:* One of the most popular mounted bearings (referred to as *plummer blocks* in Europe), they act as integral units to support shafts parallel to the mounting surface. Pillow block bearings may be a one-piece design or a split design (see Figures 5.32 and 5.33). The one-piece design slides over the shaft and is then locked into position. The split pillow block is made up of a base and a cap with a horizontal split joint. This

FIGURE 5.33
Split pillow block bearing.

FIGURE 5.34
Two-bolt flange bearing.

FIGURE 5.35
Four-bolt flange bearing.

design allows the bearing to be mounted without disturbing other components mounted on the same shaft. The base and cap are accurately machined and use dowels for exact alignment.

2. *Flange bearings:* Flange bearings are used when the mounting surface is perpendicular to the shaft. They may be mounted on vertical or horizontal supports. The flange bearing and the pillow block bearing can use the same bearing element and are identically rated. Flange bearings may be two-bolt mounting (see Figure 5.34) or four-bolt mounting (see Figure 5.35).
3. *Take-ups:* Frame-mounted bearings called *take-ups* are used to provide a way to adjust bearing position (see Figure 5.36). They allow movement of the bearing and shaft. Take-up units are used to adjust the center distance between two shafts. They are commonly found in conveyor applications where belt-tightening is required.

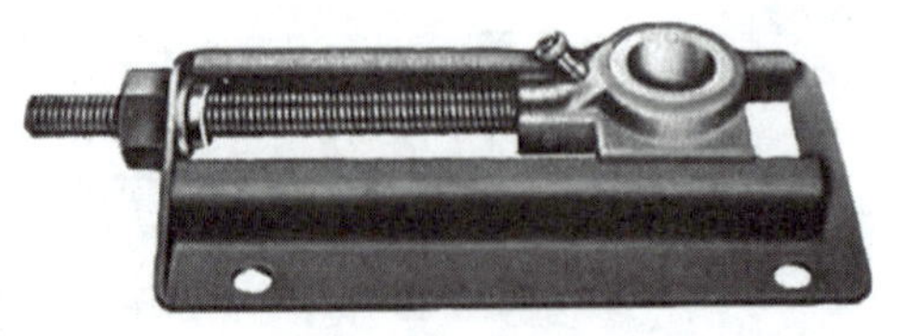

FIGURE 5.36
Bearing and take-up frame.

Premounted bearings may be *rigid* or *self-aligning* types. Rigid bearings will tolerate very little misalignment and must be accurately aligned. Self-aligning bearings are more expensive but are designed to compensate for minor misalignment, making them more desirable for most applications.

Premounted bearings may also be *expansion* or *nonexpansion* types. Expansion bearings allow axial shaft movement that may occur as the shaft heats and expands. Nonexpansion bearings limit shaft movement and keep the shaft accurately positioned. This makes them very useful as thrust bearings for vertical or horizontal applications.

5-16 TROUBLESHOOTING BEARINGS

Pending bearing failure can be detected by periodic inspections of three areas:

1. *Bearing temperature:* Monitoring bearing temperature provides an indication of problems before bearing failure occurs. Increased bearing temperature can be caused by either too little or too much lubrication. Inadequate lubrication may allow metal-to-metal contact, generating friction and heat. Excessive lubrication generates friction and heat because of the extra resistance created by too much grease. Increased bearing temperature may also be caused by misalignment, overloading, or overspeeding.
2. *Vibration:* Increased vibration may be caused by a loose bearing, a loose mounting, an unbalanced load, or contaminants in bearing lubrication.
3. *Change in the bearing's normal running sound:* A whine or whistle may indicate a damaged bearing caused by shock loading or overloading. Knocks or rattles may indicate a loose mounting. Grinding or growling may indicate the presence of contaminants (usually dirt or sand) inside the bearing.

Seven of the more common causes for bearing failure are as follows:

1. *Lubrication problems:* These problems are caused by too much or too little lubrication, contamination with moisture or trash, incorrect oil viscosity, or incorrect oil type.

2. *Contamination problems:* These problems are caused by incorrect or faulty seals, dirty grease fittings, internal metallic breakdown, contaminated lubricant, and incorrect assembly procedures.
3. *Misalignment problems:* These problems are caused by angular misalignment, parallel misalignment, bent shafts, or the flexing of the mounting or frame.
4. *Overloading problems:* Overloaded bearings tend to flake, looking similar to fatigue failure.
5. *Poor installation techniques:* Any of the previously discussed problems can be caused by incorrect installation.
6. *Electric pitting problems:* This problem is caused by an electric current passing through the bearing. A pit is burned into the bearing by the arcing of the electric current. This is commonly found in motor or generator bearings or may be caused by welding on some part of the machine in which the bearing is operating. Be careful where you place the ground clamp when welding.
7. *Fatigue failure:* All bearings eventually experience fatigue failure after a length of service, depending on speed, load, and type of service. Fatigue failure starts as small cracks in the race. The number of cracks increases until a piece of metal breaks off. This deterioration continues until the bearing fails completely.

QUESTIONS

1. What is a bearing?
2. The three types of loads a bearing must deal with are ______________, ______________, and ______________.
3. Any bearing using a sliding action rather than a rolling action is called a ______________ bearing.
4. What is a class II plain bearing?
5. Antifriction is another name used for this type of bearing.
6. What are the three basic types of rolling-element bearings?
7. Name the three basic types of loads that bearings are required to handle.
8. What is the dynamic load rating of a bearing?
9. What is a sealed bearing?
10. What are the two basic types of bearing seals?
11. When is it all right to spin a bearing with compressed air to clean it?
12. A common method of mounting a bearing is the press fit, or ______________ fit mounting.
13. The three most common types of premounted bearings are ______________, ______________, and ______________.
14. What do you call premounted bearings that are designed to compensate for minor misalignment?
15. When troubleshooting a bearing, what does increased vibration usually indicate?

CHAPTER 6

Seals, Gaskets, and Packing

6-1 INTRODUCTION

The earliest seals were leather strips wrapped around the end of wagon axles to hold grease in place. With the Industrial Revolution, better sealing materials became more critical. Shafts were sealed with stuffing boxes that used packing made of natural fibers (see Figure 6.1). As new and better machines were developed, seals were required to handle higher speeds and loads. High temperature and wear capabilities of asbestos packing made it the most widely used packing in the early 1900s replacing most natural-fiber packings. The young, booming automobile industry began to require a new seal for high-speed applications that was simple in design. *Lip seals* fulfilled this need and became widely used. The first lip seals were made of leather locked in a light sheet-metal ring (see Figure 6.2). Since then, leather has been replaced by synthetic oil-resistant elastomers that come in a variety of shapes and special designs.

Nonleak mechanical seals suited for most applications have now replaced the old stuffing box and packing system (see Figure 6.3).

Modern seals can be classified into two different types: *static* and *dynamic*. **Static seals** are used where there is little or no movement between mating surfaces. Gaskets and O-rings are the most common examples of static seals.

Dynamic seals are used to seal moving surfaces. Rotating or reciprocating shafts require some type of dynamic seal. Lip seals and mechanical seals are two widely used dynamic seals.

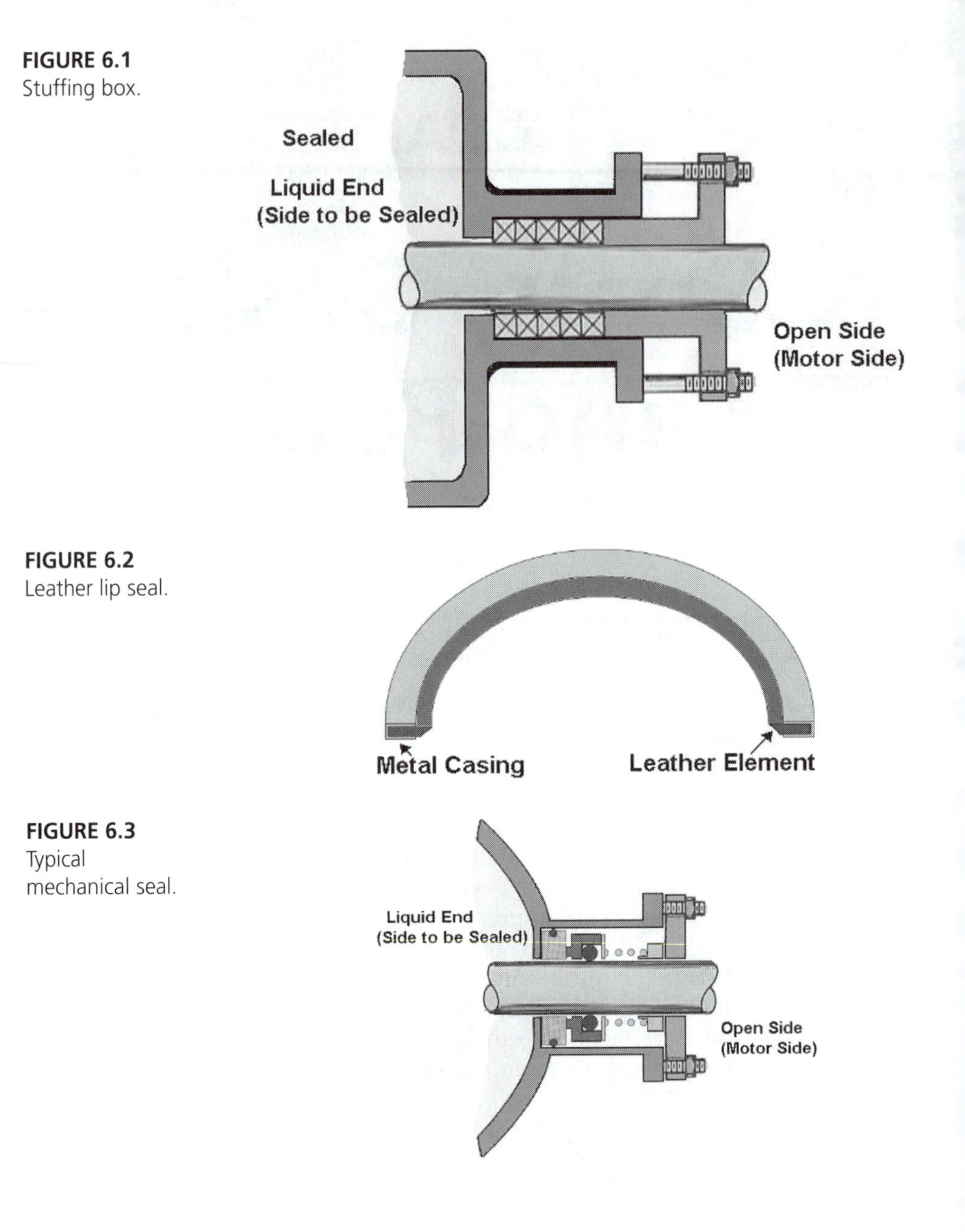

FIGURE 6.1
Stuffing box.

FIGURE 6.2
Leather lip seal.

FIGURE 6.3
Typical mechanical seal.

6-2 GASKETS

A gasket is an elastic material placed between two metal machine parts to create a seal. Gaskets may be used to seal something *in* (such as liquids, gases, and so on) or to seal something *out* (such as moisture, dirt, or other contaminants). Most gaskets are designed to perform both functions.

Gasket materials come in a wide variety, depending on the application and media in which the gasket must operate. The manufacturer's recommended gasket should be used in most applications.

For good gasket sealing, you should follow the three guidelines when installing new gaskets:

1. Clean and check mating surfaces for warping, surface damage, and surface finish to make sure the gasket contact area is smooth and will seal properly.
2. Use the correct gasket for your application.
3. Torque all bolts evenly.

Proper torque and even tightening is important for good gasket installation. Always follow the manufacturer's recommendation for the correct torque values. Follow these steps when installing a gasket:

- Line up the gasket and mating surfaces.
- Install all bolts finger-tight.
- Tighten the bolts in at least three steps using the correct pattern. For example, if the final bolt torque is 60 ft-lb, first torque to 20 ft-lb, then to 40 ft-lb, and finally to 60 ft-lb. (Note: Some applications require more than three steps; always use manufacturer's recommendations.)

The bolt-tightening sequence is also important. In general, all mating surfaces should be tightened evenly. Figure 6.4 shows a typical flange-fastening sequence. In this example, the three-step method for achieving a final torque of 60 ft-lb is to tighten bolt 1 to 20 ft-lb, then bolt 2 to 20 ft-lb, and so forth, until all bolts are tightened to 20 ft-lb. Then you start again with bolt 1 tightened to 40 ft-lb, bolt 2 to 40 ft-lb, and so forth.

For a rectangular shape, a spiral-fastening sequence should be used. Figure 6.5 shows the tightening sequence for this type of shape. The idea is to start at the middle and work your way outward to the edges. Again, the finished torque should be accomplished in at least three steps. Bolts should always be retorqued after a break-in period to ensure the proper torque is maintained.

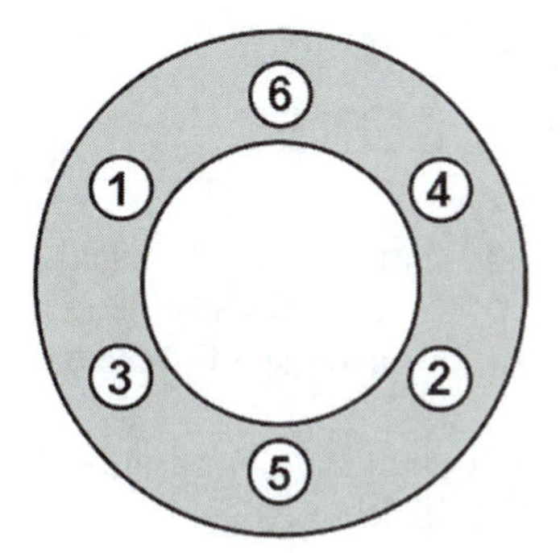

FIGURE 6.4
Bolt-tightening sequence for flanges.

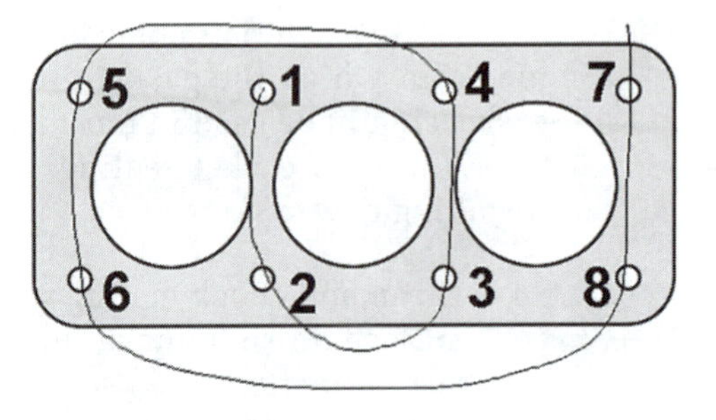

FIGURE 6.5
Use a spiral sequence for a rectangular shape.

Gasket failure is typically a result of installation mistakes, damaged gaskets or surface, dirt, or contaminants, incorrect gasket materials for the application, or, sometimes, poor design. Gaskets, in general, should never be reused. They lose resilience and may not seal properly when reused.

6-3 ELASTOMERIC STATIC SEALS

Elastomeric static seals made from a flexible polymer, called an **elastomer,** are now widely used in modern industrial applications. They come in a broad variety of shapes, but the most common seal is the *O-ring.* O-rings are molded in sheets and then separated. Any excess rubber flash is removed by cold tumbling the O-rings with an abrasive in a large rotating drum, similar to rock tumbling. O-rings are identified by type of material, inner diameter (ID), and cross-section diameter. For an O-ring to seal properly it must be compressed. The minimum amount of compression should be at least 0.006 in. (0.15 mm) regardless of the O-ring diameter. Typically, the compression should be about 20% of the cross-section diameter and should never exceed 35%.

In addition to O-rings, rectangular or square-shaped rings are also widely used for higher-pressure applications. O-ring failure usually occurs as a result of installation mistakes, a damaged O-ring or grooves, poorly prepared sealing surfaces, or incorrect O-ring material for the application.

6-4 STUFFING BOXES

Dynamic seals are used to seal rotating or reciprocating shafts. The stuffing box is one of the first widely used dynamic seals. A simple stuffing box consists of three main parts (see Figure 6.6):

1. The packing chamber, or the packing box, is the stationary housing for the packing rings.

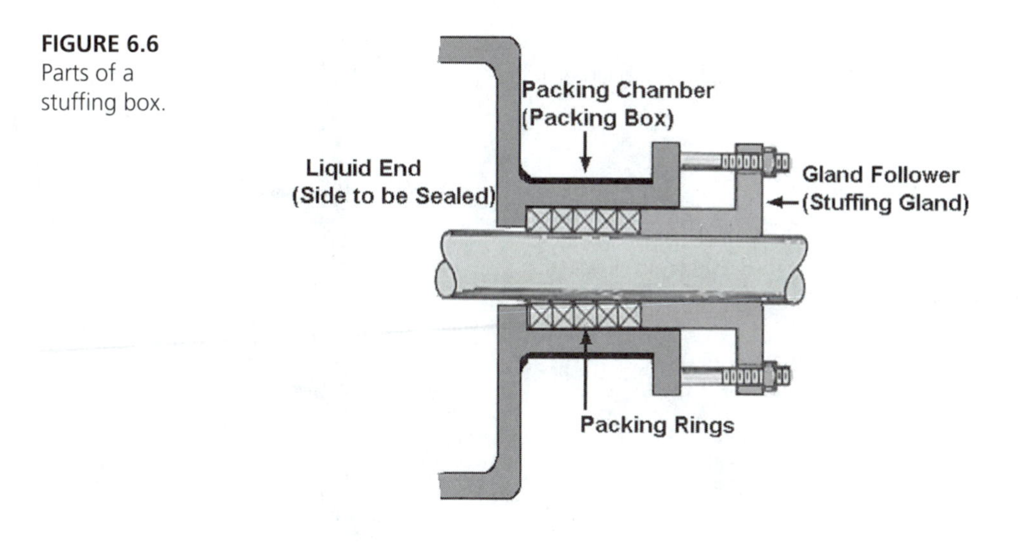

FIGURE 6.6
Parts of a stuffing box.

2. The packing rings are made of soft, compressible material that does the actual sealing.
3. The gland follower (sometimes called the *stuffing gland*) is used to compress the packing rings, thus reducing leakage.

The packing chamber and the gland follower can be any type of metal construction but are more commonly cast iron or bronze.

Packing rings come in a wide variety of materials. Older packings were made of cotton impregnated with asbestos, graphite, and grease. Newer packing materials use natural or synthetic fibers with graphite or Teflon. Lead or other metals are also used for higher-temperature applications.

Packing cross sections may be square, rectangular, or round. They may be braided, twisted, plaited, or laminated for increased life. Because packing rings are made of soft, pliable material, they usually compress and wear quickly leading to increased leakage. To decrease leakage, the gland follower can be tightened; however, there *must* be some leakage for good maintenance and long shaft life. Because the rubbing action of the shaft on the packing creates friction and generates heat, a small amount of leakage is absolutely necessary to assist in lubrication and to cool the shaft to ensure maximum wear. It also ensures that the gland follower is not too tight. The heat and wear are kept to a minimum by using hardened polished shafts (and sometimes shaft sleeves) and embedding lubricant in the packing rings.

Most stuffing boxes use at least five rings for good sealing and long life. A **lantern ring** is often installed in the middle of the packing (see Figure 6.7). A lantern ring is a specially shaped spacer (usually made of metal) that allows the introduction of an outside fluid. This provides a seal to prevent air from being drawn in. It also provides positive lubrication for the primary packing rings. Lantern rings also permit the injection of a clean,

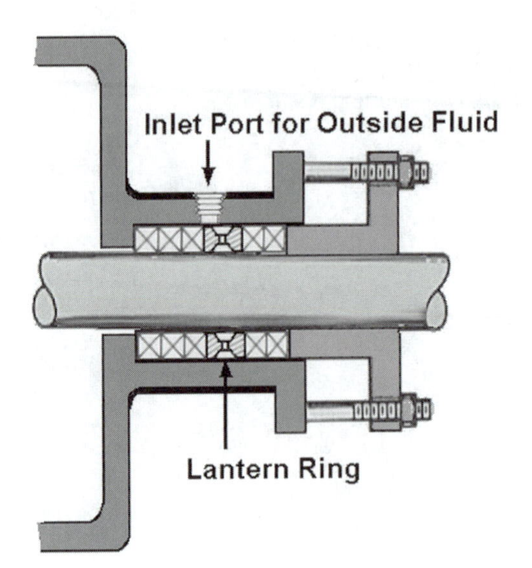

FIGURE 6.7
Stuffing box with lantern ring.

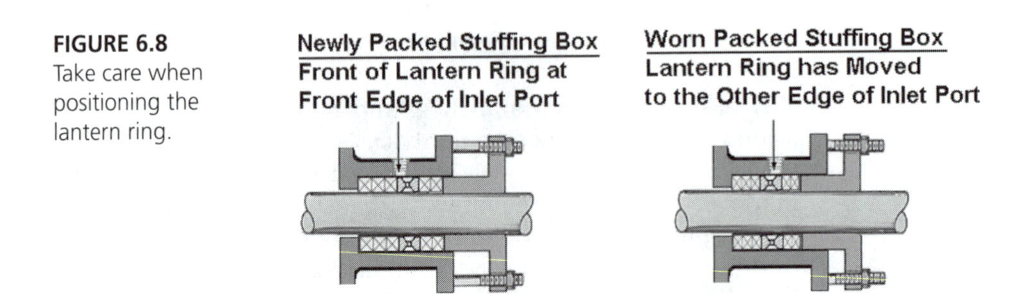

FIGURE 6.8
Take care when positioning the lantern ring.

cool fluid for pumps handling chemical or abrasive materials. This eliminates an otherwise necessary chemical leakage for cooling. With abrasive pumpage, premature damage to the shaft is avoided with injection of this clear liquid.

The lantern ring requires careful positioning to ensure that it is aligned with the inlet port (see Figure 6.8). Because the lantern ring moves as the packing wears and is compressed, it should be aligned with the front edge of the inlet port. As wear occurs, the lantern ring becomes more centered and then continues to move to the other edge of the inlet. Careful positioning helps the lantern ring stay operational for the life of the packing.

Using bolts with springs to tension the gland follower can reduce required periodic adjustment of the gland follower. The springs continue to compress the packing as it wears, keeping an even pressure on the gland follower, which is on the packing (see Figure 6.9).

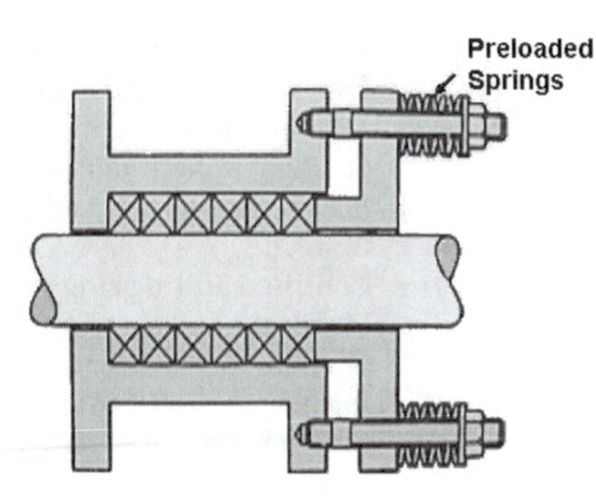

FIGURE 6.9
Stuffing box with preloaded springs.

6-5 STUFFING BOX PACKING INSTALLATION

Proper installation of the packing is important for low maintenance and long life of the packing and the shaft. When removing the old packing, look for signs of problems. If the inner diameter of the old packing is charred and dry and the shaft is scored and discolored, the packing may not be getting enough lubrication. Increase the controlled leakage for better lubrication. Wear on the outside of the packing rings indicates that the rings are rotating with the shaft. This is usually caused by shaft corrosion, causing rings to stick. An incorrect size of packing rings can also cause this problem by either not contacting the stuffing box or fitting too tightly on the shaft.

After removing the old packing, you should clean the stuffing box, making sure that all parts are clear and clean. Inspect the shaft or shaft sleeve for excessive scoring and other damage. Replace or repair these as necessary.

Precut and presized packing-ring kits can be purchased from the manufacturer. Most packing, however, is generally bought in rolls and must be cut to size for the application. The packing shaft or a mandrel of the same diameter should be used as a guide in cutting the rings to size. A butt joint cut, as illustrated in Figure 6.10, should be used for most pump applications.

For value stems and most reciprocating applications, a 45° skive cut should be used instead of a butt joint cut for better sealing.

The following are the steps required to complete the installation:

- The installation of the first packing ring is most critical. Always keep the packing square with the shaft and gently seat it at the bottom of the

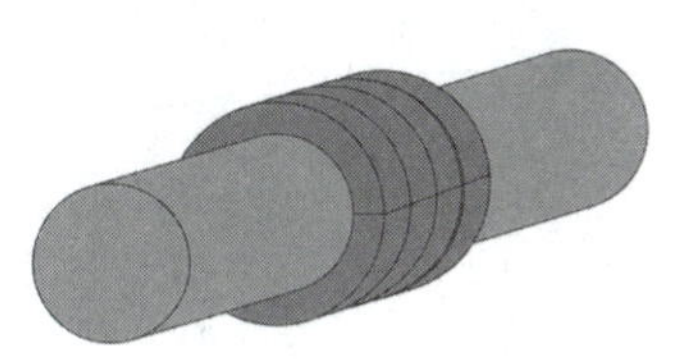

FIGURE 6.10
Butt joint packing cut.

stuffing box. An old sleeve or bushing is helpful for fully seating the first few rings. Do not use the gland follower to push multiple rings. The critical bottom rings will not seat properly, and the top rings may be crushed or damaged by excessive pressure.

- Rotate the second packing ring joint 120° and firmly seat it against the first ring.
- Continue adding rings, staggering each joint 120°.
- Install the lantern ring as previously discussed, making sure it is aligned with the front edge of the inlet port.
- Install the rest of the packing rings and the gland follower. The gland follower should extend out of the stuffing box about one-third of the total packing depth. Initial adjustment of the gland follower is made by tightening the follower while rotating the shaft by hand.
- Tighten the bolts evenly.
- Flood the pump and start the motor.
- Adjust the gland follower to allow a slight controlled leakage. After a few hours of operation, the gland follower will probably need retightening.
- Periodically recheck the stuffing box for proper controlled leakage.

6-6 AUTOMATIC OR MOLDED PACKING

This type of molded packing is called a*utomatic* because internal air or fluid pressure forces the lips of this type of packing to seal against some hard surface, usually a cylinder. The most common shapes of this molded packing are cup, U-shape, V-shape, and flange. Automatic packing is typically used in reciprocating applications.

Cup packing is the oldest style of automatic packing (see Figure 6.11). Leather was originally used for slow-speed applications, but molded rubber replaced leather as technology improved the reliability of molded packing. This type of automatic packing is generally used for slow-speed, lighter-duty applications. Cup packing is considered to be an unbalanced seal because fluid force operates only on one lip. In order to seal properly, cup seals require backplates or follower plates for support. Proper torque is important when installing and tightening follower plates; they must be tight enough to flare the cup to seal initially but not so tight that they create excessive friction. The cup

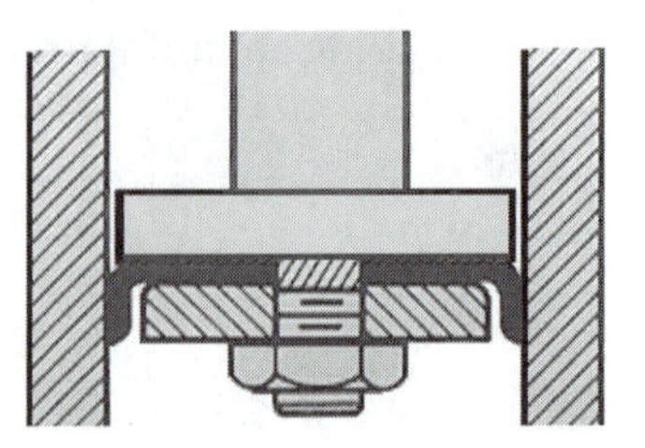

FIGURE 6.11
Automatic flange cup packing.

seal is created by fluid pressure, not excessive mechanical pressure produced by a follower plate that is too tight.

The U-shaped cup, or U-cup, was one of the first-balanced molded seals, because it seals on the inside and the outside diameter surfaces. Like all cup packing, U-cups are pressure-activated. They are commonly found in low-to-moderate-pressure hydraulic and pneumatic applications. To perform its automatic sealing well, the U-cup needs proper back and side support to prevent the lips from collapsing until fluid pressure becomes high enough to provide hydraulic support. Traditionally, a natural fiber, rubber, or metal ring called a **pedestal ring** provided this internal support. Today's pedestal rings are typically made of nylon or Teflon.

Figure 6.12 shows a single U-cup application. Figure 6.13 shows U-cups in a double-acting cylinder application. Notice that the U-cups are back to back if correctly installed. This configuration requires complete disassembly to replace U-cups, which may be a disadvantage for some applications. Figure 6.14 shows a face-to-face arrangement, in which both of the U-cups can be accessed from one side. This compromise has the disadvantage of trapping pressure between the two U-cups. It also makes it very easy to install U-cups backward. Always check your specific application to ensure that the U-cups face the right direction. When U-cups are installed backward, they do not seal properly. Fluid pressure may also crush improperly installed U-cups and cause rapid wear.

V-shaped (also called V-ring) packing is a modern-design molded packing that is primarily used for higher pressures, although it can also be found in all pressure ranges in different applications. Like U-cup packing, V-shaped

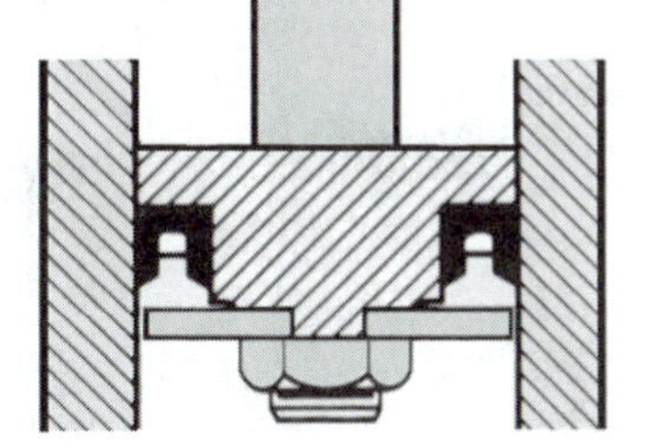

FIGURE 6.12
Single U-cup packing.

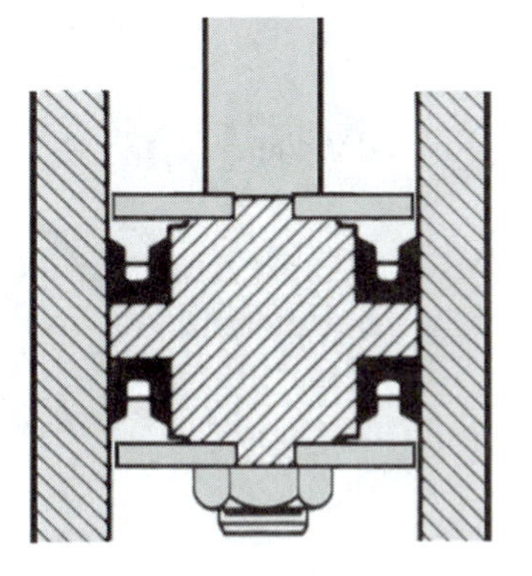

FIGURE 6.13
Back-to-back U-cups in a double-acting cylinder.

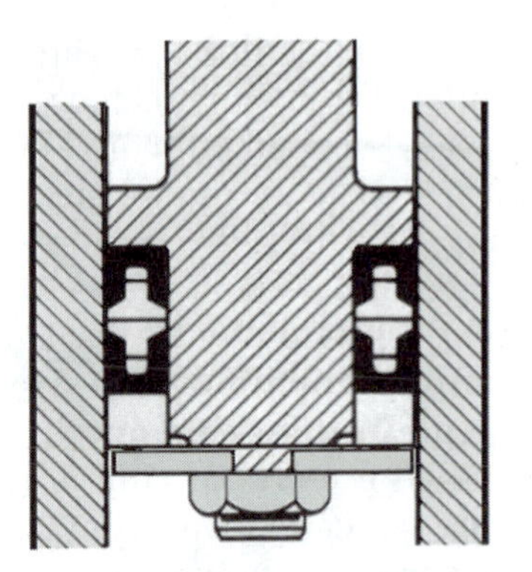
FIGURE 6.14
Face-to-face U-cups in a double-acting cylinder.

packing requires a gland for support. However, because a smaller cross section is used, there is much less pressure on the gland ring. This provides a more compact installation with greater dependability. Properly installed V-shaped packing lasts longer and performs better than any other type of molded packing. Most applications use 45° angle packing as a standard. V-shaped automatic packing has the unique advantage of allowing gland-ring adjustment when excessive wear occurs, extending the normal life of this type of packing. The number of V-rings required depends on the rod size, operating pressure, and ring material. Sometimes, a wiper or seal is used to prevent external contaminants from damaging V-shaped packing. V-shaped packing used for piston seals is *inside packed.* The type used as a shaft or rod seal is *outside packed.* In general, if the packing moves, it is considered to be an inside-packed application. If the packing stays stationary, it is an outside-packed application. Figure 6.15 illustrates an inside-packed V-ring, and Figure 6.16 shows an outside-packed V-ring.

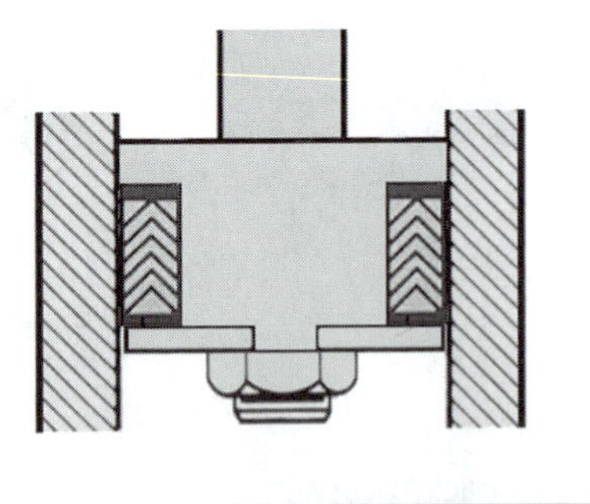
FIGURE 6.15
Inside-packed V-ring.

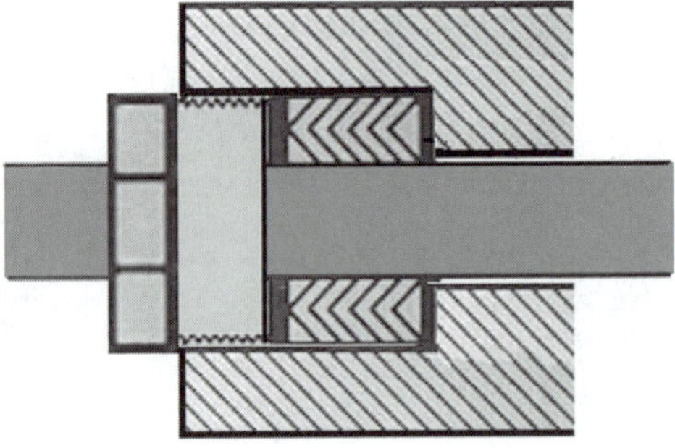
FIGURE 6.16
Outside-packed V-ring.

6-7 RADIAL LIP SEALS

Radial lip seals maintain a flexible interference fit with rotating or reciprocating shafts (see Figure 6.17). Radial lip seals are primarily used to seal in lubricant, but they can also be used to seal out contaminants or a combination of both. When used to seal in lubrication, the lip of the seal should be installed facing the lubricant side. If used for keeping contaminants out, the lip should be installed facing out. Dual-element lip seals are used to seal in lubricants and seal out contaminants (see Figure 6.18).They provide the best protection and are widely used in industry.

In some applications, radial lip seals rely on the elasticity of the seal material to maintain pressure on the shaft to seal it. Many other applications use a **garter spring** to provide additional force to help maintain lip contact with the shaft and extend the operational life. Figure 6.19 shows a typical radial lip seal with a garter spring.

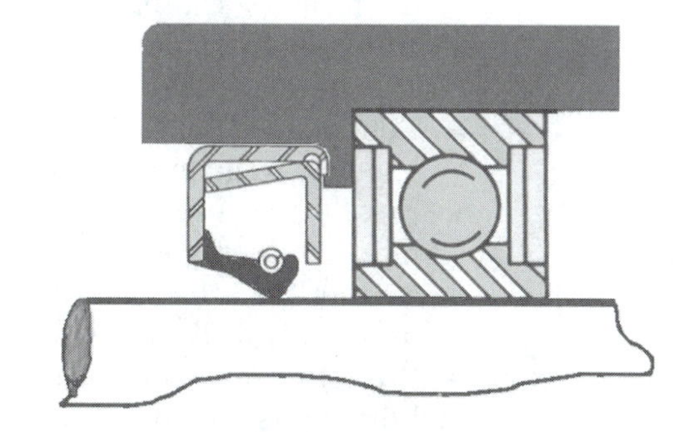

FIGURE 6.17
Radial lip seal and bearing.

FIGURE 6.18
Dual-element lip seal.

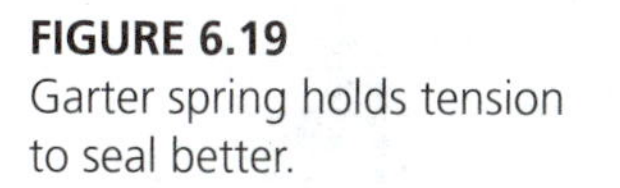

FIGURE 6.19
Garter spring holds tension to seal better.

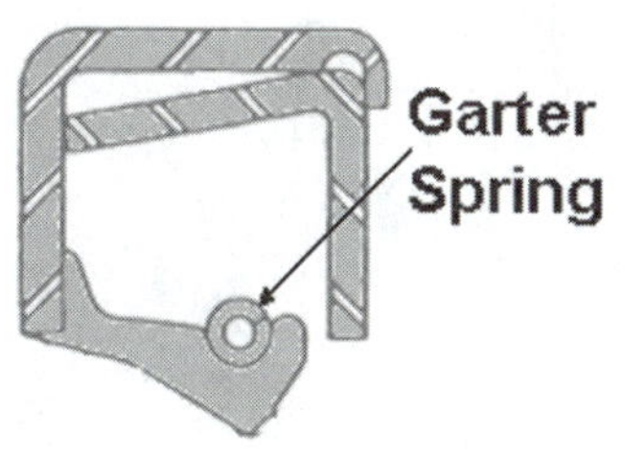

6-8 INSTALLING RADIAL LIP SEALS

Correctly installed radial lip seals give excellent service and long life. The following checklist will help you make sure your installation is a good one:

- Make sure the replacement seal is correct for the application. Check the size, material, and type of seal.
- Check for bad bearings, excessive runout, or endplay.
- Before sliding the seal on the shaft, repair any nicks or scratches that might damage the seal.
- Cover the sharp edges of the keyway or shaft end with a sleeve or tape to protect the lip seal from damage.
- Inspect the lip seal for cuts or other damage. Lubricate it with a thin coat of lubricant.
- Determine which way the seal should face. For retention of lubricant, the primary sealing lip should face the lubricant. For exclusion of contaminants, the primary sealing lip should face the source of contaminants (see Figure 6.20).
- Gently slide the seal on the shaft until it is lined up squarely and contacts with an interference fit.
- Press the seal in, using the correct adapter (see Figure 6.21). If you must fabricate the adapter, make it a ten-thousandths of an inch smaller than the seal-housing bore. Do *not* use a punch or screwdriver or directly strike the seal with a hammer.
- Press the seal squarely until it bottoms out in the housing. If the housing is a through-bore type, use an adapter with a shoulder

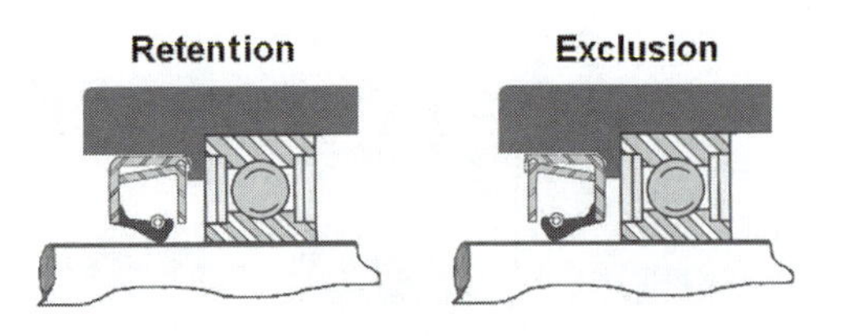

FIGURE 6.20
A lip seal may be installed for retention of lubricant or exclusion of contaminants.

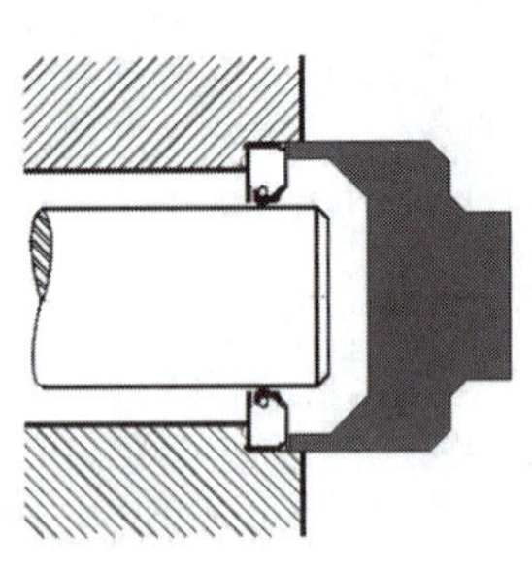

FIGURE 6.21
Use the correct adapter to press in the seal.

FIGURE 6.22
For through bores, use an adapter with shoulders.

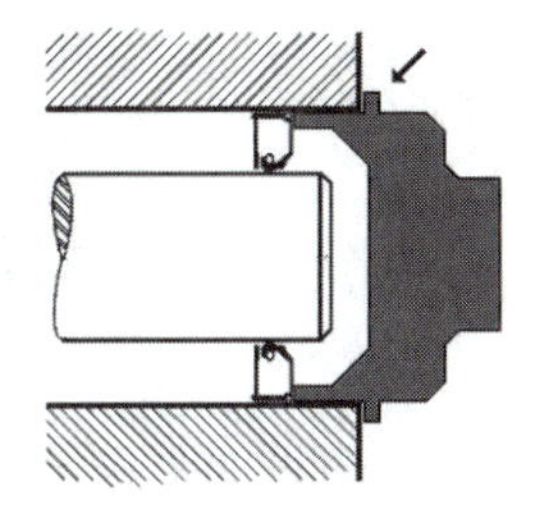

to ensure a properly installed seal; square it with the shaft (see Figure 6.22).

- Check the seal for any damage that might have occurred during installation. If equipment is to be painted, cover the seal and shaft area to protect them from paint.

6-9 MECHANICAL SEALS

Mechanical seals are used in most pumps to prevent leakage of the fluid being pumped. Packing boxes (stuffing boxes) were originally used to perform this function, but they have been for the most part replaced because of the superiority of the mechanical seal. Advantages of mechanical seals over conventional packing are as follows:

- Zero leakage of fluid
- Capable of higher pressures
- Better suited for chemical sealing
- Little or no shaft wear
- Less maintenance
- No need for adjustment
- Less friction
- More efficient

A mechanical seal must seal at the following three points:

1. A static seal between the housing and the stationary part of the seal.
2. A static seal between the shaft and the rotating part of the mechanical seal.
3. A dynamic seal between the rotating seal face and the stationary seal face.

Figure 6.23 shows the components of a basic mechanical seal. Note that the stationary seal is locked in the housing with an O-ring to seal and prevent rotation. The rotating seal is sealed on the shaft with another O-ring. The dynamic seals formed by the mating faces of the stationary and rotating seal are precision-lapped for a flatness of typically three light bands and a surface

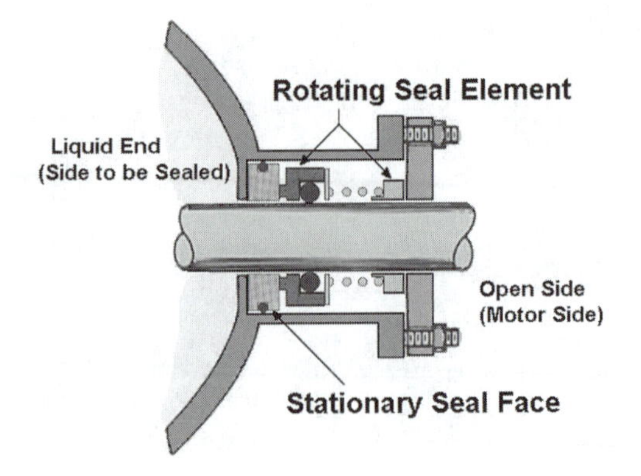

FIGURE 6.23
Parts of a mechanical seal.

finish of 5 micro-inches. The driving member is locked on the shaft and positions the spring assembly and rotating face. The spring assembly rotates with the shaft and provides pressure to keep the mating faces together during periods of shutdown or when there is insufficient hydraulic pressure. As wear takes place between the mating faces, the rotating face must move along the shaft to maintain contact with the stationary face. The O-ring must be free to move along the shaft to make this work properly.

The stationary seal member is usually made of ceramic, stainless steel, or some other hard material, depending on the application. The rotating face member typically has a carbon/graphite base and is the wear member of the seal. The springs and other metal parts are plated steel or stainless steel. Many different types of materials are used for components involved in chemical applications.

Each basic component of a mechanical seal must perform its function properly for a good seal. Fluid and spring pressure in the seal chamber forces the faces together and provides a thin film of liquid between them for lubrication. The faces, selected for low frictional qualities, are the only rubbing parts. These basic components are a part of every mechanical seal. The form, shape, type, and design vary greatly depending on the service and manufacturer. The basic theory, however, remains the same.

Mechanical seals may be mounted *internally* or *externally.* The internal, or *inside,* seal, which is more common, has a rotating member mounted inside the seal housing on the fluid side of the pump. With an external, or *outside,* mechanical seal, the rotating member is mounted outside the seal housing away from the fluid side of the pump (see Figure 6.24).

The springs and rotating element are not in contact with the fluid being pumped, thus reducing corrosion problems and preventing product accumulation in the springs. The external mechanical seal is usually easier to install, adjust, and maintain. The disadvantage of an external seal is that fluid pressure tends to push sealing faces apart. With internal seals, fluid pressure pushes sealing faces tighter.

Seals are available in two types: single or double. The internal and external seals just described are single seals, in which only one sealing element is used. Most pumps use single internal seals. Double seals have two mechanical

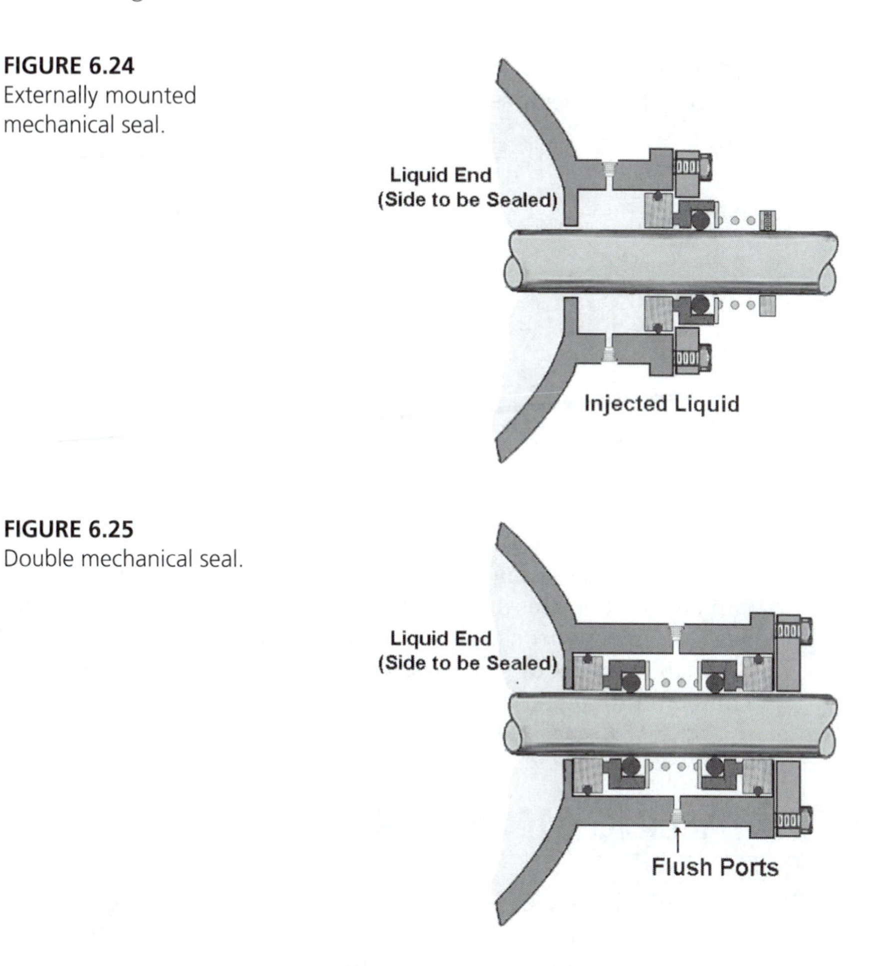

FIGURE 6.24
Externally mounted mechanical seal.

FIGURE 6.25
Double mechanical seal.

seals mounted back to back in the seal housing and are typically used for specialized applications (see Figure 6.25). The seal housing is pressurized with a clear liquid from an external source. This liquid is circulated through the double-seal chamber at 1/4 to 1 gal/min to cool and lubricate the mechanical seals. Double mechanical seals are used for pumping solutions that have solids or abrasives or that are toxic or extremely corrosive. The external-source fluid must be compatible with the pumped material.

Mechanical seals may be classified as *balanced* or *unbalanced.* This does not refer to weight distribution but to the rotating face configuration. Figure 6.26 shows an internal unbalanced seal. The fluid pressure as well as the spring pressure act on the rotating member, forcing it tighter against the stationary member. This type of configuration is the most widely used type for pumps that handle clear liquids up to 100 lb/in^2. An unbalanced external seal is good for pressures up to about 35 lb/in^2.

Figure 6.27 shows an inside-balanced mechanical seal. Note that the rotating member has a step design on its front edge. This step allows fluid to counteract the total force on the back of the rotating member. This opposition

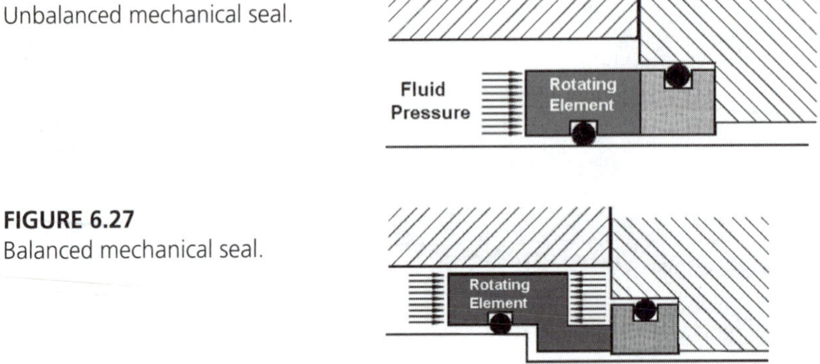

FIGURE 6.26
Unbalanced mechanical seal.

FIGURE 6.27
Balanced mechanical seal.

reduces the force of the fluid pressure pressing the rotating member against the stationary member. Lighter-contact pressure increases lubrication and seal life. It also permits sealing of higher pump pressures; single internal balanced seals can be used for pressures up to 2,000 lb/in^2. They are also used extensively for sealing light hydrocarbons, which tend to vaporize easily. Balanced external seals are capable of pressures of more than 150 lb/in^2. That is a great increase compared to the 35 lb/in.^{2} limits for the unbalanced outside mechanical seal.

6-10 INSTALLING MECHANICAL SEALS

Mechanical seals are available in many different designs. Installing these seals can be extremely simple or extremely complicated. This section covers some basic information that applies to installing most mechanical seals. Always refer to the manufacturer's recommendations as the ultimate authority for a specific application. When installing a new mechanical seal, you should follow these guidelines:

- Make sure the mechanical seal is correct for the application. Verify the part number; check the size and chemical compatibility of the seal and the fluid pumped. Check the type of seal for pressure limitations, temperature limitations, and so on.
- Check for bad bearings, excessive runout, or endplay.
- Check the shaft and housing for burrs, nicks, scratches, or other damage. Repair or replace the parts as necessary.
- Cover the sharp edges of the keyway or shaft end with a sleeve or tape to protect the mechanical seal from damage.
- Mount the rotating member on the shaft. Location of this part is critical because its location determines the amount of force applied to the seal face. Too little force will result in a leaking seal. Too much force will result in premature seal wear and failure.

6-11 LABYRINTH SEALS

Seals can be two types: *contacting* and *noncontacting*. All the seals discussed so far have been the contacting type, or those that physically touch the moving part to complete the seal. Noncontacting-type seals do not rub or make physical contact with moving parts. Labyrinth seals are the noncontacting type. Labyrinth seals provide a series of small chambers separated by ribbed protrusions whose clearance with the shaft is very small. Each of the multiple ribs creates a pressure drop, so the lubricant is retained by back pressure created by this type of seal.

Labyrinth seals that are used in industrial bearings utilize grease to exclude contaminants and moisture from the bearing. Because of their noncontacting construction, these seals cannot retain a liquid as contacting seals do; however, they do retain grease under normal conditions.

Labyrinth seals are also used to seal gases in compressors and steam turbines where a small leakage is acceptable. The leakage rate is directly proportional to the radial clearance between the labyrinth seal and the shaft. It is inversely proportional to the number of teeth. That is, a smaller radial clearance means less leakage and adding more teeth means less leakage. The advantage of labyrinth seals in this type of application is that their all-metal construction can handle very high temperatures [up to 932°F (500°C)]. They are also very efficient because they have no rubbing friction. Labyrinth seals usually have long lives with no maintenance required. Common materials used for labyrinth seals are brass, bronze, aluminum, or other materials, depending on the application.

One of the simplest labyrinth seal designs is shown in Figure 6.28. This straight-through design creates backpressure to seal. It is not as efficient as the alternating-teeth design (shown in Figure 6.29) but is easier to manufacture

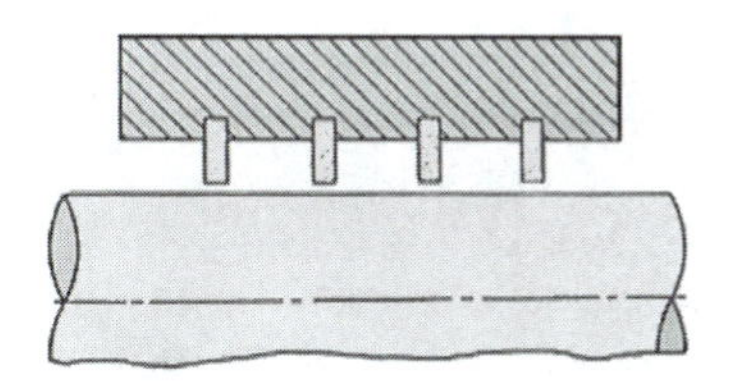

FIGURE 6.28
Labyrinth seal.

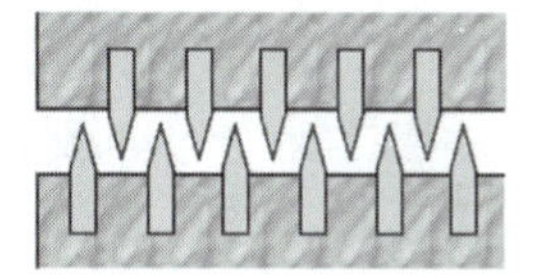

FIGURE 6.29
Alternating-teeth labyrinth seal.

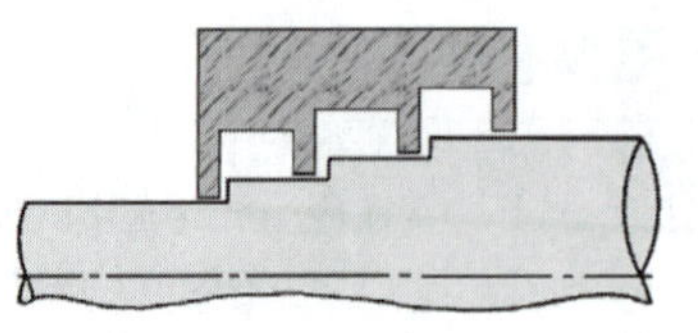

FIGURE 6.30
Stepped labyrinth seal.

and maintain. Figure 6.30 shows a simple stepped labyrinth seal. It is more effective than a straight labyrinth seal, and the assembly is much simpler than it is for labyrinths with alternating teeth.

QUESTIONS

1. What are the two main classifications of modern seals?
2. In general, how should you correctly torque a rectangular shape?
3. Name two typical causes of gasket failure.
4. The O-ring is considered to be the most common ______________ ______________ seal.
5. What are the most common causes of O-ring failure?
6. What are the three main parts of a simple stuffing box?
7. Why do stuffing box applications require some leakage?
8. What is the function of the lantern ring?
9. What is automatic packing?
10. Name three advantages of mechanical seals.

CHAPTER 7

Belt Drives

7-1 INTRODUCTION

Belt drives have been used for hundreds of years to transmit mechanical power. It has been only recently, however, that anything other than flat belts made of leather has been used. Flat belts could not handle very much power, and pulleys had to be fairly large to get enough wrap to prevent slippage. As the world became more industrialized, the power and speed of machinery needed to increase. Better belt drives were required to meet these increased demands. These power-transmission belts are either *friction* drive or *positive* drive (see Figure 7.1). Flat belts rely on friction for their gripping action and require a good bit of tension to prevent slippage of the belt. In 1917, the first V-belt was patented. V-belts are a friction-type drive (see Figure 7.2) but require much less tension than flat belts because they use the mechanical advantage of one of the simple machines we have already studied: the *wedge*. The wedge design of the V-belt multiplies the gripping force with less belt tension. Less belt tension means increased bearing and belt life. Compared to flat belts, V-belt drives can handle high loads and are much more compact.

FIGURE 7.1
Industrial belt drive.

FIGURE 7.2
V-belt drive.

FIGURE 7.3
Timing belt drive.

The first *synchronous,* or *timing,* belt (see Figure 7.3) was a positive-type drive introduced in the 1940s. Synchronous drives use the engagement of the belt teeth with the teeth on a pulley (sometimes called a *sprocket* because it has teeth). This type of belt drive does not depend on friction for its transmission of power. In a positive drive, there is no designed slippage of the belt at all. Synchronous drives require even less tension than V-belt drives and are capable of transmitting more power with smaller pulleys and belts.

V-belts are the most widely used type of belt drive in industry today. There are many types of specialized V-belts, but we discuss only the main types used in industry.

7-2 V-BELT STANDARDIZATION

Standardization is very important to modern industry. Henry Ford recognized this when he started using production-line techniques to manufacture the Model 'T' automobile. Standardization made the finished goods cheaper and easier to repair by the consumer. With the popularity of the production line, standardization became necessary to increase interchangeability of parts. In the United States, the American National Standards Institute (ANSI) establishes many standards used in industry. The Rubber Manufacturers' Association (RMA) writes industrial power-transmission-belt standards for

TABLE 7.1
Standardized industry V-belts.

Belt Section	Industry Standard, Description	Width *W*, in inches	Thickness *T*, in inches
3L		3/8	1/32
4L	FHP, single	1/2	5/16
5L		21/12	3/8
3V		3/8	5/16
5V	Narrow or wedge	5/8	17/32
8V		1	9/12
A		1/2	5/16
B	Classical	21/32	3/32
C	multiple	7/8	7/12
D		11/4	3/4
E		11/2	29/32

ANSI. All manufacturers of these belts contribute to and agree upon each of the standards that is written. With the increase of world trade, the international aspect of standardization has become increasingly important. The International Organization for Standardization (ISO) has created industrial standards that are supported by most manufacturing nations in the world. ISO standards do not officially govern U.S. industry, but many U.S. industries have adopted these ISO standards to be more competitive with international companies. ISO and ANSI review their standards on a five-year cycle to ensure that they are current with industry knowledge and practices (see Table 7.1).

7-3 CLASSICAL V-BELTS

The classical V-belt was the first widely used, standardized, industrial V-belt. This industry workhorse has been in use for more than 60 years and is still not ready to be retired. Some industry mechanics also refer to this belt as a *conventional,* or *standard,* V-belt. Classical V-belt drives are typically designed for up to 7:1 speed ratios, up to an 8,000-rev/min maximum speed, and up to a 300-hp operation. There are five belt sizes identified by a letter (A, B, C, D, and E) indicating the belt cross section followed by numbers indicating belt length. The A size is the smallest, the B size is next largest, and so on. The E sizes are being phased out and should be used only for replacement on existing drives not for new drives. Figure 7.4 shows the cross-section dimensions and relative size for each classical belt.

FIGURE 7.4
Classical V-belts.

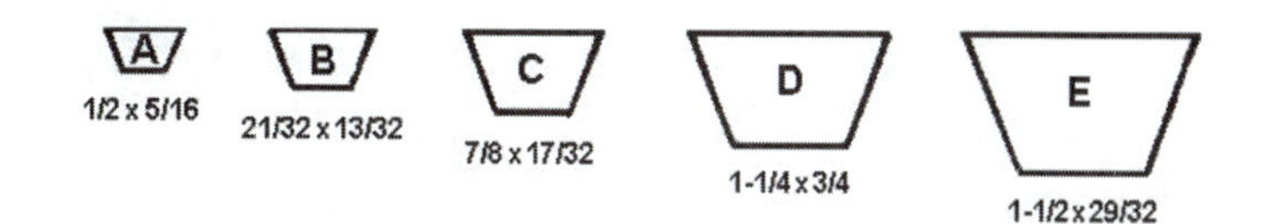

The number following the letter designates the length of the inside circumference of the V-belt in inches. For example, a belt marked *B90* has cross-sectional dimensions of 21/32 in. × 7/16 in. (from Figure 7.4) with an inside circumference of 90 in. Classical V-belts with an inside circumference greater than 210 in. are measured by their *pitch length.* **Belt pitch length** is the approximate circumference of the tensile cord of the V-belt located just above the center of the cross section.

Another important number, called the **match number,** is stamped on a V-belt. Match numbers provide a way to match exactly the V-belts used in a multiple V-belt drive. A match number of 50 means that the V-belt is exactly the length it was designed to be. As the match number increases or decreases from the standard number of 50, the V-belt circumference increases or decreases. Each digit represents a change of 1/16 in. For the B90 belt, if the match number is 52, the inside circumference of the V-belt is 90 in. plus 2/16 in. long. If the match number is 49, the V-belt length is 90 in. minus 1/16 in. Match numbers are used when selecting single V-belts for multiple-groove *sheaves.* (*Sheave* is the U.S. term used for a V-belt pulley. In ISO, the term *pulley* is applied to all belt drives.) Matching exactly the length of all V-belts in a multiple V-belt drive ensures that all V-belts pull equally and that the drive is more efficient.

7-4 NARROW-GROOVE V-BELTS

This second-generation V-belt was introduced around 1959 and are sometimes called "wedge" belts. It can handle more power, uses smaller-diameter sheaves, and is more economical than a classical V-belt. Three narrow-groove standard V-belts replace all five sizes of the classical V-belts in power capability (see Figure 7.5). This is a great advantage to industry because fewer V-belts have to be kept in inventory for repairs, which lowers the cost of inventory and

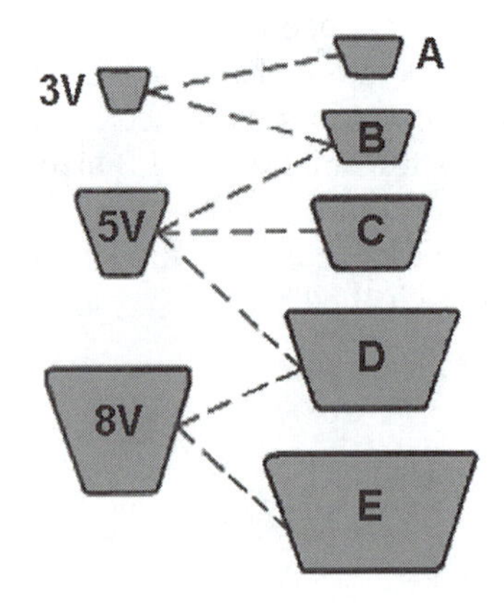

FIGURE 7.5
Three narrow-groove, V-belts can do the work of five of the old classical V-belts.

FIGURE 7.6
Narrow-groove industrial V-belts.

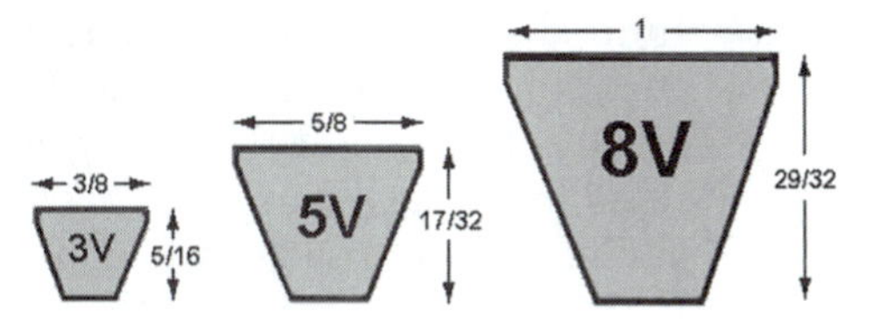

reduces recordkeeping associated with larger inventories. The cross-sectional sizes of the narrow-groove V-belts are represented by 3V, 5V, and 8V (from the smallest to the largest).

Figure 7.6 shows that the first digit of the narrow-groove V-belt number indicates the nominal top width in 1/8-in. increments. As with the classical V-belt, a number that represents the circumference of the V-belt in inches and tenths of inches follows this prefix. Unlike the classical V-belt numbering system, these numbers designate the *outside* circumference, not the inside. The last digit represents tenths of inches. For example, a narrow-groove V-belt marked 3V375 has cross-section dimensions of 3/8 in. × 5/16 in. with an outside circumference of 37.5 in.

Narrow-groove V-belts have no match numbers because newer technology enables manufacture of these V-belts to an exact size. This is achieved through processing innovations in cord treatment, material compounding, and belt curing.

You should never use narrow-groove V-belts with classical sheaves, or vice versa. The angle of the sides of the sheaves where the V-belt rides is not the same. If the correct sheave is not used, the drive will experience lower-horsepower capabilities, V-belt slippage, shortened V-belt life, and sheave damage.

These V-belts also require more tension than classical V-belts.

7-5 COGGED V-BELTS

Cogged V-belts are improved versions of standard classical and narrow-groove belts. These modified V-belts should not be confused with synchronous belts.

Cogged V-belts have notches molded in the underside of a standard V-belt (see Figure 7.7). These notches offer greater flexibility and better heat dissipation. Greater flexibility means that smaller sheaves can be used for areas with

FIGURE 7.7
Cogged V-belt.

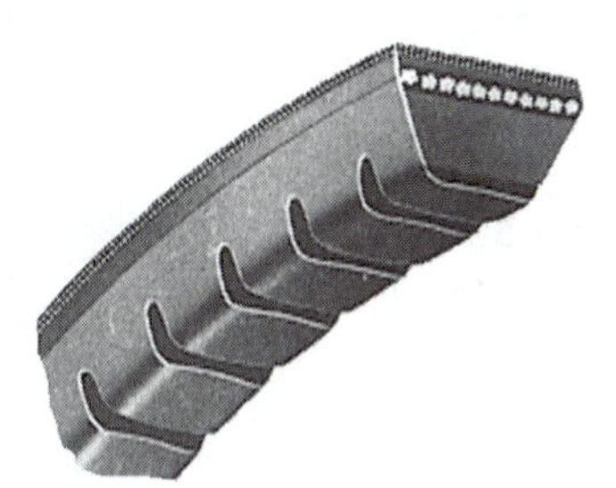

limited space to mount the drive. Improved heat dissipation increases V-belt life. This design can increase the horsepower capacity of the V-belt by as much as 30% because of the better gripping power of the V-belt. Cogged V-belts at rated loads can be up to 98% efficient. Typically, they can also improve energy savings of up to 5% over conventional V-belts. Although they are generally more expensive initially than conventional V-belts, the increased efficiency and longer life of cogged V-belts can make them the best economical choice. Cogged V-belts are usually designated by the classical or narrow-grove cross-sectional prefix followed by the letter *X* and then the length code. For example, a classical B90 V-belt in its cogged version is labeled BX90, and a narrow-groove 3V375 belt becomes 3VX375.

Cogged V-belts can be used to replace conventional V-belts without replacing existing drive sheaves. The belts are fully interchangeable, and by replacing conventional V-belts with cogged V-belts you get all the benefits of the newer technology of cogged belts.

CAUTION: *You should never mix conventional V-belts and cogged V-belts in a multiple-V-belt drive.*

7-6 BANDED V-BELTS

A banded V-belt is a multiple V-belt in which individual classical or narrow-groove V-belts are attached with a common backing (see Figures 7.8 and 7.9). A common back ensures that all belts are exactly the same length and share equally in the transmission of power. This design is used where vibration, shock loads, or misalignment causes belts to turn over, whip, or jump off multiple-groove sheaves. Banded V-belts are commonly found in two-, three-, four-, and five-belt bands. They are not identified by a universally used number or designation; for ordering purposes, they are referred to, for example, as a two-band B90 banded V-belt.

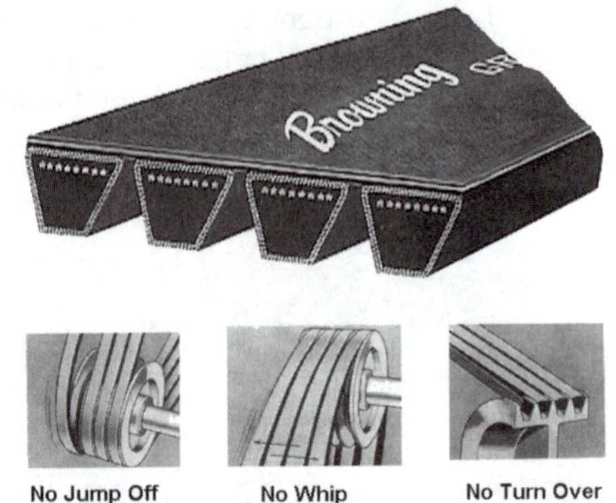

FIGURE 7.8
Banded V-belt.

FIGURE 7.9
Advantages of banded V-belts.

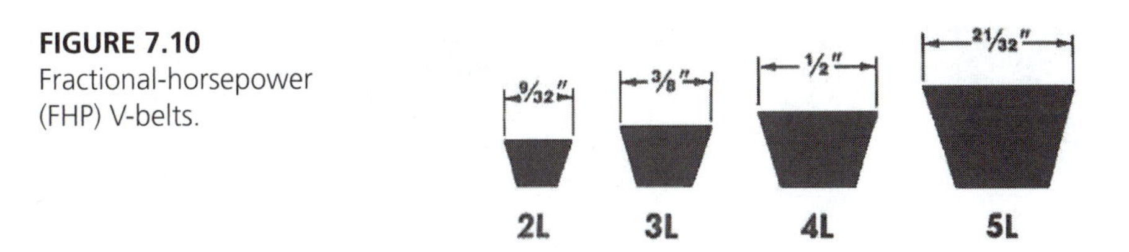

FIGURE 7.10
Fractional-horsepower (FHP) V-belts.

7-7 LIGHT-DUTY OR FRACTIONAL-HORSEPOWER (FHP) V-BELTS

These V-belts are designed for light-duty use with smaller-diameter sheaves. They are intended for infrequent or intermittent use rather than continuous use. Single V-belts are available in four common sizes: *2L, 3L, 4L,* and *5L.*

Figure 7.10 shows that the first number of the FHP V-belt is the approximate measurement of the top width of the belt in 1/8-in. increments. The *L* indicates light duty; the largest belt in this series is not capable of even 3 hp. A number that represents the circumference of the V-belt in inches and tenths of inches follows the prefix. As with the narrow-groove V-belts, these numbers designate the outside circumference, with the last digit representing tenths of inches. For example, a 3L335 FHP V-belt has cross-sectional dimensions of 3/8 in. × 7/32 in. with an outside circumference of 33.5 in. Notice that the cross-sectional dimension of this V-belt is almost the same as a 3V narrow-groove V-belt (3/8 in. × 5/16 in.). These V-belts are *not* interchangeable! The horsepower capacity of the industrial-duty narrow-groove V-belt is much greater. Comparing these two V-belts would be like comparing a domestic cat to a tiger. They look alike and are from the same family, but they are definitely not the same.

7-8 RIBBED V-BELTS

Ribbed V-belts are a technological update of the old flat belts that have integrally molded ribbing on the contact side. Friction between the ribs and sheave grooves provide traction for these belts in addition to the wedging action of conventional belts.

Because the ribs are relatively small, V-ribbed belts do not have as much mechanical advantage because they have less wedging action than conventional V-belts do; therefore, they require approximately 20% more belt tension than conventional V-belts. V-ribbed belts have high lateral stability, but they have almost no side flexibility, which means that they require more accurate alignment than conventional V-belts do. V-ribbed belts offer the power transmission capability of conventional V-belts with the flexibility of flat belt

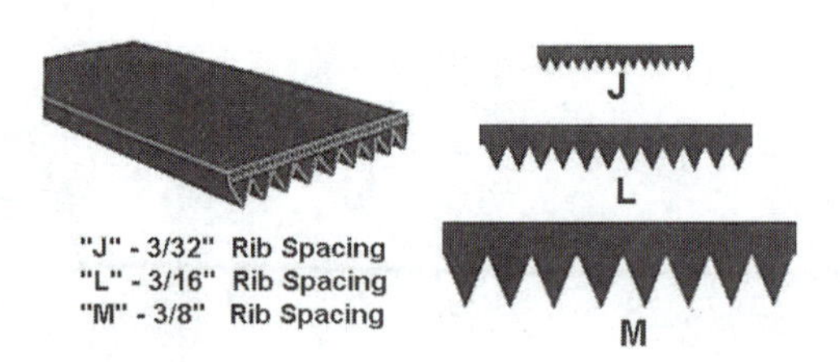

FIGURE 7.11
Ribbed V-belts.

drives. They are capable of handling up to 2,000 hp and have possible ratios of up to 40:1. V-ribbed belts also provide improved performance on 1/8-turn, 1/4-turn, and serpentine drives.

V-ribbed belts are manufactured in three sizes, J, L, and M, with J being the smallest and M being the largest. With this series, the prefix is the pitch length of the belt in inches and tenths of inches, the middle letter is the cross-sectional size, and last digit is the number of ribs the belt has (see Figure 7.11). These belts usually contain 6, 10, or 16 ribs, but special belts may have over 100 ribs. For example, a V-ribbed belt marked 460J6 has a pitch length of 46.0 in. with six of the J-size ribs.

7-9 VARIABLE-SPEED V-BELTS

Variable-speed V-belts are wider than conventional V-belts to handle the higher-horsepower requirements demanded of them (see Figure 7.12).

These special belts have top widths typically ranging from 7/8 in. to 3 in. and can handle up to 100 hp with a single V-belt. Drive ratios of up to 10:1 are possible.

A variable-speed V-belt is similar to a conventional V-belt drive except that it is wider (for higher-horsepower capabilities) and it operates on sheaves with movable sides. Conventional V-belt drives operate with solid sheaves that deliver a fixed ratio; to change the drive ratio, you must change the sheaves. Adjusting the movable sides in and out can change variable-speed drive ratios,

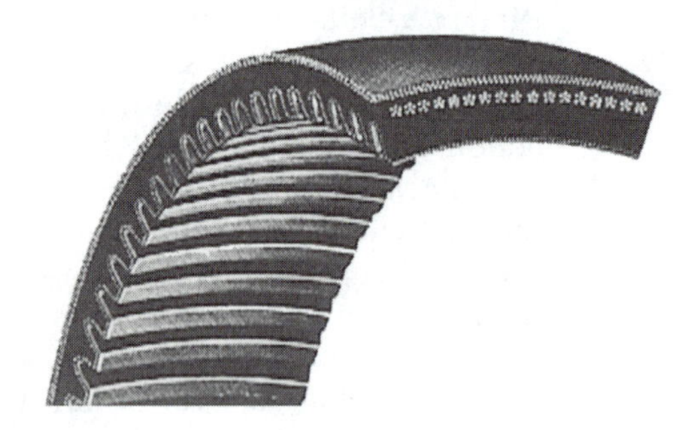

FIGURE 7.12
Variable-speed V-belt.

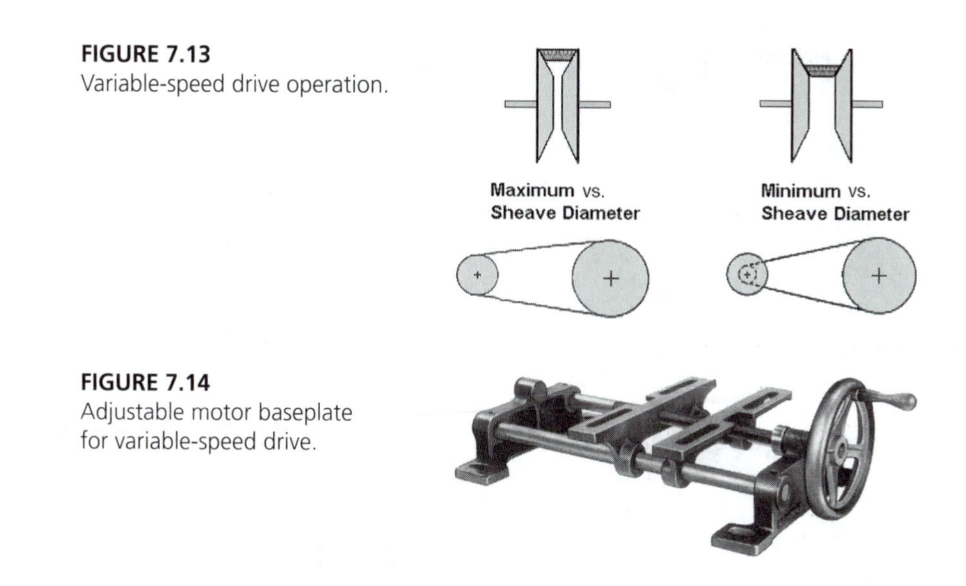

FIGURE 7.13
Variable-speed drive operation.

FIGURE 7.14
Adjustable motor baseplate for variable-speed drive.

which changes the effective pitch diameter of the sheave (see Figure 7.13). A typical variable-speed drive consists of a V-belt, a movable side, a spring-loaded sheave, and a sheave mounted on an electric motor with an adjustable baseplate (see Figure 7.14).

As with the other V-belt numbering systems we have studied, variable-speed V-belt numbers tell us the physical dimensions of the V-belt. Variable-speed belts have a four-digit prefix followed by a V and then the pitch-length number. The first two digits of the prefix give the top width of the belt in 1/16 in. increments. The next two digits indicate the angle of the groove in which the V-belt is designed to operate. The letter *V*, indicating a V-belt, follows next. The last digits indicate the pitch length of the V-belt in inches and tenths of inches. For example, a variable-speed V-belt marked 1422V540 has a top width of 14/16 in., operates on a 22° side-angle sheave, and has a pitch length of 54.0 in.

7-10 REPLACING A V-BELT

Before replacing an old V-belt with a new one, you should first check the condition of the sheaves. Damaged, worn, or dirty sheaves will substantially reduce V-belt life. Nicks and gouges can cut the V-belt. Dirt in the grooves can abrade the V-belt, and oil can attack the belt materials. Worn grooves will allow the belt to bottom out, slip, and become damaged.

CAUTION: *Never use damaged or worn sheaves with new V-belts. Repair or replace them.*

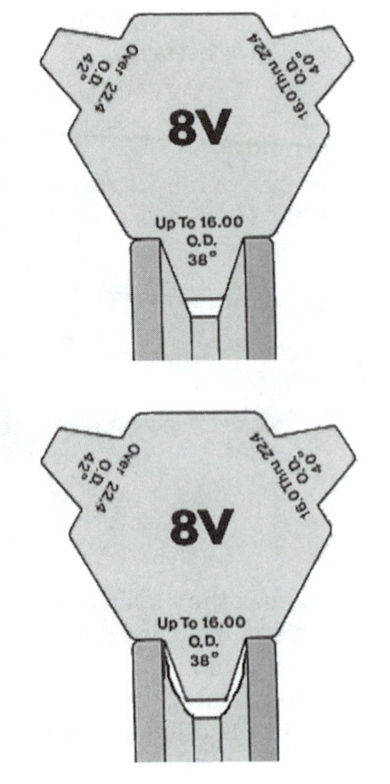

FIGURE 7.15
Sheave groove gauge with good sheave.

FIGURE 7.16
Badly worn sheave.

When checking the sheave for wear you should follow these guidelines:

- Select the correct sheave groove gauge and template for the sheave size (see Figure 7.15).
- Insert the gauge in the groove and look for voids that indicate dishing or other uneven and abnormal wear (see Figure 7.16).
- Place a new belt in the sheave groove. The top of the belt should be flush with the outer diameter of the sheave. If the belt top is below the outer diameter of the sheave, the groove is worn. An exception to this is the *combination sheave* (see Figure 7.17).

Combination sheaves are sometimes used with *A*- or *B*-section belts. This interesting design saves money in production and inventory because one sheave can be used to run *A*- or *B*-type V-belts. This sheave is designed with

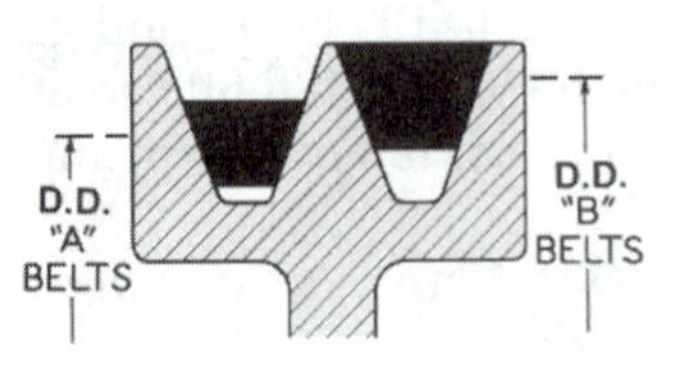

FIGURE 7.17
Combination groove sheave.

a deeper groove so *B*-section V-belts will be flush with the top of the groove, but smaller *A*-section V-belts will run farther down the groove. (Note: The effective pitch diameter depends on whether you use the sheave with an A-section or B-section V-belt.) In either case, the V-belt should never contact the bottom of the groove when operating. If the bottom of the groove is shiny, the V-belt is riding too far down. If the sheaves need to be replaced, refer to Section 7.11, "Sheave Replacement," for sheave removal and installation.

The next step is to identify the correct replacement V-belt for the drive. If the old V-belt has worked well, all you need to do is get the number from the old V-belt and replace it. On multibelt drives, do not mix new and used V-belts. Always replace the complete set of V-belts with new V-belts, even if only one or two seem worn or damaged. Used V-belts are usually worn in the cross section and stretched. A new V-belt will ride higher in the sheave, travel faster, and be at a much higher tension. If the drive is tensioned correctly for the old V-belt, the new V-belt cord center may become ruptured, allowing the new V-belt to stretch. Shortly after this occurs, it will cease to accept its full share of the load, leaving the drive underbelted. Thus, the new V-belt will be wasted. V-belts of different brands should not be mixed for the same reasons. If the V-belts have match numbers, be sure they are all the same.

Banded V-belts are always a good choice for multibelt replacement. A banded V-belt set will generally greatly increase the service life over a multibelt drive with single V-belts. Sheave condition is very important to long life for any V-belt, but it is especially critical for banded V-belt applications. Worn or damaged sheaves will destroy banded V-belts very quickly. If the center distance of the drive cannot be adjusted to properly tension the replacement V-belts, then a different-length V-belt must be used. Belts should generally be replaced by the same type of V-belt unless the drive tension cannot be adjusted satisfactorily using the exact replacement (the belt is still slack with take-up fully extended, or vice versa). If the belt numbers are illegible, then belt length can be calculated by the equation

$$L = 2C + 1.57\,(D + d) + \frac{(D - d)^2}{4C}$$

where L = length of the belt
C = center distance at the center of adjustment
D = diameter of the large sheave
d = diameter of the small sheave

Long center distances are generally not recommended for V-belts because the excess vibration of the slack side of the belt shortens belt life. Generally, center distances should not be greater than three times the sum of the sheave diameters nor less than the diameter of the larger sheave.

7-11 SHEAVE REPLACEMENT

If sheaves are worn or damaged beyond repair, they should be removed and replaced with new sheaves before the V-belts are replaced. There are three common types of sheaves:

1. Bored-to-size, or press-fit, sheaves
2. Taper-lock-type compression bushing
3. Quick-disconnect-type (QD) compression bushing

They are all removed and installed differently.

A press-fit sheave is a simple sheave with a straight hole bored in the center. Press-fit sheaves are typically used for light-duty applications with FHP V-belts. The hub and the sheave are pressed on the shaft. A set screw and a keyway lock the sheave in place. The set screw usually aligns with the keyway of the shaft (see Figure 7.18).

To replace the press-fit sheave, follow these steps:

- Loosen the set screw.
- Use a puller or press to remove the sheave.
- Before installing the press-fit sheave, make sure that the groove for the V-belt is as close to the supporting bearing as possible. This reduces the overhung load and increases bearing life. The hub should be facing the end of the shaft, if possible.
- Check for a bent or damaged shaft. Dressing the shaft with a file or emery cloth may be necessary.
- Tap the sheave onto the shaft using a soft-face hammer, being careful to avoid sheave damage.
- If the sheave is too difficult to install, remove the sheave and heat it in a bearing heater or oven. Do not heat the pulley to more than 350°F (177°C). Overheating can weaken the metal.
- After heating the sheave, quickly place it on the shaft and align it. Do not install the V-belt until the sheave has completely cooled.

FIGURE 7.18
Press-fit sheave.

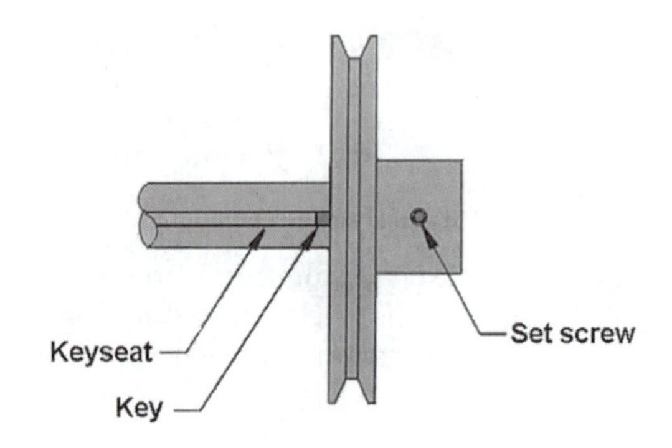

- Tighten the set screw.
- Install V-belt.

Both taper-lock and quick-disconnect (QD) sheaves are compression-type sheaves. They use the mechanical advantage of the principle of the wedge to tightly squeeze the shaft, allowing no slippage.

Most industrial sheaves are compression-type mountings. This type of mounting is much easier to install and remove than the press-fit is, and it holds the shaft more securely. It also has the ability to accommodate slightly undersize or oversize shafts without gripping loss. Compression-type bushings are the same bushings used on chain sprockets and other mechanical devices. An advantage of these bushing systems is that less inventory is needed in the storeroom because bushings for different shaft sizes can be stocked to fit several different sheaves and sprockets.

Taper-lock sheaves use bushings that are tapered and have no flange (see Figure 7.19). They are assembled with set screws that cause the bushing to squeeze the shaft as they are tightened. To remove this sheave and bushing, additional threaded holes are provided for set screws to force separation between the sheave and bushing. QD (quick-disconnect) sheaves use tapered bushings in a manner similar to that of taper-lock bushings (see Figure 7.20). They can be easily identified because QD bushings have a characteristic flange that taper-lock bushings do not have. Instead of set screws, QD bushings use hex-head bolts to hold the sheave and squeeze the shaft. Like taper-lock bushings, they also have jack-bolt holes to force separation of the sheave and bushing. Another type of bushing is now being marketed called a *split-taper bushing.* It is very similar to the QD bushing. Figure 7.21 shows the split-taper bushing and how to assemble and disassemble it.

7-12 BELT-DRIVE ALIGNMENT

New sheaves and V-belts are no guarantee that a belt drive will have a long life. Accurate alignment and proper belt tension are key factors for long life of a belt drive (see Figure 7.22). The following steps should be used to ensure good alignment for long-drive life for any V-belt (belt tension is discussed later):

- Use a dial indicator to check shaft runout and sheave wobble. Replace or straighten shaft or sheave as necessary.
- Using a machinist's level, make sure both shafts are leveled horizontally. This ensures vertical angular alignment.
- Use a straightedge or string to be sure the shafts are parallel. Correct horizontal angular alignment first; then correct horizontal parallel alignment (see Figure 7.23). Generally, the recommended alignment should be within 1/16 in. for each 12 in. of center distance.
- Repeat steps 1 through 3 to verify alignment. Correct as necessary.

TAPER-LOCK® BUSHING

> **WARNING**
> Disconnect power supply to the machine before removing or installing sheaves.

HOW TO MOUNT THE TAPER-LOCK BUSHING

Look at the bushing and the hub. Each has a set of half-holes. The threaded holes in the hub are the mates to the non-threaded holes in the bushing. Insert the bushing in the hub and slide it onto the shaft. Align the holes *(not the threads).* Start the setscrews into the holes that are threaded in the hub only. Do not tighten the setscrews yet.

Align both edges of the sheave with the edges of its mating sheave. See section 7-12 "Belt Alignment." Tighten the screws alternately and evenly. This will wedge the bushing inward and cause it to contract evenly and grip the shaft.

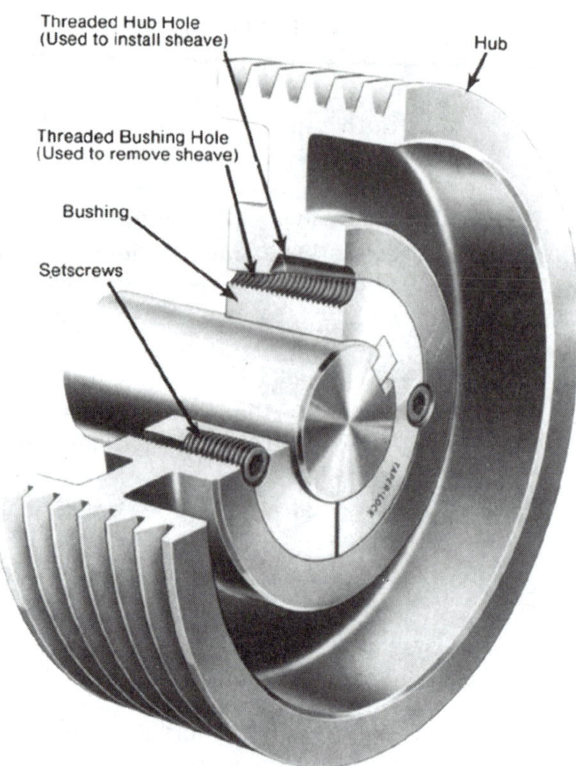

Taper-Lock: TM Reliance Electric Company

HOW TO REMOVE THE TAPER-LOCK BUSHING

Remove all the setscrews. Place two of the setscrews in the holes that are threaded in the bushing only. Turn the setscrews alternately and evenly. This will unlock the grip and permit easy removal of the assembly with no shock to the bearings or machinery.

FIGURE 7.19
Taper-lock bushing instructions.
Courtesy of Dodge/Rockwell Automation.

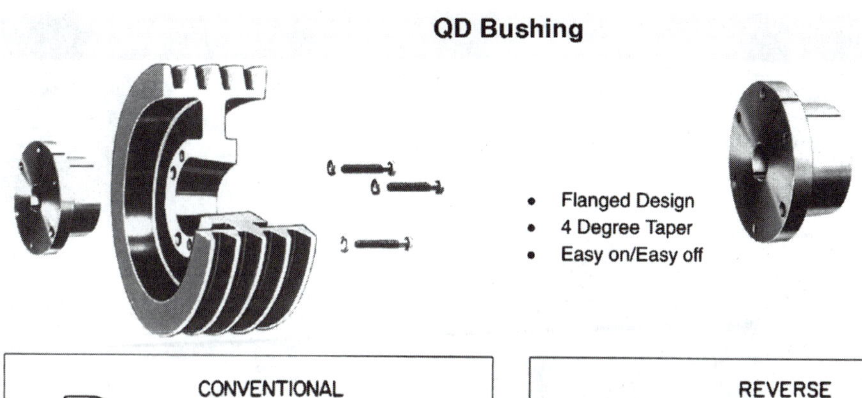

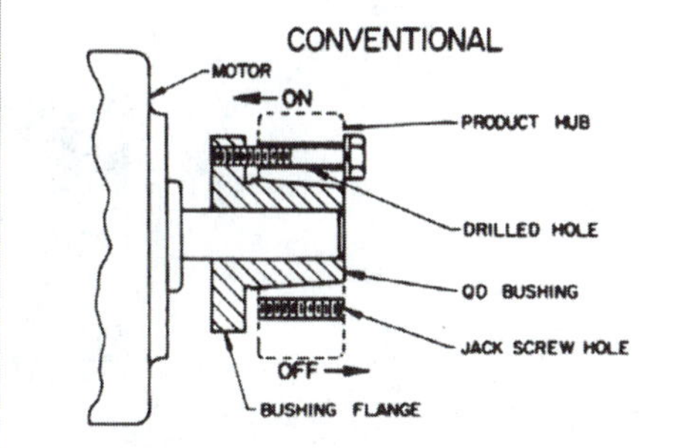

REVERSE
MOTOR
OFF
PRODUCT HUB
BUSHING FLANGE
JACK SCREW HOLE
QD BUSHING
ON
THREADED HOLE

Conventional Mounting

Easy On

- Place bushing in product
- Align clearance holes in product with threaded holes in bushing.
- Install screws and lockwashers thru clearance holes, finger tight.
- Slide assembly onto shaft, flange first.
- Locate assembly on shaft for proper drive alignment.
- Tighten cap screws alternately and evenly to specified torque.

Easy Off

- Remove cap screws and install in product threaded holes.
- Alternately and evenly tighten screws until bushing grip is released.

Reverse Mounting

Easy On

- Place bushing in product
- Align clearance holes in product with threaded holes in bushing.
- Install screws and lockwashers thru clearance holes, finger tight.
- Slide assembly onto shaft, flange outward.
- Locate assembly on shaft for proper drive alignment.
- Tighten cap screws alternately and evenly to specified torque.

Easy Off

- Remove cap screws and reinstall in flange threaded holes.
- Alternately and evenly tighten screws until bushing grip is released.

IMPORTANT! Do not use lubricants or anti-seize compounds on tapered bore or bushing surfaces. For complete installation instructions, refer to the sheet packaged with each bushing.

FIGURE 7.20
QD bushing instructions.
Courtesy of Emerson Electric.

WARNING
Disconnect power supply to the machine before removing or installating sheaves.

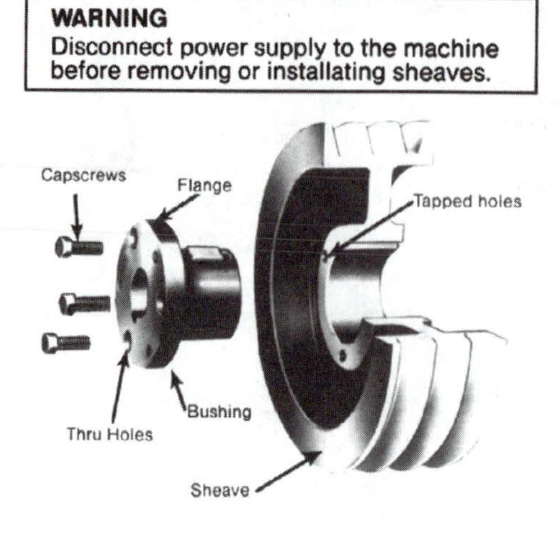

HOW TO MOUNT SPLIT TAPER BUSHING SHEAVES

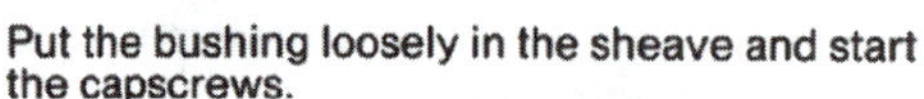

Put the bushing loosely in the sheave and start the capscrews.

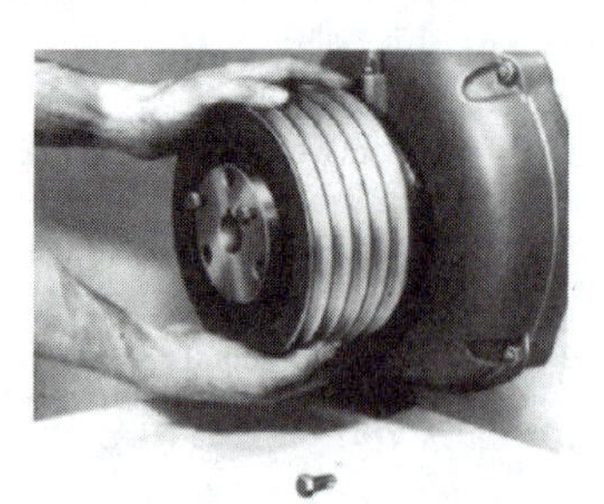

Place the assembly on the shaft. Align both edges of the sheave with the edges of its mating sheave (i.e. the sheave on the driven shaft).

Tighten the capscrews according to the instructions furnished with the bushing.

HOW TO REMOVE SPLIT TAPER BEARING SHEAVES

Remove all capscrews.

Put two of the capscrews in the tapped holes in the flange of the bushing. Turn the bolts alternately and evenly until the sheave has loosened.

Remove the sheave/flange assembly from the shaft.

FIGURE 7.21
Split taper bushing instructions.
Courtesy of Browning Manufacturing.

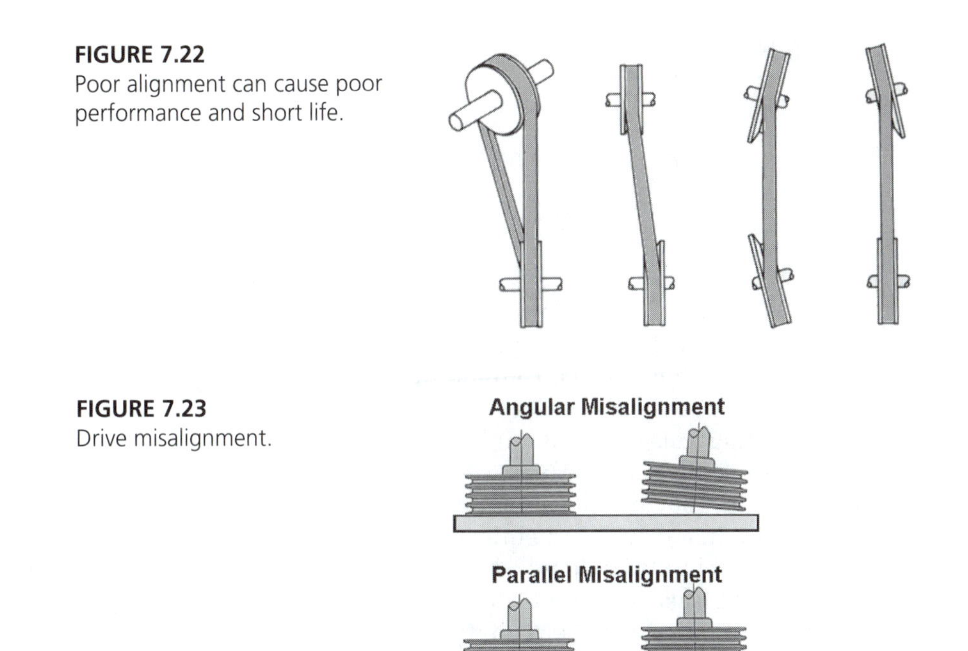

FIGURE 7.22
Poor alignment can cause poor performance and short life.

FIGURE 7.23
Drive misalignment.

7-13 V-BELT INSTALLATION AND TENSIONING

When installing V-belts, you should never force a belt on with a pry bar or screwdriver (see Figure 7.24). V-belts should also never be *run on* (a practice where you start the belt on a too-tight sheave and then jog the motor to fully install the V-belt). These practices place excessive strain on the V-belt, damage the belt cover, and break the belt cords. Broken cords weaken the V-belt and may cause it to turn over or jump off the sheave when it is under a load. The center distance should also be reduced so that V-belts can be slipped on by hand without forcing them (see Figure 7.25).

FIGURE 7.24
V-belts should be installed without forcing.

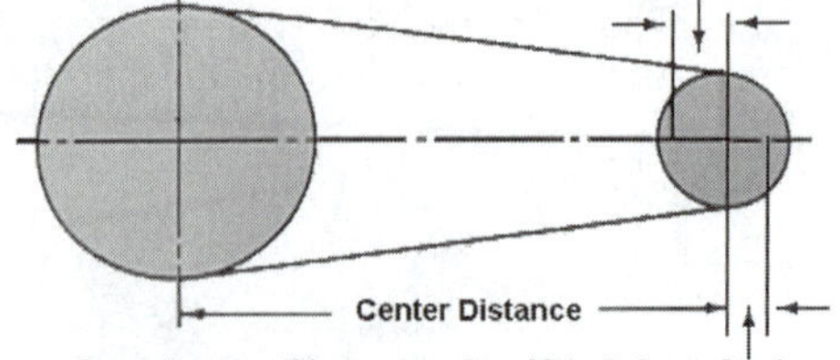

FIGURE 7.25
Adjust the center distance of the V-belt.

Belt tension is an important factor for good belt life. Belt drives must have proper tension to produce the wedging action in the sheave. Drives that are too loose can cause accelerated wear of belts, slippage, and heat and sometimes a squealing noise. You should never use belt dressing on V-belts to attempt to correct a slippage problem. Although belt dressings may temporarily cure a problem, they are detrimental to the life of the belt. Belt dressings increase traction in a drive by chemically softening the rubber compound of the V-belt. Soft, gooey rubber does not slip as much and does not make noise; however, the belt is sacrificed for a quick fix. Tightening a belt too much can cause separation of the tension cords in the load-carrying part of the belt, resulting in belt failure. Highly tensioned belts can also cause bearing failure and possibly bend the shafting. *Ideal tension is the lowest tension at which the belt will not slip under peak load conditions.* In the past, with older belt drives, mechanics were taught to tension belts by feel. To obtain the correct tension, the mechanic would strike the belt with the edge of his or her hand and see how much spring the belt had. With too little tension, the belt would feel loose and dead. Too much tension would cause the belts to have no spring. Belts today are available with different materials and construction; therefore, the "feel method" is no longer reliable enough to check belt tension. Manufacturers' manuals give the proper belt tension. Check the tension on all belts every 1,000 h of use and adjust when necessary. The following steps should be followed when installing and tensioning V-belts:

- Check sheaves for damage and runout. (See Section 7–10, "Replacing a V-Belt," for details.)
- Align the drive initially and recheck it periodically (see Section 7–12, "Belt-Drive Alignment").
- Reduce center distance so that V-belts can be installed on sheaves.
- Tighten V-belts to correct tension (see Figure 7.26). The ideal tension is the lowest tension at which the belt will not slip. Always use manufacturers' recommendations for tensioning.
- Install belt guards and run the drive for a short period to allow V-belts to seat in grooves.
- Remove belt guards, check alignment, and retension V-belts. Do not overtension V-belts; this shortens belt and bearing life.
- Recheck the V-belt tension after 24 to 48 h of operation.

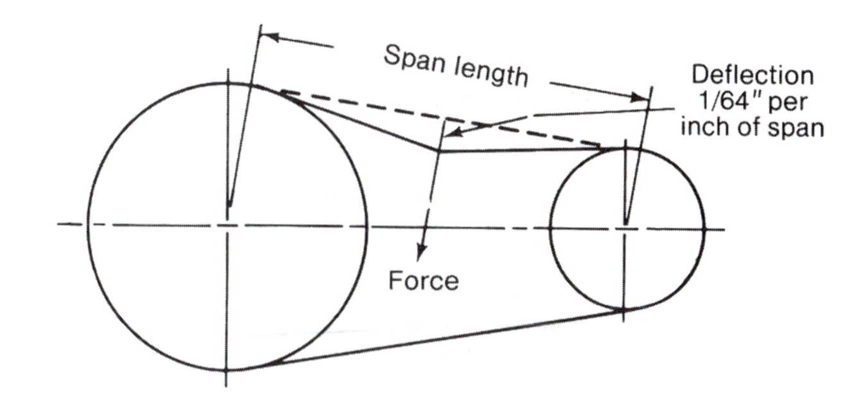

FIGURE 7.26
V-belt deflection-force tables.

Classical V-Belts

Cross Section	Smallest Sheave Diameter Range	RPM Range	Belt Deflection (Force Pounds)			
			Standard Belts		Cogged Belts	
			Used Belt	New Belt	Used Belt	New Belt
A, AX	3.0–3.6	1000–2500 2501–4000	3.7 2.8	5.5 4.2	4.1 3.4	6.1 5.0
	3.8–4.8	1000–2500 2501–4000	4.5 3.8	6.8 5.7	5.0 4.3	7.4 6.4
	5.0–7.0	1000–2500 2501–4000	5.4 4.7	8.0 7.0	5.7 5.1	9.4 7.6
B, BX	3.4–4.2	860–2500 2501–4000			4.9 4.2	7.2 6.2
	4.4–5.6	860–2500 2501–4000	5.3 4.5	7.9 6.7	7.1 7.1	10.5 9.1
	5.8–8.6	860–2500 2501–4000	6.3 6.0	9.4 8.9	8.5 7.3	12.6 10.9
C, CX	7.0–9.0	500–1740 1741–3000	11.5 9.4	17.0 13.8	14.7 11.9	21.8 17.5
	9.5–16.0	500–1740 1741–3000	14.1 12.5	21.0 18.5	15.9 14.6	23.5 21.6
D	12.0–16.0	200– 850 851–1500	24.9 21.2	37.0 31.3		
	18.0–20.0	200– 850 851–1500	30.4 25.6	45.2 38.0		

Narrow-Groove V-Belts

Cross Section	Smallest Sheave Diameter Range	RPM Range	Belt Deflection (Force Pounds)			
			Standard Belts		Cogged Belts	
			Used Belt	New Belt	Used Belt	New Belt
3V, 3VX	2.2–2.4	1000–2500 2501–4000			3.3 2.9	4.9 4.3
	2.65–3.65	1000–2500 2501–4000	3.6 3.0	5.1 4.4	4.2 3.8	6.2 5.6
	4.12–6.90	1000–2500 2501–4000	4.9 4.4	7.3 6.6	5.3 4.9	7.9 7.3
5V, 5VX	4.4–6.7	500–1749 1750–3000 3001–4000			10.2 8.8 5.6	15.2 13.2 8.5
	7.1–10.9	500–1740 1741–3000	12.7 11.2	18.9 16.7	14.8 13.7	22.1 20.1
	11.8–16.0	500–1740 1741–3000	15.5 14.6	23.4 21.8	17.1 16.8	25.5 25.0
8V	12.5–17.0	200– 850 851–1500	33.0 26.8	49.3 39.9		
	18.0–22.4	200– 850 851–1500	39.6 35.3	59.2 52.7		

*Note: For above 3000 fpm, reduce the deflection force by 20%.

7-14 V-BELT MAINTENANCE

As with any mechanical device, good maintenance is essential to obtain maximum life for a belt drive. You should check the belt tension and alignment every 1,000 h of use under normal operating conditions. It is necessary to check more often if a problem is suspected or if operating conditions are severe. Improper belt tension, either too much or too little, may reduce belt life up to 90%.

Sheaves should be inspected when checking belts. Look for wear, nicks, and cracks on all sheaves. At the same time, check for loose or missing bolts, and inspect the shafts for wear or possible bearing failure.

Belts should always be kept clean. Grease and oil damage belts, so check them carefully. If belts need to be cleaned, use a clean dry cloth or soap and water. Never use gasoline or flammable cleaners to remove oil and grease. Find the source of any leaking oil, and *stop the leak.*

In addition, always follow these guidelines:

- Never use belt dressing on V-belts because it is no substitute for proper maintenance of belt drives.
- Make sure that belt guards are in place. Check them for damage and loose or missing bolts.
- *Never* operate a belt drive without the proper guards in place.

Proper storage is important to ensure maximum belt life. Always store belts away from damp floors, high-heat areas, or sunlight. Keep them in their original packaging and stack them flat to prevent stretching. If belts are hung on pegs, use at least two pegs to prevent distortion. With proper storage, the shelf life of belts will be several years with no loss of service life.

V-belts stretch with use. When replacing belts on a multibelt drive, replace all of them for maximum belt life and drive efficiency (see Table 7.2).

TABLE 7.2
V-Belt troubleshooting guide.

Type of Failure	Cause of Failure	Corrective Action
Belt stretch beyond take-up	Misaligned drive, unequal work done by belts	Realign and retension drive.
	Belt's tensile member broken from improper installation	Replace all belts with new matched set properly installed.
	Insufficient take-up allowance	Check take-up.
	Greatly overloaded or underdesigned drive	Redesign.
Relatively rapid failure; no visible reason	Tensile members damaged through improper installation	Replace with all new matched set, properly installed.
	Worn sheave grooves (check with groove gauge)	Replace sheaves.
	Underdesigned drive	Redesign.
Sidewalls soft and sticky; low adhesion between cover plies; cross section swollen	Oil or grease on belts or sheaves	Remove source of oil or grease. Clean belts and grooves with cloth moistened with alcohol.
Sidewalls dry and hard; low adhesion between cover plies; bottom of belt cracked	High temperatures	Remove source of heat. Ventilate drive better.
Belt turnover	Excess lateral belt whip	Use banded belt.
	Foreign material in grooves	Remove material-shield drive.
	Misaligned sheaves	Realign the drive.
	Worn sheave grooves (check with groove gauge)	Replace sheave.
	Tensile member broken through improper installation	Replace with new matched set properly installed.

Continued

TABLE 7.2
Continued.

Type of Failure	Cause of Failure	Corrective Action
	Incorrectly placed flat idler pulley	Carefully align the flat idler on the slack side of the drive as close as possible to the driver sheave.
Deterioration of rubber compounds used in belt	Belt dressing	Never use dressing on V-belts. Clean with cloth moistened with alcohol. Tension drive properly to prevent slip.
Extreme cover wear	Belts rub against the belt guard or other obstruction	Remove obstruction or align drive to give needed clearance.
Spin burns on belt	Belts slip under starting or stalling load	Tighten drive until slipping stops.
Bottom of belt cracked	Too small sheaves	Redesign for larger sheaves.
Broken belts	Object falling into or hitting drive	Replace with new matched set of belts. Provide shield for drive.
Improper driven speeds	Design error	Use correct sheave sizes.
	Belt slip	Retension drive until belt stops slipping.
Belt noises	Belt slip	Retension drive until it stops slipping.
Hot bearings	Worn grooves; belts bottoming and will not transmit power until overtensioned	Replace sheaves. Tension.
	Improper tensioning	Retension drive.
	Sheaves too small; motor manufacturer's sheave diameters not followed	Redesign drive.
	Underdesigned bearing or poor bearing maintenance	Observe recommended bearing design and maintenance.
	Sheaves out too far on shaft	Place sheaves as close as possible to bearings. Remove any obstruction preventing this.
	Belts slipping and causing heat buildup	Retension drive.

7-15 SYNCHRONOUS BELT DRIVES

All the belt drives we have discussed so far have been *friction*-type drives. In this section, we discuss **positive belt drives.** Synchronous belts, or *timing belts,* as most industrial mechanics call them, do not rely on friction for the driving

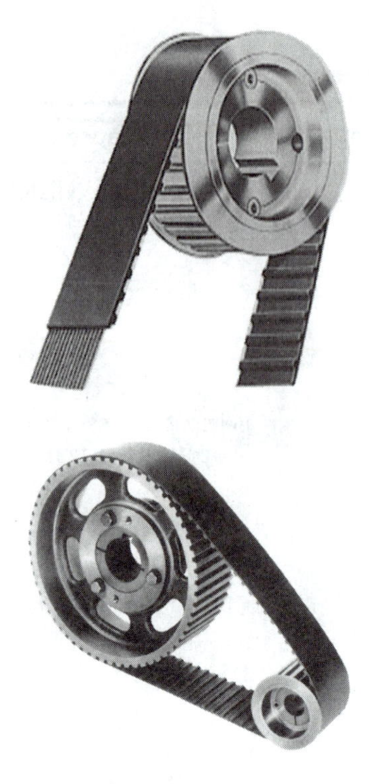

FIGURE 7.27
Synchronous belt drive.

FIGURE 7.28
Synchronous belt drives require no lubrication.

force (see Figure 7.27). They transmit power by the positive engagement of belt teeth with pulley teeth, much like a chain drive. Because of the construction of timing belts, there is no slippage or speed variation. Timing belt drives are often used to replace chain drives in timing operations, which is how they got their name. An advantage of timing belt drives over chain drives is that there is no metal-to-metal contact. This provides a quieter, low-maintenance drive, requiring no lubrication (see Figure 7.28). Unlike friction belts, timing belts do not get their tensile strength from their cross-sectional area. Timing belts are relatively thin, using steel-cable tension cords and rugged wear-resistant teeth. In some belts, fiberglass is also used for tension cords, typically encased in a durable, flexible backing of neoprene, which protects the tensile members from dirt, oil, moisture, and abrasive wear. The teeth are molded as part of the backing and are protected by a facing of tough nylon duck. Most timing belt drives are designed so that there are at least six teeth engaged with the small pulley. When six or more teeth engage, the tooth strength of the belt exceeds the tensile strength.

Timing belts are made in five standard tooth pitches for industrial use.

Code	Description	Stock Widths	Pitch
XL	Extra light	1/4 in. 5/16 in. 3/8 in.	1/5 in.
L	Light	1/2 in. 3/4 in. 1 in.	3/8 in.
H	Heavy	3/4 in. 1 in. 1 1/2 in. 2 in. 3 in.	1/2 in.
XH	Extra heavy	2 in. 3 in. 4 in.	7/8 in.
XXH	Double extra heavy	2 in. 3 in. 4 in. 5 in.	1 1/5 in.

Timing belt numbers are made up of three components. The prefix is the belt pitch length in inches and tenths of inches. The middle letter is the tooth pitch code letter. The last part of the number is the belt width in inches multiplied by 100. For example, a timing belt marked 900H300 has a belt pitch length of 90.0 in.; the *H* tells us the tooth pitch is ½ in.; and the belt width is 3 in.

7-16 TIMING BELT PULLEYS

Various types of pulleys are used on positive belt drives. Pulleys are usually solid for small diameters and may have spokes or be webbed for larger diameters. The type available is governed by the pitch diameter and manufacturer's preference. In general, small-diameter pulleys do not use taper bushings but have an integral hub with a keyway and set screw for a press fit (see Figure 7.29). The larger pulleys generally use industry-standard tapered bushings (see Figure 7.30). Table 7.3 shows the minimum pulley diameters.

FIGURE 7.29
Timing belt pulley with keyway and set screw.

FIGURE 7.30
Timing belt pulley with taper-lock bushing.

TABLE 7.3
Minimum pulley diameter.

Pitch	Speed Rev/min	Recommended Minimum*	
		Pitch Diameter (in.)	Number of Grooves
1/5 in. (XL)	3,500	0.764	12 XL
	1,750	0.700	11 XL
	1,160	0.637	10 XL
3/8 in. (L)	3,500	1.910	16 L
	1,750	1.671	14 L
	1,160	1.432	12 L
1/2 in. (H)	3,500	3.183	20 H
	1,750	2.865	18 H
	1,160	2.546	16 H
7/8 in. (XH)	1,750	7.241	26 XH
	1,160	6.685	24 XH
	870	6.127	22 XH
1 1/4 in. (XXH)	1,750	10.345	26 XXH
	1,160	9.549	24 XXH
	870	8.753	22 XXH

*Smaller-diameter pulleys can be used if a corresponding reduction in belt service life is satisfactory.

7-17 INSTALLING TIMING BELTS

If you are only replacing belts, you should check the old pulleys (see Figure 7.31). Damaged, worn, or dirty pulleys substantially reduce belt life. Worn teeth cause belt wear and damage. Nicks and gouges can cut the belt. Dirt or debris on the teeth and in the grooves can abrade the belt, and oil can attack the belt materials. Debris and dirt should be cleaned off with a stiff brush. Always keep pulleys clean of oil and grease. If pulleys are damaged or worn, replace them. Use the same procedures given for V-belt sheave replacement (see Table 7.4).

Alignment is more critical for timing belt drives than for conventional V-belt drives. Belt alignment should be checked when belt maintenance is performed,

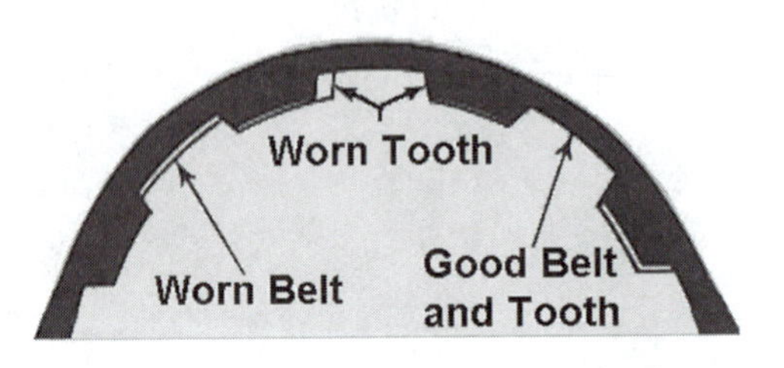

FIGURE 7.31
Inspect timing belt pulley for wear.

TABLE 7.4
Timing belt troubleshooting guide.

Type of Failure	Cause of Failure	Corrective Action
Excessive edge wear	Misalignment or nonrigid centers	Check alignment and/or reinforce mounting.
	Bent flange	Straighten the flange.
Jacket wear on pressure-face side of belt tooth	Excessive overload and/or excessive belt tightness	Reduce installation tension and/or increase drive load-carrying capacity.
Excessive jacket wear between belt teeth (exposed tension members)	Excessive installation tension	Reduce installation tension.
Cracks in neoprene backing	Exposure to excessive low temperature (below −30°F)	Eliminate the low-temperature condition or consult the factory for proper belt construction.
Softening of neoprene backing	Exposure to excessive heat (+200°F) and/or oil	Eliminate the high-temperature and oil condition or consult the factory for proper belt construction.
Tensile or tooth shear failure	Small or subminimum diameter sprocket	Increase sprocket diameter or use the next-smaller pitch.
	Acid or caustic atmosphere	Refer to the factory for best construction.
Excessive sprocket tooth wear (on pressure face and/or outside diameter)	Excessive overload and/or excessive belt tightness	Reduce installation tension and/or increase drive load-carrying capacity.
	Insufficient hardness of sprocket material	Surface-harden sprocket or use harder material.
Unmounting of flange	Incorrect flange installation	Reinstall the flange correctly.
	Misalignment	Correct alignment.
Excessive drive noise	Misalignment	Correct alignment.
	Excessive installation tension	Reduce tension.
	Excessive load	Increase drive load-carrying capacity.
	Subminimum sprocket diameter	Increase sprocket diameters.
Tooth shear	Fewer than six teeth in mesh (TIM)	Increase TIM or use the next-smaller pitch.
	Excessive load	Increase drive load-carrying capacity.
Apparent belt stretch	Reduction of center distance or nonrigid mounting	Retention drive and/or reinforce mounting.

The common causes of premature failure of "timing" belt drives are shown here. There are other less apparent causes of drive failure: excessive reverse bend, subminimum diameter idler, variable center (as caused by rubber-mounted motors and mountings, etc.). However, when a drive is correctly designed—with full consideration to proper application factor for the service conditions—premature failure should not be encountered.

when timing belts are replaced, and when sheaves are removed or installed. The procedure for timing belt alignment is the same as for conventional V-belts. (See prior sections for detailed alignment procedures.) When installing belts, you should never force the belts onto a pulley with a screwdriver or pry bar. Doing so can rupture the facing fabric, damage the belt teeth, or break the tensile cords. Always reduce the center distance so that the belts can be easily installed on the pulley without forcing them. To assure smooth operation and prevent premature failure, belts should be protected against sharp bending during handling and storage. They should not be subjected to extreme heat, low temperatures, or high humidity.

7-18 TENSIONING TIMING BELTS

The timing belt should be installed with a snug fit, neither too tight nor too loose. The belt's positive grip eliminates the need for initial tension. Consequently, a belt installed with a snug fit (i.e., not too taut) has a longer life, wears on bearings less, and has a quieter operation. Preloading, often the cause of premature failure, is not necessary. A belt in either the 7/8- or 1 1/4-in. pitch can usually be installed slightly slack (because of its deeper tooth section) unless shock loads or reversals are abnormally high.

When torque is unusually high, a loose belt may "jump teeth" on starting. In such cases, the tension should be increased gradually until satisfactory operation is attained.

To tension a drive properly, follow these guidelines:

- Apply a force at the midpoint of the span between the two pulleys. Deflect the belt 1/64 in. for each inch of span length.
- Installation tension should be regulated so that the value of this applied force equals the value of f given in the following formula:

$$f = \frac{T + (s/L)K}{16}$$

where s = the span distance (in.)
T = the tension (lb) found in Table 7.5
K = the constant from Table 7.5
L = the length of the belt

- If the deflecting force is less than that given in the formula, the belt is too loose. If the deflecting force is greater than that given in the formula, the belt is too tight.

TABLE 7.5
Timing belt tension.

		Synchronous Belt-Deflection Forces									
Belt Pitch	**Values**	**1/4**	**3/8**	**1/2**	**3/4**	**1**	**1 1/2**	**2**	**3**	**4**	**5**
1/5	*T*										
	K	.85	1.7	2.7	4.7	6.7					
3/8	*T*		11.4	17.1	27.9	39.3	61.4	84.3	132.1		
	K		8.5	9.9	17.0	24.0	37.0				
1/2	*T*			39.3	66.0	93.3	145.3	200	313	444	574
	K			17.0	32.0	46.0	71.0	95	152	210	265
7/8	*T*							227	356	504	652
	K							190	305	440	568
1 1/4	*T*							556	873	1238	1599
	K							310	500	710	920

QUESTIONS

1. What is a friction belt drive?
2. What is a positive belt drive?
3. What is a match number?
4. Why should you use a narrow-groove V-belt instead of a classical V-belt in a new installation?
5. Will narrow-groove V-belts use the same sheaves as the classical V-belts when you are replacing them?
6. What is a banded V-belt?
7. Where would you use fractional-horsepower (FHP) V-belts?
8. What is the main advantage of using V-ribbed belts over conventional V-belts?
9. Name the three common types of mounting sheaves.
10. When should belt dressing be used?
11. Tensioning V-belts is an important part of good maintenance. What is the ideal V-belt tension?
12. How often should you check V-belt tension and alignment under normal operation conditions?
13. For most applications, how many teeth of a timing belt should engage the small pulley?
14. What is the recommended minimum pulley pitch diameter for a 3/8-in. (*L*) timing belt drive running at 1750 rev/min?

CHAPTER 8

Chain Drives

8-1 INTRODUCTION

A chain drive is probably the most widely used positive type of drive. It can be found in photocopiers, vending machines, and even in heavy construction equipment. A simple example is a bicycle (see Figure 8.1). The chain transfers mechanical power from the pedals to make the rear wheels move. If the chain breaks or comes off, you cannot ride the bike. Metal chains were used in medieval times, but the roller chain of today was first widely used in the nineteenth century. Brunel's famous ship, the *SS Great Britain,* used an early form of roller chain to transfer mechanical power to drive the propeller shaft. Demand for chains increased along with the demand for automobiles, airplanes, and countless other applications in the industrialized twentieth century (see Figure 8.2). In this chapter, we cover only chain drives commonly used in industrial applications.

FIGURE 8.1
Bicycle chain drive.

8-2 TYPES OF ROLLER CHAINS

The roller chain is the most common kind of chain drive found in industrial drive units. Industrial roller chains are used to transfer power in many different types of equipment and machinery. Because it is a positive drive (one in which

FIGURE 8.2
Small industrial chain drive.

FIGURE 8.3
Rolling contact means less friction.

there is no slip or creep), a roller chain is also often used for timing or synchronizing different parts of machinery. A roller chain is also a very efficient means of power transmission because the rollers that come into contact with the sprocket teeth rotate on bushings. This allows the chain to make a rolling contact instead of a sliding contact, creating less friction (see Figure 8.3). Less friction means higher efficiency.

Roller chain standards for dimensions and specifications have been established by the American National Standards Institute (ANSI). These standards ensure that a replacement chain can be interchanged with the old chain in both size and performance even if it is not from the same manufacturer. If two roller chains bear the same identification number, they can be interchanged without any serious problems. A benefit of this standardization is that the choice of replacement chains can be based on economic and availability factors, rather than on a specific manufacturer's product.

There are two types of roller chain construction: *riveted* and *cottered*. Bradded pins hold the side plates in riveted chain. Cotter keys to hold the side plates in cottered chain.

Standard *single-strand* roller chain is probably the most common chain to be standardized by ANSI (see Figure 8.4). This roller chain consists of overlapping plates connected by pins. Rollers or bushings fit over the pins between the side plates. There are two different types of side plates: a *pin link* and a *roller link*. The pin-link plates always lap on the outside of the roller-link plates. Pin links connect roller links to each other. Therefore, a strand of roller

FIGURE 8.4
Roller chain parts.
Courtesy of Emerson Power Transmission Corp.

FIGURE 8.5
Connecting links.
Courtesy of Emerson Power Transmission Corp.

chain is an assembly of alternating pin links and roller links. A length of chain is made into a continuous strand by using a special type of pin link called a *connecting,* or *master,* link (see Figure 8.5). The connecting link's plate is held in place by a spring clip or cotter pins. These spring clips or cotter pins are necessary for ease of assembly or disassembly. The necessity of having one pin and one roller link to mate means that chain strands usually have an even number of pitches—that is, an equal number of pin links and roller links.

Sometimes because of fixed-shaft center distances or other reasons, a special link called an *offset link,* or *half-link,* is desirable (see Figure 8.6). It is installed to produce a strand of chain with an odd number of links. An offset link is a combination of a pin link and a roller link. Half the offset link attaches to the pin link; the other half attaches to the roller link. By using an offset link, a strand of chain can be assembled with an odd number of links. The chain can be increased or decreased by half the distance of the full link.

FIGURE 8.6
Offset link.
Courtesy of Emerson Power Transmission Corp.

Heavy series chains are similar to standard chains, but they have side plates that are thicker than the corresponding pitches (sizes) of the standard series. In general, the thickness of heavy series chains is the same as the next larger pitch, or size, in the standard series. This offers greater resistance to fatigue failures on more demanding applications. This type of chain is usually used as replacement chain when the standard series chain is not durable enough for extreme conditions or heavy shock loading. Heavy series chain runs on standard series sprockets and uses the same horsepower ratings. This chain is usually designated by the letter *H* following the chain number and is usually manufactured in 3/4-in. (number 60) and larger pitches.

Standard series multistrand chain attaches single-strand standard roller chains together to make double- and triple-strand roller chain. It is used when greater capacity is required, but there is not enough room for the larger-diameter sprockets needed for larger single-strand roller chain drives. Each additional strand proportionally increases the horsepower capacity of the drive. However, the single-strand rating is not multiplied by the number of strands but rather by a lesser multiplier:

Number of Strands	Strand Factor
2	1.7
3	2.5
4	3.3

For example, a single-strand number 60 chain drive with a small sprocket having 21 teeth running at 400 rev/min is rated at 10.1 base hp. A number 60–2 (two-strand) chain drive is rated at (10.1×1.7), or 17.17 hp instead of 20.2 hp.

Multistrand roller chain is usually indicated by a dash followed by a number equal to the number of strands at the end of the chain number. For example, 3/4-in.-pitch triple roller chain is designated by the number 60–3. (The –3 indicates a triple-strand chain.)

8-3 SELF-LUBRICATING CHAIN

Roller chain usually requires external lubrication (lubrication requirements are discussed in greater detail later). The exception to this is a sintered-bushing roller chain. This unique chain, interchangeable with the same sizes of ANSI standard chains, uses sintered, oil-impregnated bushings that release lubricant to all bearing surfaces during operation and then reabsorb it when the drive is idle. This virtually eliminates wear and power loss between sprocket, chain, and chain components. Self-lubricating chain solves many problems that arise when lubrication maintenance is difficult or impossible to perform (see Figure 8.7). Another advantage is that the chain is clean running and ideal for use in food processing, packaging, and textile applications. Its ever-present oil film also

FIGURE 8.7
Self-lubricating roller chain.
Courtesy of Emerson Power Transmission Corp.

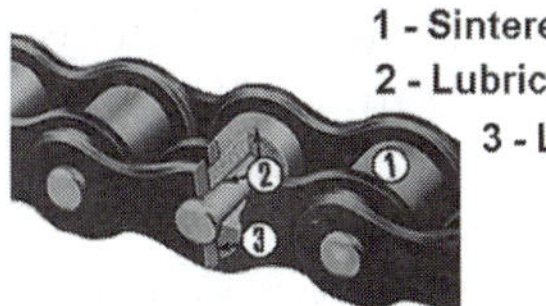

protects internal parts against corrosion encountered in outdoor applications. The major disadvantage of this type of roller chain is its lack of strength and capacity compared to standard roller chain.

8-4 ROLLER CHAIN IDENTIFICATION

Roller chain identification is a simple matter of measuring the pitch. Roller chain pitch is determined by measuring the center distance between two pins. The physical size and power capability of a roller chain increases as the pitch increases. In general, the standard industry number is determined by measuring the roller chain pitch. The first digits denote the total number of 1/8-in. increments in the pitch. For example, a roller chain with a pitch of 5/8 in. is number 50 roller chain; a pitch of 1 in., or 8/8 in., is number 80; and a pitch of 1.5 in., or 1 2/8 in., is number 120.

Each identification number has at least two digits. In general, the last digit of a standard roller chain is zero. A last digit of 1 specifies a chain that is narrower than the standard width. A last digit of 5 specifies a rollerless bushing chain.

8-5 DOUBLE-PITCH ROLLER CHAIN

Double-pitch roller chain, also called *extended-pitch chain,* is used in light-duty industrial-power transmission and conveyor applications (see Figure 8.8). It can also be found in many agricultural applications. Longer distances between sprocket centers are not unusual. Double-pitch chain can be run only at low or moderate speeds. The chain's strength depends on the strength of its sidebars and pins rather than how many it has.

Specifications and dimensions for double-pitch roller chain can be found in ANSI Standard B29.3. These dimensions are the same as for standard roller chain of comparable load capacity, except the pitch is doubled. Because a given length of double-pitch chain contains only half as many pitches, it is

FIGURE 8.8
Double-pitch roller chain.
Courtesy of Emerson Power Transmission Corp.

lighter and less expensive than standard roller chain. All dimensions of double-pitch roller chain are the same except pitch, offering essentially the same strength characteristics as standard roller chain.

The first digit of a double-pitch roller chain number is always the number 2, designating double-pitch chain. The second and third digits designate the total number of 1/4-in. increments in the pitch. (Because it is double-pitch, roller chain uses 1/4-in. instead of 1/8-in. increments.) If the fourth digit is zero, the chain has a standard-size roller; if the fourth digit is the number 2, the chain has large rollers. (Double-pitch chains are available with oversized rollers that extend beyond the tops of the sidebars. These chains are typically used in conveyors rather than in power-transmission drives.) As with standard series chains, the letter *H* following the four-digit number indicates heavy series chain. The letter *C* preceding the number usually indicates conveyor series double-pitch chain. For example, double-pitch roller chain with the number 2060 designates 1 1/2-in.-pitch roller chain. (The number 2 indicates double-pitch, 06 × 1/4 in. equals 1.5-in. pitch, and the last zero means it has standard rollers.) The number C2082H indicates a conveyor series, double-pitch roller chain with a pitch of 2 in., large rollers, and heavy side plates.

Double-pitch roller chain can run on standard roller chain sprockets (using only every other tooth), or it may use sprockets made especially for double-pitch roller chain. Standard sprockets are referred to as *double-cut* (two teeth per link). Sprockets made for use only with double-pitch roller chain are called *single-cut* (one tooth per link). Using standard roller chain sprockets with an odd number of teeth can reduce sprocket wear because teeth that engage rollers on the first revolution do not engage rollers on the next revolution. Because of the double pitch of the roller chain, sprocket teeth engage chain rollers only once for every two revolutions of the sprocket because the sprocket has an odd number of teeth. When using standard roller chain sprockets with an even number of teeth, the same teeth engage rollers on every revolution. Half the sprocket teeth are never used and, therefore, have no wear. After half of the teeth wear to the point of replacement, you can loosen the chain and advance the sprocket one tooth space so that the unworn teeth engage chain rollers and the worn teeth do not.

8-6 ROLLER CHAIN ATTACHMENTS

Attachments to roller chains can solve many difficult conveyor problems, such as moving the most delicate equipment or moving heavy bulky material. Attachments are available for standard and double-pitch chain. Commonly used ANSI attachments include *B1, B2, S1,* and *S2.* B1 attachments have 90° bent lugs on one side of the roller chain. If you use two parallel roller chains with B1 attachments turned inward, you can attach a series of plates between

FIGURE 8.9
B2 attachment chain.
Courtesy of Emerson Power Transmission Corp.

FIGURE 8.10
S1 attachment chain.
Courtesy of Emerson Power Transmission Corp.

the chains to form a flat conveying surface. B2 attachments have 90° bent lugs on both sides of the roller chain (see Figure 8.9). B2 attachments are used when an application calls for greater stability. S1 attachments have 180° straight lugs on one side of the roller chain; S2 attachments have 180° straight lugs on both sides of the roller chain. S1 and S2 attachments can be used as pushers, for timing, or for supporting loads. Figure 8.10 shows a typical S1 attachment roller chain. Attachments may be located as close as every pitch of chain, every foot, or whatever arrangement the application warrants. When practical, if the spacing is even, attachments should be located on pin links rather than on roller links. If the spacing is odd, attachments should be located alternately on pin and roller links. Attachments may be ordered loose and then attached in the field, but preassembled chain and attachments are usually recommended. If you order attachments preassembled into chain, the total number of attachments and the spacing required for a given number of pitches should be specified (such as every pitch, every second pitch, and so on). For example, a 10-ft number 60 standard chain with B2 attachments every second pitch requires 80 B2 attachments; number 60 chain has 16 pitches per foot and 8 for every second pitch. Therefore, every second pitch equals 8 attachments per foot and 80 attachments for 10 ft.

8-7 INVERTED-TOOTH (SILENT-CHAIN) DRIVES

The inverted-tooth chain drive was invented by Hans Renold in 1895. This chain consists of a number of thin plates, each cut in an inverted tooth, like a long U-shape. These plates are assembled side by side in a laminated fashion with pins. The projecting ends overlap slightly. The main advantage of this construction is that the chain conforms well to the shape of the sprocket, making it run much more quietly than standard roller chain (which is why it is sometimes called *silent chain*). In addition to standard link plates, the assembled chain has guide plates, which keep the chain aligned with the sprockets. These guide plates are usually located in the center of the chain but are sometimes on the outside of the chain (see Figure 8.11).

Chordal action is the vibratory motion caused by the rise and fall of the chain as it goes over a small sprocket. It is a serious limiting factor in roller chain performance. The problem is that when a roller chain enters a sprocket, the line of approach is not tangent to the pitch circle. The chain makes contact below the tangent line is then lifted to the top of the sprocket, and is dropped again as sprocket rotation continues. Because of its fixed-pitch length, the pitch line of the link cuts across the chord between two pitch points on the sprocket and remains in this position relative to the sprocket until the chain disengages. This chordal action seriously detracts from chain performance and life. The pulsations produced by chordal action generate noise and vibration and considerably limit the power capacity and speed range of a roller chain. Inverted-tooth chain drives minimize chordal action. Smooth engagement with the sprocket minimizes shock loading and stresses in the links as well as noise and vibration.

Because ANSI standards do not specify inverted-tooth chain joint design, different brands of inverted-tooth chain usually cannot be coupled with each other. However, you can interchange the complete chain with other brands of sprockets when those products have the same pitch and width as the chain (identified by the same two-letter symbol and with one or two numerical digits

FIGURE 8.11
Silent chain drive.
Courtesy of Emerson Power Transmission Corp.

as the chain). Standard silent chain numbers consist of a two-letter symbol and one or two numerical digits to indicate the pitch in 1/8-in. increments. In addition, two or three digits indicate the chain width in 1/4-in. increments. For example, in the chain number SC 302, the letters *SC* mean that it is a silent chain, the number 3 stands for 3/8-in. pitch, and the numbers 02 equal 2/4-in., or 1/2-in., width. The numbers designating silent chain and its pitch are normally stamped on the chain plates. (The width of the chain assembly is not usually included.) Chain numbers without the SC prefix are usually not made to ANSI standards and may indicate a specially made chain that is noninterchangeable.

8-8 ENGINEERED-CLASS STEEL CHAIN

Engineered-class steel chain is designed to deliver high power at low speeds and is known for its ability to operate in adverse environments. It is often used in construction equipment drives and conveyor drives located in outside environments. There are two main types of engineered-class steel chains: *straight-side bar* and *offset-side bar*. The chain workload ranges from about 8,000 to more than 400,000 lb, depending on the size and duty of the chain. Each of these two main types also has *roller* and *rollerless* chains.

The roller type of engineering-class steel chain is more common than the rollerless type because it is used primarily for power transmission. Chain with offset-side bars is used more often than straight-side bars.

8-9 GENERAL-PURPOSE CAST CHAIN

There are two common types of general-purpose cast chain: *pintle chain* and *H-type mill chain* (see Figure 8.12). **Pintle chain** is an offset-type chain usually made of malleable iron. Malleable iron is more wear-resistant than steel and performs better in hostile environments. There are several classes of pintle chain, but the most common are *class 400* and *class 700*. Class 400 is a general-duty pintle chain used in many conveyor applications and for power transmission. Class 700 has a longer pitch and is made for heavier duty.

FIGURE 8.12
General-purpose cast chain.
Courtesy of Emerson Power Transmission Corp.

H-type mill chain is basically the same design as pintle chain. It is an offset-type chain usually made of malleable iron. The advantage of H-type mill chain over pintle chain is that it has greater shock resistance and is generally better for heavier duty. It also has side bars that have flanged wear shoes on one side, making it useful for dragging or sliding applications. H-type mill chain can be found in the paper and lumber industries, many conveyor applications, and in power-transmission drives.

8-10 DRAG CONVEYOR CHAINS

There are three types of drag conveyor chains: *H-type drag chain, steel-bar drag chain,* and *combination-type drag chain.*

H-type drag chain is similar to H-type mill chain but is physically larger and can drag more material. This chain is found in all types of drag conveyor applications. Some of the more common industries using this type of drag chain are the lumber and paper industries for handling wood products and industries involved in conveying highly abrasive material.

Steel-bar drag chain is an inexpensive alternative to H-type drag chain because it is made of formed steel. The single-side-bar construction is used primarily for nonabrasive applications such as conveying sawdust, shavings, plastic, or other light materials. Double-side-bar chain is sometimes used to convey abrasive materials in light-duty or intermittent applications.

Combination-type drag chain is a heavy-duty chain that uses cast links similar to H-type drag chain. The cast links are referred to as *block links* because they are connected to each other by links of steel side bars that provide a joint similar to H-type mill chain. This combination chain with cast block links and steel side bars is assembled with larger rivets, making it stronger and more shock resistant. The overall result is a very heavy-duty chain capable of handling loads that would destroy other types of drag chain.

8-11 CHAIN SPROCKETS

A chain is only half of a chain drive. Sprockets are needed to attach the chain to the shaft so that power can be transferred. Most sprockets are made of steel or cast construction. Standard-series roller chains generally use steel sprockets. The smaller relative size and precision of steel construction makes it more economical than cast construction. Engineered-class chain and most larger chain drives usually use cast construction because the larger size makes it the economical choice. Smaller sprockets are usually solid steel construction. Because of economic considerations and weight, larger cast sprockets have spokes, and steel sprockets have flame-cut holes. Split sprockets are used with large chain drives where it would be inconvenient or impractical to disassemble the machine to slide a conventional sprocket off its shaft. Sprocket teeth may also be hardened for longer life or use in harsh conditions.

8-12 SPROCKET HUB DESIGN

There are four basic types of hubs used with chain drives:

1. Type A sprockets have no hubs (see Figure 8.13). They are flat and require some other device to be mounted. Sometimes they are mounted with flanges on the hubs of the device that they drive. Type A sprockets have a small plain bore with no keyway. They are most commonly found on friction clutches with chain drives.
2. Type B sprockets have a hub on one side only (see Figure 8.14). The flat side should be placed close to the bearings on the machine where it is mounted. This reduces the overhung load on the bearings of the machine. Type B sprockets are most commonly used in typical industrial roller chain drives.
3. Type C sprockets have hubs on both sides of the sprocket (see Figure 8.15). This type of sprocket is used with larger sprockets to increase the contact area with the shaft. More contact area increases the rigidity of the shaft and better distributes overhung load for large drives.
4. Type D sprockets have a bolt-on detachable or split hub (see Figure 8.16). Type D sprockets are not common for industrial chain drives. They are used for special applications where it would be difficult to replace type B or type C sprockets.

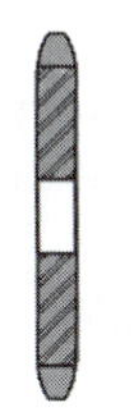

FIGURE 8.13
Type A sprocket (no hub).

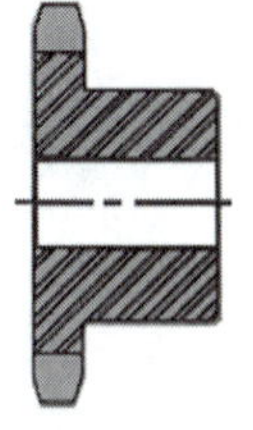

FIGURE 8.14
Type B sprocket (hub on one side).

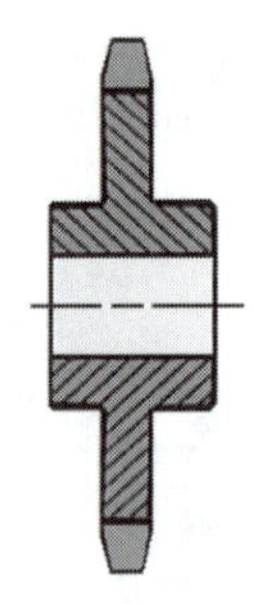

FIGURE 8.15
Type C sprocket (hub on both sides).

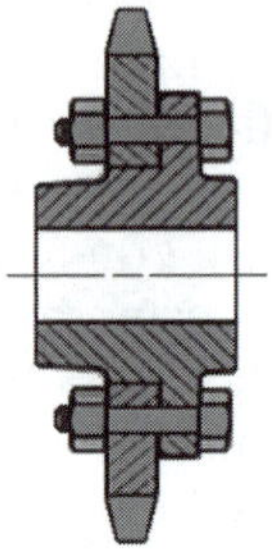

FIGURE 8.16
Type D sprocket (detachable hub).

8-13 SPROCKET MOUNTING

Type A sprockets have only a small plain bore with no keyway. Type D sprockets may be ordered with a minimum plain bore with no keyway. The sprocket is then bored to the shaft size, and the keyway is cut in the field. Most type D sprockets are factory-finish bored with a keyway. Type B and type C sprockets may be ordered with a minimum plain bore with no keyway, factory-finish bored with a keyway, or set up for an interchangeable bushing. This industry-standard interchangeable bushing system provides easy installation and removal, along with an extremely secure fit on the shaft. An advantage of these bushing systems is that less inventory is needed because bushings for different shaft sizes can be stocked to fit several different sprockets. There are two popular industry types of industrial compression bushings: *taper-lock* and *quick-disconnect* (QD). Taper-lock sprockets use **taper-lock bushings** that are tapered and have no flange (see Figure 8.17). They are assembled with set

FIGURE 8.17
Taper-lock bushing sprocket.
Courtesy of Dodge/Rockwell Automation.

FIGURE 8.18
QD bushing sprocket.
Courtesy of Dodge/Rockwell Automation.

screws that cause the bushing to squeeze the shaft as they are tightened. To remove these sprockets and bushings, additional threaded holes are provided for set screws to force separation between the sprocket and bushing. QD sprockets use tapered bushings called **QD bushings** in a manner similar to taper-lock bushings (see Figure 8.18). QD sprockets can be easily identified because they have a characteristic flange that taper-lock sprockets do not have. Instead of set screws, QD bushings use hex-head bolts to hold the sprocket and squeeze the shaft. As with like taper-locks, they also have jack-bolt holes to force separation of the sprocket and bushing.

INSTALLING AND REMOVING TAPER-LOCK BUSHINGS

To install a taper-lock bushing:

Step 1. Place the bushing in the sprocket.

Step 2. Apply oil to both the thread and the point of set screws. Place the screws loosely in pull-up holes.

Step 3. Make sure the bushing is free in the sprocket. Slip the assembly onto the shaft and align-locate.

Step 4. Tighten the screws alternately and progressively until they are pulled up tight.

Step 5. Tap the large end of the bushing (use the hammer and soft block to avoid damage). Tighten the screws further according to the manufacturer's recommendation included in box. Fill the other holes with grease to exclude dirt.

To remove a taperlock bushing:

Step 1. Remove both set screws.

Step 2. Apply oil to both the thread and the point of one set screw. Insert this screw in the tapped removal hole. (Note one setscrew is not used.)

Step 3. Tighten the inserted screw until the bushing is loosened in the sprocket.

8-14 STANDARD ROLLER CHAIN DRIVE SELECTION

For long life and ease of maintenance, the following steps should be used when selecting a standard roller chain drive:

- Determine the revolutions per minute and diameter of the high-speed shaft.
- Determine the total horsepower to be transmitted.
- Apply service-factor multipliers based on drive duty and shock loading from Table 8.3.
- Select the chain pitch and number of teeth in the small sprocket from Table 8.1.
    a. Be sure that the small sprocket will accommodate the high-speed shaft diameter.
    b. If the high-speed shaft diameter exceeds the maximum bore in the selected small sprocket, either increase the number of teeth in the sprocket or select the next-larger-pitch chain (see Table 8.2).
- Determine the required ratio:

$$\frac{\text{Rev/min of high-speed shaft}}{\text{Rev/min of slow-speed shaft}} = \text{ratio of drive}$$

- Multiply the number of teeth in the small sprocket by the ratio to obtain the number of teeth in the large sprocket. To avoid ordering special sprockets, ratios are sometimes adjusted to change the speed slightly. If it is not possible to adjust the ratio, a different small sprocket can sometimes be selected, which then changes the number of teeth of the larger sprocket.

SERVICE FACTORS

A service factor is applied to the horsepower ratings for other-than-normal duty drives, taking into consideration source of power, nature of the load, and load inertia strain or shock. Average hours per day of continuous service should also be considered. Normal duty drives are those with relatively little shock or load variation. When in doubt about the correct service factor, consult the manufacturer.

The type of load can be divided into the following three classifications:

Smooth: Running load is fairly uniform. Starting and peak loads may be somewhat greater than the running load but occur infrequently.

Moderate shock: Running load is variable. Starting and peak loads are considerably greater than running load and occur frequently.

Heavy shock: Starting loads are extremely heavy. Peak loads and overloads occur continuously and are of maximum fluctuation.

TABLE 8.1
Basic horsepower ratings.

60 3/4" Pitch Single-Strand Roller Chain — HORSE POWER RATINGS

No.	Small Sprocket RPM																					
Teeth	10	25	50	75	100	125	150	175	200	250	300	350	400	500	600	700	800	900	1000	1200	1400	1600
9	0.15	0.33	0.62	0.89	1.16	1.42	1.67	1.91	2.16	2.64	3.12	3.58	4.04	4.94	5.82	6.68	7.54	8.38	9.21	8.77	6.96	5.70
10	0.16	0.37	0.70	1.00	1.30	1.59	1.87	2.15	2.43	2.96	3.49	4.01	4.53	5.53	6.52	7.49	8.44	9.39	10.3	10.3	8.15	6.67
11	0.18	0.41	0.77	1.11	1.44	1.76	2.07	2.38	2.69	3.28	3.87	4.44	5.02	6.13	7.23	8.30	9.36	10.4	11.4	11.9	9.4	7.70
12	0.20	0.45	0.85	1.22	1.58	1.93	2.28	2.62	2.95	3.60	4.25	4.88	5.51	6.74	7.94	9.12	10.3	11.4	12.6	13.5	10.7	8.77
13	0.22	0.50	0.92	1.33	1.73	2.11	2.49	2.86	3.22	3.93	4.64	5.33	6.01	7.34	8.65	9.94	11.2	12.5	13.7	15.2	12.1	9.89
14	0.24	0.54	1.00	1.44	1.87	2.28	2.69	3.09	3.49	4.26	5.02	5.77	6.51	7.96	9.37	10.8	12.1	13.5	14.8	17.0	13.5	11.1
15	0.25	0.58	1.08	1.55	2.01	2.46	2.90	3.33	3.76	4.59	5.41	6.21	7.01	8.57	10.1	11.6	13.1	14.5	16.0	18.8	15.0	12.3
16	0.27	0.62	1.16	1.66	2.16	2.64	3.11	3.57	4.03	4.92	5.80	6.66	7.52	9.19	10.8	12.4	14.0	15.6	17.1	20.2	16.5	13.5
17	0.29	0.66	1.24	1.78	2.31	2.82	3.32	3.81	4.30	5.25	6.20	7.12	8.03	9.81	11.6	13.3	15.0	16.7	18.3	21.6	18.1	14.8
18	0.31	0.70	1.31	1.88	2.45	2.99	3.53	4.05	4.58	5.59	6.59	7.57	8.54	10.4	12.3	14.1	15.9	17.7	19.5	22.9	19.7	16.1
19	0.33	0.75	1.39	2.00	2.60	3.17	3.74	4.30	4.85	5.92	6.99	8.02	9.05	11.1	13.0	15.0	16.9	18.8	20.6	24.3	21.4	17.5
20	0.35	0.79	1.47	2.11	2.75	3.36	3.96	4.55	5.13	6.26	7.38	8.48	9.57	11.7	13.8	15.8	17.9	19.8	21.8	25.7	23.1	18.9
21	0.36	0.83	1.55	2.23	2.90	3.54	4.17	4.79	5.40	6.59	7.78	8.94	10.1	12.3	14.5	16.7	18.8	20.9	23.0	27.1	24.8	20.3
22	0.38	0.87	1.63	2.34	3.05	3.72	4.39	5.04	5.68	6.94	8.19	9.40	10.6	13.0	15.3	17.5	19.8	22.0	24.2	28.5	26.6	21.8
23	0.40	0.92	1.71	2.45	3.19	3.90	4.60	5.28	5.96	7.28	8.59	9.85	11.1	13.6	16.0	18.4	20.8	23.1	25.4	29.9	28.4	23.3
24	0.42	0.96	1.79	2.57	3.35	4.09	4.82	5.53	6.24	7.62	8.99	10.3	11.6	14.2	16.8	19.3	21.7	24.2	26.6	31.3	30.3	24.8
25	0.44	1.00	1.87	2.68	3.50	4.27	5.04	5.78	6.52	7.96	9.40	10.8	12.2	14.9	17.5	20.1	22.7	25.3	27.8	32.7	32.2	26.4
26	0.46	1.05	1.95	2.80	3.65	4.45	5.25	6.03	6.81	8.31	9.80	11.3	12.7	15.5	18.3	21.0	23.7	26.4	29.0	34.1	34.2	28.0
28	0.50	1.13	2.12	3.04	3.95	4.82	5.69	6.53	7.37	8.99	10.6	12.2	13.8	16.8	19.8	22.8	25.7	28.5	31.4	37.0	38.2	31.3
30	0.54	1.22	2.28	3.27	4.26	5.20	6.13	7.04	7.94	9.67	11.4	13.1	14.8	18.1	21.4	24.5	27.7	30.8	33.8	39.8	42.4	34.7
32	0.57	1.31	2.45	3.51	4.56	5.57	6.57	7.55	8.52	10.4	12.3	14.1	15.9	19.4	22.9	26.3	29.7	33.0	36.3	42.7	46.7	38.2
35	0.63	1.44	2.69	3.86	5.03	6.14	7.24	8.31	9.38	11.4	13.5	15.5	17.5	21.4	25.2	29.0	32.7	36.3	39.9	47.1	53.4	43.7
40	0.73	1.67	3.11	4.46	5.81	7.09	8.37	9.59	10.8	13.2	15.6	17.9	20.2	24.7	29.1	33.5	37.7	42.0	46.1	54.4	62.5	53.4
45	0.83	1.89	3.53	5.07	6.60	8.05	9.50	10.9	12.3	15.0	17.7	20.4	23.0	28.1	33.1	38.0	42.9	47.7	52.4	61.7	70.9	63.7
	TYPE A									TYPE B									TYPE C			

80 1" Pitch Single-Strand Roller Chain — HORSEPOWER RATINGS

No.	Small Sprocket RPM																					
Teeth	10	25	50	75	100	125	150	175	200	250	300	350	400	500	600	700	800	900	1000	1200	1400	1600
9	0.34	0.78	1.45	2.08	2.71	3.31	3.90	4.48	5.05	6.17	7.28	8.36	9.43	11.5	13.6	15.6	17.6	17.0	14.5	11.0	8.76	7.17
10	0.38	0.87	1.63	2.33	3.03	3.70	4.37	5.02	5.66	6.91	8.16	9.38	10.6	12.9	15.2	17.5	19.7	19.9	17.0	12.9	10.3	8.40
11	0.42	0.97	1.80	2.58	3.36	4.10	4.84	5.56	6.28	7.66	9.04	10.4	11.7	14.3	16.9	19.4	21.9	23.0	19.6	14.9	11.8	9.69
12	0.47	1.06	1.98	2.84	3.69	4.51	5.32	6.11	6.89	8.41	9.93	11.4	12.9	15.7	18.5	21.3	24.0	26.2	22.3	17.0	13.5	11.0
13	0.51	1.16	2.16	3.10	4.03	4.91	5.80	6.66	7.52	9.16	10.8	12.4	14.0	17.1	20.2	23.2	26.2	29.1	25.2	19.2	15.2	12.5
14	0.55	1.25	2.34	3.35	4.36	5.33	6.29	7.22	8.14	9.92	11.7	13.4	15.2	18.6	21.9	25.1	28.4	31.5	28.2	21.4	17.0	13.9
15	0.59	1.35	2.52	3.61	4.70	5.74	6.77	7.77	8.77	10.7	12.6	14.5	16.4	20.0	23.6	27.1	30.6	34.0	31.2	23.8	18.9	15.4
16	0.63	1.45	2.70	3.87	5.04	6.15	7.26	8.34	9.41	11.5	13.5	15.6	17.6	21.5	25.3	29.0	32.8	36.4	34.4	26.2	20.8	17.0
17	0.68	1.55	2.88	4.13	5.38	6.57	7.75	8.88	10.0	12.3	14.5	16.6	18.7	22.9	27.0	31.0	35.0	38.9	37.7	28.7	22.7	18.6
18	0.72	1.64	3.07	4.40	5.72	6.99	8.25	9.48	10.7	13.1	15.4	17.6	19.9	24.4	28.7	33.0	37.2	41.4	41.1	31.2	24.8	20.3
19	0.76	1.74	3.25	4.66	6.07	7.41	8.74	10.0	11.3	13.8	16.3	18.7	21.1	25.8	30.4	35.0	39.4	43.8	44.5	33.9	26.9	22.0
20	0.81	1.84	3.44	4.93	6.41	7.83	9.24	10.6	12.0	14.6	17.2	19.7	22.3	27.3	32.2	37.0	41.7	46.3	48.1	36.6	29.0	23.8
21	0.85	1.94	3.62	5.19	6.76	8.25	9.74	11.2	12.6	15.4	18.2	20.9	23.5	28.8	33.9	39.0	43.9	48.9	51.7	39.4	31.2	25.6
22	0.90	2.04	3.81	5.46	7.11	8.66	10.2	11.8	13.3	16.2	19.1	22.0	24.8	30.3	35.7	41.0	46.2	51.4	55.5	42.2	33.5	27.4
23	0.94	2.14	4.00	5.73	7.46	9.08	10.7	12.3	13.9	17.0	20.1	23.1	26.0	31.8	37.4	43.0	48.5	53.9	59.3	45.1	35.8	29.3
24	0.98	2.24	4.19	6.00	7.81	9.58	11.3	13.0	14.6	17.8	21.0	24.1	27.2	33.2	39.2	45.0	50.8	56.4	62.0	48.1	38.2	31.2
25	1.03	2.34	4.37	6.27	8.16	9.98	11.8	13.5	15.2	18.6	21.9	25.2	28.4	34.7	40.9	47.0	53.0	59.0	64.8	51.1	40.6	33.2
26	1.07	2.45	4.56	6.54	8.52	10.4	12.3	14.1	15.9	19.4	22.9	26.3	29.7	36.2	42.7	49.1	55.3	61.5	67.6	54.2	43.0	35.2
28	1.16	2.65	4.94	7.09	9.23	11.3	13.3	15.3	17.2	21.0	24.8	28.5	32.1	39.3	46.3	53.2	59.9	66.7	73.7	60.6	48.1	39.4
30	1.25	2.85	5.33	7.64	9.94	12.1	14.3	16.4	18.5	22.6	26.7	30.7	34.6	42.3	49.9	57.3	64.6	71.8	78.9	67.2	53.3	43.6
32	1.34	3.06	5.71	8.21	10.7	13.0	15.3	17.6	19.9	24.3	28.6	32.9	37.1	45.4	53.5	61.4	69.2	77.0	84.6	74.0	58.7	48.1
35	1.48	3.37	6.29	9.00	11.7	14.3	16.9	19.4	21.9	26.8	31.6	36.3	40.9	50.0	58.9	67.6	76.3	84.8	93.3	84.7	67.2	55.0
40	1.71	3.89	7.27	10.4	13.6	16.5	19.5	22.4	25.3	30.9	36.4	41.8	47.2	57.7	68.0	78.1	88.1	99.0	108	103	82.1	67.2
45	1.94	4.42	8.25	11.8	15.4	18.8	22.2	25.5	28.7	35.1	41.4	47.5	53.6	65.6	77.2	88.7	100	111	122	123	98.0	80.2
	TYPE A							TYPE B										TYPE C				

Lubrication Note :
- TYPE A: Manual or Drip
- TYPE B: Bath or Disc
- TYPE C: Oil Stream

Multiple Strand Chain HP Factors:

Single Strand	1.0
Double Strand	1.9
Triple Strand	2.8

Courtesy of Dodge/Rockwell Automation.

TABLE 8.2
Table for recommended small sprocket.

Rev/Min of Small Sprocket	Chain Size and No. of Teeth for Design HP											
	1/4	1/3	1/2	3/4	1	1–1/2	2	3	4	5	6	7–1/2
951–1000	35 17	35 17	35 17	35 17	35 17	35 17	35 17	35 23	40 17	40 17	40 20	40 24
901–950	35 17	35 17	35 17	35 17	35 17	35 17	35 17	35 24	40 17	40 18	40 21	40 25
851–900	35 17	35 17	35 17	35 17	35 17	35 17	35 17	35 24	40 17	40 18	40 22	50 17
801–850	35 17	35 17	35 17	35 17	35 17	35 17	35 18	40 17	40 17	40 19	40 23	50 17
751–800	35 17	35 17	35 17	35 17	35 17	35 17	35 19	40 17	40 17	40 20	40 24	50 17
701–750	35 17	35 17	35 17	35 17	35 17	35 17	35 20	40 17	40 18	40 22	40 25	50 17
651–700	35 17	35 17	35 17	35 17	35 17	35 17	35 22	40 17	40 19	40 23	50 17	50 18
601–650	35 17	35 17	35 17	35 17	35 17	35 18	35 23	40 17	40 20	40 24	50 17	50 19
551–600	35 17	35 17	35 17	35 17	35 17	35 19	40 17	40 17	40 21	50 17	50 17	50 21
501–550	35 17	35 17	35 17	35 17	35 17	35 21	40 17	40 17	40 23	50 17	50 18	50 22
471–500	35 17	35 17	35 17	35 17	35 17	35 22	40 17	40 17	40 24	50 17	50 19	50 24
441–470	35 17	35 17	35 17	35 17	35 17	35 23	40 17	40 17	50 17	50 17	50 20	50 25
411–440	35 17	35 17	35 17	35 17	35 17	40 17	40 17	40 21	50 17	50 18	50 22	60 17
381–410	35 17	35 17	35 17	35 17	35 17	40 17	40 17	40 22	50 17	50 19	50 23	60 17
351–380	35 17	35 17	35 17	35 17	35 17	40 17	40 17	40 24	50 17	50 21	50 24	60 18
321–350	35 17	35 17	35 17	35 17	35 17	40 17	40 18	50 17	50 18	50 22	60 17	60 20
301–320	35 17	35 17	35 17	35 17	35 17	40 17	40 19	50 17	50 19	50 23	60 17	60 21
281–300	35 17	35 17	35 17	35 17	35 17	40 17	40 19	50 17	50 20	50 25	60 18	60 22
261–280	35 17	35 17	35 17	35 17	35 17	40 17	40 19	50 17	50 22	60 17	60 19	60 23
241–260	35 17	35 17	35 17	35 17	35 17	40 17	40 19	50 18	50 23	60 17	60 20	80 17
221–240	35 17	35 17	35 17	35 17	35 17	40 19	40 24	50 19	50 25	60 19	60 22	80 17
201–220	35 17	35 17	35 17	35 17	35 17	40 20	50 17	50 21	60 17	60 20	60 24	80 17
181–200	35 17	35 17	35 18	35 17	35 17	40 22	50 17	50 23	60 18	60 22	60 26	80 17
161–180	35 17	35 17	35 20	40 18	35 17	40 24	50 17	50 25	60 20	60 24	80 17	80 17
151–160	35 17	35 17	35 21	40 18	35 17	50 17	50 18	60 17	60 21	60 25	80 17	80 17
141–150	35 15	35 15	35 22	40 18	40 15	50 17	50 19	60 17	60 22	80 17	80 17	80 18
131–140	35 15	35 15	35 23	40 18	40 15	50 17	50 20	60 18	60 23	80 17	80 17	80 19
121–130	35 15	35 15	40 13	40 18	40 15	50 17	50 22	60 19	60 25	80 17	80 17	80 20
111–120	35 15	35 15	40 13	40 18	40 15	50 18	50 22	60 20	80 15	80 17	80 18	80 22
101–110	35 15	35 15	40 15	40 18	50 15	50 19	60 15	60 15	80 15	80 16	80 19	100 15
91–100	35 15	35 15	40 15	40 15	40 17	50 15	50 18	60 17	80 15	80 18	100 15	100 15
81–90	35 15	35 15	40 15	40 15	50 15	50 15	50 18	60 18	80 15	80 15	80 16	100 15
71–80	35 15	35 17	40 15	40 16	50 15	50 18	60 14	80 15	80 15	80 16	100 15	100 15
61–70	35 15	35 19	40 15	40 19	50 15	60 15	60 16	80 15	80 15	80 18	100 15	100 15
51–60	35 17	40 13	40 15	50 13	50 15	60 15	60 19	80 15	80 17	100 15	100 15	100 16
46–50	40 13	40 13	40 16	50 13	50 16	60 14	80 13	80 13	100 13	100 13	100 14	100 17
41–45	40 13	40 13	40 18	50 16	60 13	60 16	80 13	80 14	100 13	100 13	100 16	120 13
35–40	40 13	40 14	50 13	50 16	60 13	60 19	80 13	80 17	100 13	100 14	100 18	120 14
30–35	40 13	40 16	50 13	50 18	60 15	80 13	80 13	80 19	100 13	100 16	120 13	120 14
23–29	40 14	50 13	50 16	60 14	60 19	80 13	80 17	100 13	100 16	120 14	120 15	120 18
17–22	50 13	50 14	60 13	60 19	80 13	80 17	100 13	100 17	120 13	120 16	140 13	140 16
12–16	50 15	60 13	60 18	80 13	80 16	100 13	100 16	120 14	120 18	140 15	140 17	160 15
8–11	60 14	60 18	80 13	80 18	100 13	100 17	120 14	140 13	140 17	160 15	160 18	180 18
5–7	80 13	80 13	100 13	100 13	100 18	120 17	140 14	180 15	180 14	200 13	200 15	240 13

Notes:

1. Apply service factor to obtain design horsepower. Select small sprocket based upon design horsepower and rev/min on this chart.
2. Sprocket selections are recommended minimum. Larger sizes may be selected if required to obtain desired ration, etc.
3. To use this chart for double- or triple-strand chain, divide the design horsepower by the following factors: Double-strand: 1.9, Triple-strand: 2.9

Courtesy of Dodge/Rockwell Automation.

TABLE 8.3
Service factors.

	Type of Power Source		
Type of Load	**Internal Combustion Engine—Hydraulic Drive**	**Electric Motor or Turbine**	**Internal Combustion Engine—Mechanical Drive**
Smooth	1.0	1.0	1.2
Moderate shock	1.2	1.3	1.4
Heavy shock	1.4	1.5	1.7

- Determine the center distance.

  Center distance in pitches = center distance in inches ÷ pitch length

  For the average application, a center distance of 30 to 50 pitches of the chain represents good practice. For pulsating loads, centers as short as 20 pitches of chain may be desirable. To permit the sprocket teeth to clear, the center distance must at least be slightly greater than one-half the sum of the outside diameters of the two sprockets.
- Determine the chain length. You can find the approximate length of the chain in inches by the formula:

$$L = 2C + 1.57(D + d) + \frac{(D - d)^2}{4C}$$

where D = pitch diameter of the large sprocket (in.)
d = pitch diameter of the small sprocket (in.)
C = proposed drive center distance (in.)

GENERAL RECOMMENDATIONS ON SPROCKET SIZES

Unless speeds are low, it is not advisable to use less than 15 teeth in the smaller sprocket. When ratios are low, relatively large sprockets may be used, giving less chain pull, lower bearing loads, and less joint articulation. If, on the other hand, ratios and speeds are high, it may be necessary to use a relatively small number of teeth in the high-speed sprocket.

Ratios over 7:1 are generally not recommended for single-width roller chain drives. Very slow-speed drives (10 to 100 rev/min) are often practical, with as few as 9 or 10 teeth in the small sprocket, allowing ratios up to 12:1. In all cases where ratios exceed 5:1, the designer should consider the possibility of using compound drives to obtain maximum service life.

EXAMPLE OF STANDARD ROLLER CHAIN DRIVE SELECTION

A chain drive is required for a tumbling barrel for metal stampings that is to be operated at 170 rev/min. The tumbling barrel is to be driven by a speed reducer with an output speed of approximately 400 rev/min. Power to the speed reducer is supplied by a 5-hp electric motor. Starting loads are heavy peak loads and overloads occur continuously. The drive requires a center distance of 30 in.

Step 1. The high-speed shaft revolves at 400 rev/min.

Step 2. The input electric-motor horsepower is 5. Because of heavy peak loads and continuous overloads, the load should be rated as *heavy shock*. From Table 8-3, the service factor is 1.5 × 5 hp = 7.5 design hp.

Step 3. From the Table 8-2, look up 7.5 hp at 400 rev/min. The chain size is number 60 roller chain with a 17-tooth sprocket.

Step 4. 400 rev/min divided by 170 rev/min gives a ratio of 2.35:1.

Step 5. Multiplying number of small-sprocket teeth (17) times the ratio (2.35:1) gives a large sprocket with 40 teeth.

Step 6. The center distance in pitches equals center distance in inches (30) divided by pitch length (no. 60 chain pitch is 6/8, or 0.75 in.): 30/0.75 = 40 pitches. This is between 30 and 50 pitches.

Step 7. Approximate chain length is

$$L = 2C + 1.57(D + d) + \frac{(D - d)^2}{4C}$$

$$L = 2(30) + 1.57(9.56 + 4.08) + \frac{(9.56 - 4.08)^2}{4(30)}$$

$$L = 60 + 21.41 + \frac{30.03}{120}$$

$L = 81.66$ in. of no. 60 roller chain

8-15 INSTALLING ROLLER CHAIN DRIVES

Lubrication and accurate alignment are two key factors for long life of a chain drive. Lubrication is discussed later. The following steps should be included in any installation of a chain drive:

- Using a machinist's level, make sure that both shafts are leveled horizontally (see Figure 8.19). Chain drives should not be used if

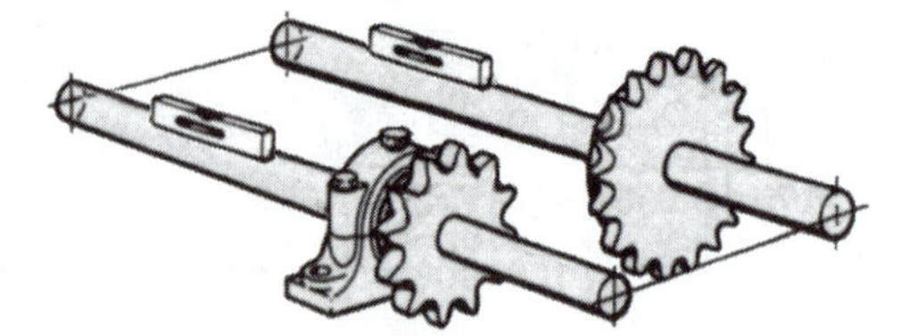

FIGURE 8.19
Shafts must be level.

shafts are not horizontal and sprockets are not vertical. Vertical alignment of sprockets is necessary to minimize wear on chain side bars when they come in contact with sprocket teeth.

- Make sure that the shafts are parallel by using a feeler bar or tape measure. The distance between shafts on both sides of the sprocket must be equal.
- Repeat steps 1 and 2 until both level and alignment are correct.
- Mount the sprockets on the shafts and align them axially. This may be accomplished by two methods:
  a. If the shafts are squared with the machine frame, you can use a tape measure to make sure that the distances between the sprockets and the frame are the same.
  b. You can also use a straightedge or taut wire along the finished sides of the sprockets (see Figure 8.20). If the shafts have any end float, lock them in their running position before aligning the sprockets.
- Check for defective or warped sprockets (see Figure 8.21). Rotating the sprocket on the shaft, use a dial indicator to check sprocket runout. If a dial indicator is not available, a ruler may be used to measure the changes in distance between the outer edge of the sprocket and a fixed point as the sprocket is rotated.
- Install the chain on the sprockets, bringing the loose ends together on one sprocket. Connect the chain by using a master or offset link.
- If necessary, adjust the chain tension by changing the shaft center distance. Chain tension may vary depending on chain size, speed, and application. For most applications, a chain sag of 2% of shaft centers is appropriate. One method of checking chain sag is to place a straightedge on top of the two sprockets, deflecting the middle of the chain downward and measuring the distance between the straightedge and the chain. For example, a drive with a 50-in. center distance should deflect no more than 1 in. (2% of 50 in. = 1 in.). After connecting the chain together, rotate it around the sprocket until the tightest spot is found. Adjust it at the tight spot of the chain. (It is better for a chain to be too loose than too tight.)

FIGURE 8.20
Using a straight to align sprockets.

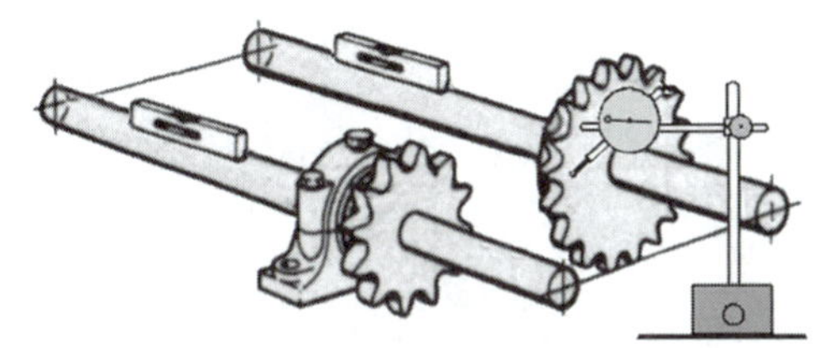

FIGURE 8.21
Check for warped sprockets.

- Make sure that all protective guards are in place and the lubrication devices are operational.
- After 100 to 200 h of operation, check alignment and tension. Seating of chain joints usually causes the chain to loosen. After the initial drive break-in period, chain drives give excellent service with few problems if kept properly lubricated.

8-16 LUBRICATING CHAIN DRIVES

Following are the basic types of chain lubrication found in industry:

- *Manual lubrication:* Oil is applied periodically with a brush or oilcan, preferably once every 8 hours of operation. The volume and frequency should be sufficient to prevent discoloration of the chain. (A brownish, rusty discoloration of a chain indicates a lack of lubrication.)
- *Drip lubrication:* Oil is dripped on the link plate edges by a drip lubricator. Oil dripping on the center of the chain may not lubricate the joint areas.
- *Bath or disk lubrication:* Bath lubrication submerses the lower strand of the chain in a sump of oil in the drive housing. The oil level should always be above the pitch line of the chain at its lowest point. With disk lubrication, the chain operates above the oil level. The disk picks up oil from the sump and deposits it onto the chain.
- *Oil-stream lubrication:* With oil-stream lubrication, the oil is pressurized by a pumping system. A continuous stream of oil is supplied to the inside of the chain loop and directed at the slack strand.
- *Oil-mist lubrication:* Oil-mist lubrication is a relatively new lubrication method in which oil is atomized into a mist through which the chain runs.

TABLE 8.4
Maintenance guide for roller chain.

Type of Maintenance	When to Do	What to Do
Lubrication—type I (manual)	Every 8 h of operation	Apply oil to chain and sprocket contact surface with brush or spout can.
Lubrication—type II (drip)	Continuous	Apply oil between link plate edges from a drip lubricator.
Lubrication—type III (bath)	Continuous	Oil level maintained in enclosed case. Lower strand of chain submerged to the pitch circle when operating.
Lubrication—type IV (oil stream)	Continuous	Oil applied inside chain loop on lower strand by a circulation pump.

Continued

TABLE 8.4
continued.

Type of Maintenance	When to Do	What to Do
Checking oil level (types III and IV)	Initial inspection—24 h Second inspection—100 h Third inspection—500 h Periodically thereafter	The oil level should be checked when the drive is idle. Add oil if necessary while drive is idle.
Checking oil flow (type IV)	Initial inspection—24 h Second inspection—100 h Third inspection—500 h Periodically thereafter	Check oil flow to be sure the pump is operating.
Changing oil (types III and IV)	Initial oil change after 500 h 2500 hours thereafter	Drain the case and refill the case with fresh oil. If at any time the lubricant is found to be contaminated, the casing should be flushed and the chain carefully cleaned. When the chain is thoroughly dry, immerse the chain in hot oil and reinstall.
Set screw and bushing capscrew check	Initial inspection—24 h Second inspection—100 h Third inspection—500 h Periodically thereafter	Check set screws and/or bushing capscrews for looseness. Retighten to recommended torques, if necessary.
Chain sag	Initial inspection—24 h Second inspection—100 h Third inspection—500 h Periodically thereafter (New chain will elongate considerably at first. Check tension often at first.)	Adjust so that the chain sag on the slack side is equivalent to 2% of the center distance.
Cleaning of chain	Necessary only if chain running in dusty, dirty atmosphere	Remove the chain and clean it with a good cleaning solution. The chain should be dipped in hot oil before reinstalling it.
Checking for chain elongation	The chain should be replaced after it is elongated in excess of 2 1/2%. Check the length after first 1000 h and periodically thereafter	Remove the chain and measure it. If the actual length is 2 1/2% longer than the calculated length for the number pitches being measured, replace it with new chain.
Checking for sprocket wear	Initial inspection—24 h Second inspection—100 h Third inspection—500 h Periodically thereafter	If teeth have a hooked appearance, replace with new sprockets.
Checking alignment	Initial inspection—24 h Second inspection—100 h Third inspection—500 h Periodically thereafter	If wear is apparent on the inner surface of the roller link sidebars and on the sides of the sprocket teeth, there is misalignment. Align the sprockets to correct problem.

8-17 MAINTAINING ROLLER CHAIN DRIVES

The average roller chain drive should have a service life of 15,000 h. The following are necessary for the long life of a chain drive (see Tables 8.4 and 8.5):

- A correctly engineered drive for the application
- Good, uncontaminated lubrication
- Proper maintenance

The following are necessary for good maintenance of a chain drive:

- After installing or replacing a drive, frequently inspect it for any obvious visible problems or any possible interference with the chain. Unusual mechanical sounds may indicate a problem. If a chain is rubbing or slapping against anything, stop the machine and make the necessary corrections.
- Never install a new chain on badly worn sprockets. Worn sprockets should always be replaced to ensure that the new chain is not damaged or stretched by the old sprockets. Many new chains last only a fraction of their useful life because of premature wear. If the old sprockets cannot be replaced, the life of a new chain may be increased by reversing the old sprocket on the shaft to bring a new set of working tooth surfaces into use. Be sure to check alignment if this is done to ensure that the sprocket runs true in its new position.
- Check the chain frequently for excessive slack. New chain almost always stretches and should be checked after 100 to 200 h of operation. If the chain has been stretched, the pitch length of the chain will be greater. You can check this visually by lifting the chain away from the large sprocket, making sure that the chain is in mesh with the sprocket teeth. Excessive clearance is conclusive evidence that the chain has elongated in pitch, and no amount of tension adjustment will keep it properly meshed with the sprocket teeth. Continued operation will quickly destroy the sprocket teeth.
- Make sure that the drive is reasonably clean and no foreign material has become packed in the sprocket teeth. If this happens, it may cause the chain to ride high on the sprocket teeth, possibly elongate the chain, and cause abnormal wear of the sprocket teeth.
- Include alignment checks of the chain drive in your preventative maintenance. Misalignment is the problem if the sides of sprocket teeth or inside surfaces of the chain-link plates show wear. Alignment is very important and should be remedied as soon as possible.
- Periodically check lubrication in all chain drives. Lubricant is a chain's lifeblood; lack of proper lubrication causes short chain life.

Always use the correct type and viscosity of lubricant. With manual lubrication, follow the correct lubrication schedule and apply the oil properly. For drip lubrication, check the rate of feed and apply the oil correctly. For bath or disk systems, inspect the oil level and check that there is no sludge. Drain, flush, and refill the system at least once a year. For force-feed systems, inspect the oil level in the reservoir and check the pump drive and delivery pressure. Check that there is no clogging of the piping or nozzles. Drain, flush, and refill the reservoir at least once a year.

- If you do not lubricate roller chains properly, the joints will have a brownish, rusty color and the pins of the connecting link of the chain, when removed, will be discolored (light or dark brown). Also, the pins will be roughened, grooved, or galled. Properly lubricated chains do not have the brownish color at the joints, and the connecting link pins are brightly polished with a very high luster.
- If the chain is exposed to dusty environments, clean the chain periodically. Clean a chain as follows: (1) Remove the chain from the sprockets. (2) Wash the chain in a nonflammable safety solvent. If the chain is thickly coated and gummed up, soak it for several hours in the cleaning fluid and then rewash it in fresh fluid. After draining off the cleaning fluid, soak the chain in oil to restore the internal lubrication. (3) Remove the excess lubricant. (4) Inspect the chain for wear or corrosion. While the chain is off the sprocket, clean the sprockets and inspect them for wear or corrosion.
- Unless properly protected, the components of a chain drive deteriorate during long periods of idleness. If you plan to store a chain, remove it from the sprockets and coat it with a heavy oil or light grease. Then wrap it in heavy, grease-resistant paper. Store the chain where it will be protected from moisture and mechanical injury. The sprockets may be left in place on the shafts. Cover each with grease, and protect them from mechanical injury. Before placing the drive in service again, thoroughly clean the chain and sprockets to remove the protective grease and then relubricate the chain.

TABLE 8.5
Troubleshooting guide for roller chain.

Type of Failure	Cause of Failure	Corrective Action
Nonsymmetrical wear on rollers	1. Shafts out of parallel or not in the same plane	1. Realign shafts, sprockets, or rollers
Wear on the inside of the roller	1. Sprockets offset on shafts (misaligned) or out of parallel	1. Realign sprockets, plates side-tooth form of the sprocket teeth
Wear on the tips of the sprocket teeth	1. Chain elongated excessively 2. Improperly cut sprockets	1. Replace chain 2. Replace with correct sprocket

Continued

TABLE 8.5
continued.

Type of Failure	Cause of Failure	Corrective Action
Worn or hooked sprocket teeth	1. Unhardened sprockets	1. Increase case clearance or move fixed object
Wear on the edges or sides of link plates	1. Chain contacting case or fixed object	1. Increase case clearance or move fixed object
Excessive vibration	1. Excessive eccentricity or face out in sprocket 2. Broken or missing roller	1. Replace with properly run machined sprocket 2. Repair or replace chain
Premature elongation	1. Inadequate or contaminated lubrication or underchaining	1. Increase oil flow or redesign
Brown-red oxide in chain joints and oil	1. Inadequate lubrication	1. Improve lubrication
Chain jumping sprocket teeth	1. Excessive chain slack 2. Worn chain or sprockets 3. Heavy overload	1. Adjust the centers or idler 2. Replace the chain 3. Reduce the load, replace drive
Broken chain parts	1. Drive overloaded 2. Excessive slack causing the chain to jump teeth 3. Foreign object 4. Excess chain speed 5. Poorly fitting sprockets 6. Inadequate lubrication 7. Corrosion	1. Redesign or avoid 2. Periodically adjust center distance 3. Prevent entry 4. Redesign or avoid 5. Replace 6. Provide proper lubrication 7. Prevent or use noncorrosive chain
Excessive noise	1. Chain contacting fixed objects 2. Inadequate lubrication 3. Broken or missing rollers 4. Misalignment 5. Chain jumping sprocket teeth	1. Remove objects 2. Improve lubrication 3. Repair or replace chain 4. Check shaft and realign 5. Adjust center distance

QUESTIONS

1. Why is a chain considered a positive drive?
2. In the United States, what organization establishes the dimensions and specifications for roller chain?
3. What are two types of roller chain construction?
4. The special link used to produce a strand of chain with an odd number of links is called a(n) __________ or _________ link.
5. When should standard chains be replaced by heavy series chain?
6. Will heavy series chain run on the same sprockets as standard chain?
7. What is the major disadvantage of self-lubricating chain?
8. What is the pitch of no. 60 roller chain?

9. What is the attachment designation of a roller chain with 90° bent lugs on both sides of the roller chain?
10. What is the main advantage of an inverted tooth drive?
11. Describe 'chordal action' in a chain drive.
12. Why is chordal action a problem in a chain drive?
13. What are the two main types of engineered-class steel chain?
14. What are the advantages of malleable iron-cast chain over steel chain?
15. Name the three types of drag conveyor chain.
16. What is a type B sprocket?
17. Why do type C sprockets have hubs on both sides?
18. What is a minimum plain bore sprocket?
19. Which popular compression bushing has no flange?
20. What is the service factor using an electric motor with moderate shock loading?
21. How do you check for defective or warped sprockets?
22. Name four types of chain lubrication.

CHAPTER 9

Gears

9-1 INTRODUCTION

Gears are used to transmit rotating motion in many machines today. They were also used by the Greeks in ancient times in astronomical instruments. Archeologists have found bronze gears used by the Chinese around the year 200 B.C. The Romans used wooden gears to power grist mills. These wooden discs were fitted with pegs and were known as **lantern gears.** By the twelfth century, England had more than 5,000 water mills that used hardwood gears (see Figure 9.1). The onset of the Industrial Revolution in the eighteenth century created a need for all kinds of gears. Initially, gears were made in cast

FIGURE 9.1
Gears are important to old Mills.

FIGURE 9.2
Pinion and gear.

iron. Soon thereafter, gear-cutting machines were invented. In the nineteenth century, a normal gear had cast-iron teeth and a precision gear had cut, or machined, teeth. In the twentieth century almost all gears were precision-cut machined gears. Gears are used in many different machines in many different ways. Today, we have many different types of gears, but we are concerned only with the basic types used in modern industry.

9-2 GEAR TERMS AND FUNDAMENTALS

Gears are used to form a positive transfer of power. This means that there is no inherent slippage in a gear drive. When two gears are used to form a drive, the larger one is called the *gear* or *wheel,* and the smaller one is called the *pinion* (see Figure 9.2). Because the pinion is smaller than the gear, it does more work than the larger gear. For this reason, pinions are often made of harder materials in order to prevent them from wearing out prematurely. If the pinion drives the gear, the unit is called a *speed reducer;* if the gear drives the pinion, it is called a *speed increaser.* Gear drives are used more often as speed reducers than as speed increasers because electric motors are usually too fast to directly couple to most machinery.

When there are only two gears in a drive, they always rotate in opposite directions. In order to make gears rotate in the same direction, a third, intermediate gear, called an *idler gear,* can be inserted between the driving and driven gears (see Figure 9.3). Multiple idler gears can also be used to increase input- and output-shaft center distances.

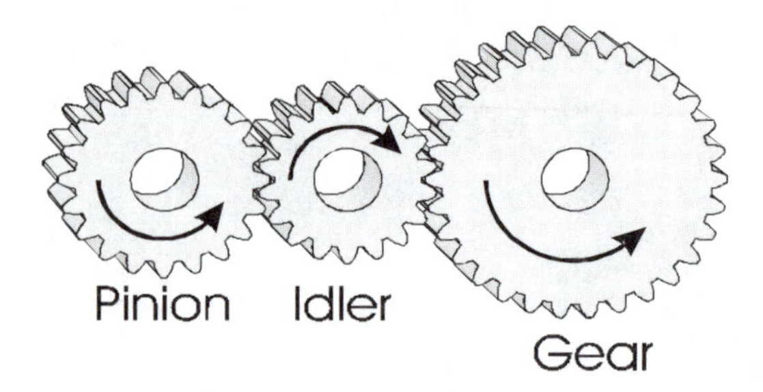

FIGURE 9.3
Idler gears are used to change direction.

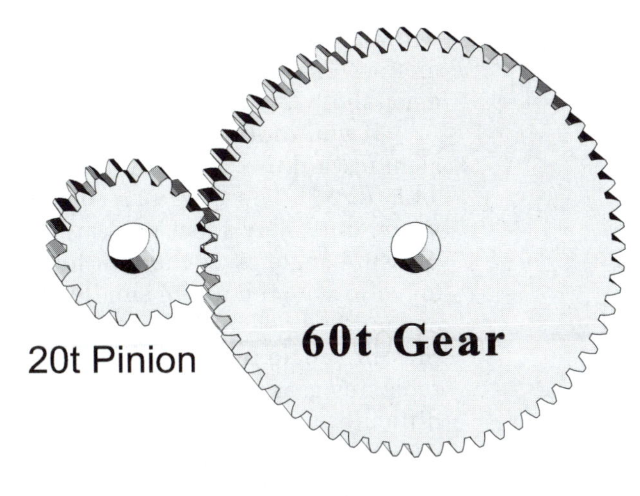

FIGURE 9.4
A ratio of 3:1.

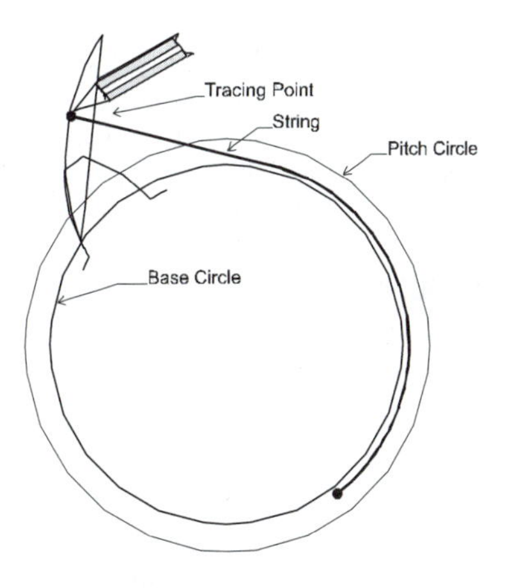

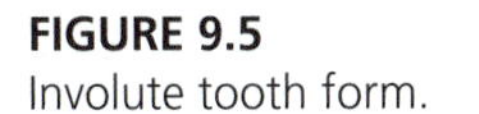

FIGURE 9.5
Involute tooth form.

A gear drive of two gears with the same number of teeth does not change the speed of the output shaft. This gear drive has a *gear ratio* of 1:1. The gear ratio is determined by comparing the number of teeth on the driven gear to the number of teeth on the driver gear. For example, if the driven gear has 60 teeth and the driver gear has 20 teeth, the ratio is 3:1 (see Figure 9.4). The driver gear rotates three revolutions every time the driven gear turns one revolution.

Most common gear teeth shapes are based on an **involute** form. An involute tooth shape is a curved line formed by the point of a tight string when it is unwrapped from a given circle, as shown in Figure 9.5.

This shape yields a gear that is easy to make and will mesh well with another gear having the same-size tooth, regardless of the number of teeth in the two gears. In a gear drive, the teeth mesh with a combined sliding and rolling motion. The involute form provides smooth tooth meshing with a minimum amount of vibration and noise.

If you look at the sides of each gear tooth, you will notice how they slant toward the center of the tooth at the top. This *pressure angle* is usually 14.5° or 20°. The 14.5° pressure angle was considered the standard for many years, but today most new applications use a 20° pressure angle. The 20° pressure angle gives the gear tooth a wider base, resulting in greater strength and allowing the use of smaller gears with fewer teeth for more compact and economical gear drives. All meshing gears in a drive must have the same pressure angle in order to operate properly (see Figure 9.6). Be careful when replacing gears, because the pressure angle on worn gears can sometimes be difficult to identify.

The shape of the gear tooth is determined by the pressure angle, but the size of the tooth is commonly determined by the *diametral* pitch. The **diametral pitch** system describes the size of the gear teeth by specifying the number of teeth for each inch of the gear's *pitch diameter* (see Figure 9.7). (Pitch diameter is determined by the point on the teeth where force is applied to rotate the gear.) Because gear teeth are located along the circumference of the gear, each inch of pitch diameter has a pitch circle diameter of π (3.1416 in.). Summed up, diametral pitch is the number of teeth per inch of pitch diameter, which translates to the number of teeth in 3.1416 in. of circumference of the gear. For example, a 4-in. pitch circle circumference gear with 20 teeth has a

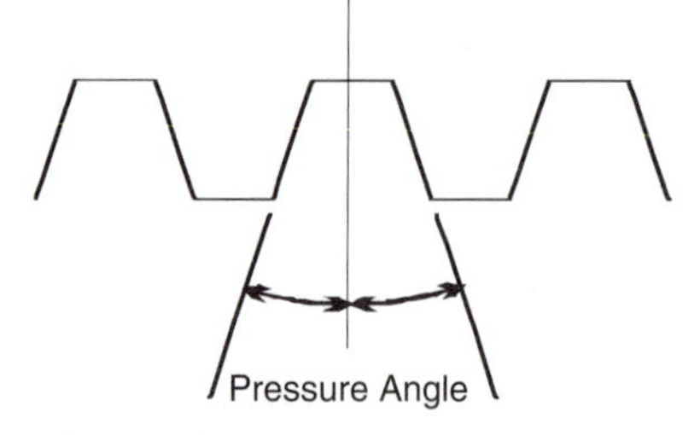

FIGURE 9.6
Gear tooth pressure angle.

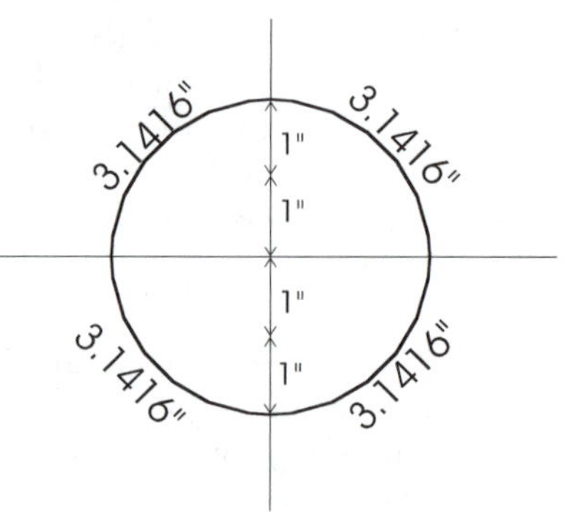

FIGURE 9.7
Diametrical pitch depends on pitch diameter.

FIGURE 9.8
5-DP gear.

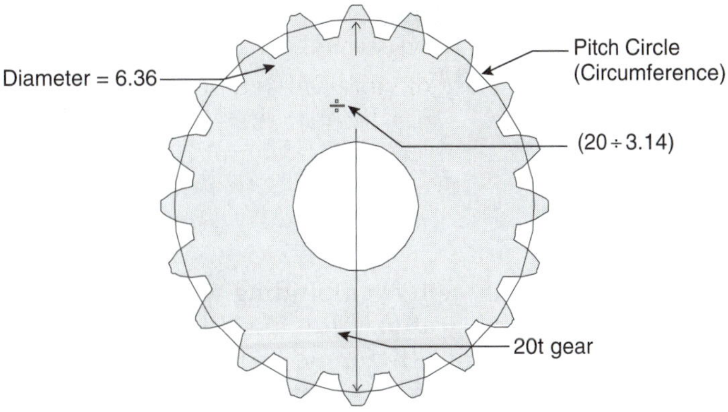

Diameter = 6.36
Pitch Circle (Circumference)
(20 ÷ 3.14)
20t gear
5-DP (Diametrical Pitch) gear

FIGURE 9.9
Gear gauge.

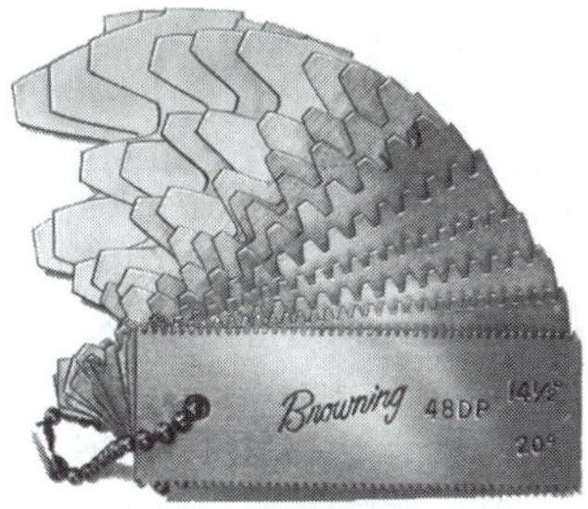

diametral pitch of 20 divided by 4, or 5 DP (see Figure 9.8). Common diametral pitch sizes range from 3 DP to 48 DP. The higher the DP number, the smaller the tooth size. Gear gauges can be used to determine diametral pitch and pressure angle (see Figure 9.9).

The metric equivalent of diametral pitch is called the *metric module system*. The metric **module** number is the ratio of the pitch diameter in millimeters to the number of teeth in the gear. For example, a 48-mm pitch-circle-circumference gear with 24 teeth has a no. 2 metric module gear-tooth size.

Face width is another important dimension of a gear. The face width is the length of the gear tooth in an axial plane; in other words, it is the thickness of the gear. Of course, the wider the gear tooth, the more power it can transmit.

When gears mesh together, there are two terms you should know and understand: *clearance* and *backlash*. **Clearance** is defined as the distance between the top of the tooth of one gear and the bottom of the meshing tooth space (see Figure 9.10). **Backlash** is the amount of *side* clearance of a gear tooth when two teeth mesh. More specifically, it is the space between meshing teeth measured at the pitch circle.

Specific backlash is obtained by machining gear teeth thinner than normal. Both backlash and clearance may also be changed by changing the center distance of the gears' shafts. The purpose of backlash is to prevent

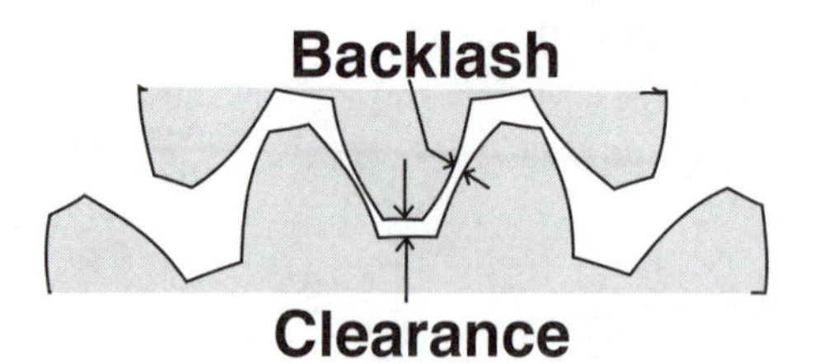

FIGURE 9.10
Backlash and clearance are important.

gear teeth from binding when meshing and to allow room between teeth for a lubrication film. Too little backlash causes overheating, noise, overload, loss of efficiency, abnormal wear, and, eventually, failure of the drive. Excessive backlash is not as detrimental as too little backlash, but noise and excessive wear can result from starting and stopping, shock loading, or drive reversing. High-speed gear drives, in general, require more backlash than slow-speed gears because of expansion from heat and problems with lubrication.

The easiest method for checking backlash is to use a feeler gauge, if the point where the gears mesh is accessible. A dial indicator can also be used to measure backlash:

- Place the dial indicator plunger near the pitch diameter on an easily accessible tooth.
- Mount the indicator so that the plunger moves parallel to the line of gear rotation.
- Lock down the other gear so it cannot move.
- Rotate the gear with the dial indicator as tightly as possible against the locked gear.
- Zero the indicator.
- Rotate the indicator gear in the opposite direction as far as it will go and read the backlash on the dial indicator.

Backlash can also be measured by inserting a soft metal wire (usually lead) into the meshing teeth. The flattened wire can then be removed and the backlash measured by measuring the compressed wire using a micrometer.

The following terms and short definitions are used when working with gears. (See Figure 9.11 for a visual reference of these terms and Table 9.1 for gear formulas.)

Addendum	The height of the tooth above the pitch circle.
Center distance	The shortest distance between the axes of two mating gears.
Chordal thickness	The thickness of a tooth measured at the pitch circle.
Circular pitch	The distance from a point on one tooth to a corresponding point on the next tooth measured on the pitch circle.

FIGURE 9.11
Basic gear terminology.

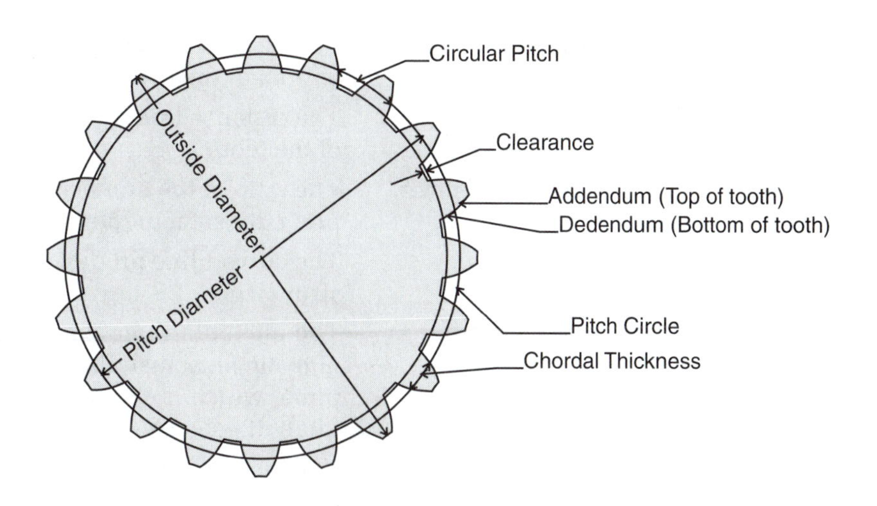

TABLE 9.1
Gear Formulas

To Find	Rule	Formula
Addendum	Divide 1 by the diametral pitch.	$A = \frac{1}{DP}$
Center distance	Divide the total number of teeth in both gears by twice the diametral pitch.	$CD = \frac{N + n}{2 \times DP}$
Circular pitch	Divide 3.1416 by the diametral pitch.	$CP = \frac{3.1416}{DP}$
Clearance	Divide 0.157 by the diametral pitch.	$Cl = \frac{0.157}{DP}$
Dedendum	Divide 1.157 by the diametral pitch.	$D = \frac{1.157}{DP}$
Diametral pitch	Divide the number of teeth by the pitch diameter.	$DP = \frac{N}{PD}$
	Add 2 to the number of teeth and divide the sum by the outside diameter.	$DP = \frac{N + 2}{OD}$
Number of teeth	Multiply the outside diameter by the diametral pitch and subtract 2.	$N = OD \times DP - 2$
	Multiply the pitch diameter by the diametral pitch.	$N = PD \times DP$
Pitch diameter	Divide the number of teeth by the diametral pitch.	$PD = \frac{N}{DP}$
	Multiply the number of teeth by the outside diameter, and divide the product by the number of teeth plus 2.	$PD = \frac{N \times OD}{N + 2}$
Outside diameter	Add 2 to the number of teeth and divide the sum by the diametral pitch.	$OD = \frac{N + 2}{DP}$
Tooth thickness	Divide 1.5708 by the diametral pitch.	$T = \frac{1.5708}{DP}$
Whole depth	Divide 2.157 by the diametral pitch.	$WD = \frac{2.157}{DP}$

Clearance	The distance between the top of one tooth and the bottom of the mating tooth space.
Dedendum	The distance from the pitch circle to the bottom of the tooth.
Diametral pitch	The ratio of the number of teeth for each inch of pitch diameter of the gear.
Involute	The curved line produced by a point of a stretched string when it is unwrapped from a given cylinder.
Module	The pitch diameter of a gear divided by the number of teeth; an actual dimension, unlike diametral pitch, which is a ratio of the number of teeth to the pitch diameter.
Linear pitch	The distance from a point on one tooth to the corresponding point on the next tooth of a gear rack.
Pitch diameter	The diameter of the pitch circle.
Tooth thickness	The thickness of the tooth measured on the pitch circle.
Whole depth	The full depth of the tooth.

FIGURE 9.12
External gear drive.

9-3 SPUR GEARS

A spur gear can be thought of as the original gear. It is one of the oldest designs and the simplest to produce. Spur gears are used to transmit power between parallel shafts. The teeth are cut straight and are parallel to the drive shafts. Spur gear teeth are of the involute design, and as previously discussed, the gear teeth mesh with a rolling, sliding motion. This produces friction and requires some form of lubrication. Because no more than two sets of teeth are in mesh at one time, the load is transferred abruptly from one tooth

FIGURE 9.13
Internal gear set.

FIGURE 9.14
Planetary, or epicyclic, gear set.

to another. This makes spur gears noisier and produces more vibration than most other gear types. It also limits spur gears to low or moderate speeds. Higher speeds cause excessive noise, vibration, and wear. Spur gears are typically constructed of steel, cast iron, bronze, or synthetic materials (such as plastic or nylon).

Spur gear drives can be classified in three major configurations: external drive, internal drive, and rack and pinion. The external gear drive is thought of as a conventional gear drive (see Figure 9.12). The gear shafts are parallel and rotate in opposite directions unless idler gears are added. Internal gear drives are the opposite of external gear drives (see Figure 9.13). They have a cylindrical ring with the teeth located *inside* the ring. A conventional gear is mounted inside this ring. Tooth strength is greater and shaft center distances can be much closer than in their external counterparts. Drive shafts rotate in the same direction. Internal gears are also used in *planetary,* or *epicyclic,* gear drives (see Figure 9.14). These versatile compact drives are widely used in transmissions to transmit power. The rack-and-pinion gear drive converts rotary power to linear power, or vice versa (see Figure 9.15). This special type of gear drive is usually limited to slower speeds.

FIGURE 9.15
Rack-and-pinion gear set.

9-4 HELICAL GEARS

Helical gears are an improvement over spur gears. They resemble spur gears, but the teeth are cut at an angle (see Figure 9.16). The angle of the helical gear tooth is called the *helix angle* (see Figure 9.17). To avoid side thrust, a helix angle should not exceed 20° in parallel shafts, although a helix angle of 30° is typically used in the British system and 45° helix angle gears can be commonly found. With these larger helix angles, thrust bearings should be used for end thrust support.

Because helical gear teeth are cut at an angle, rather than straight like spur gears, the length of a helical gear tooth is greater. This produces greater load and power capabilities. Unlike spur gears, the angled tooth design also ensures that multiple teeth mesh at the same time, which results in smoother

FIGURE 9.16
Helix angle on helical gears.

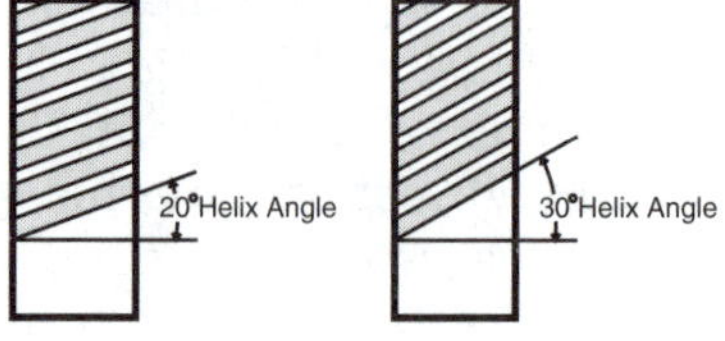

FIGURE 9.17
Different helix angles.

engagement and quieter operation than spur gears. With straight-tooth spur gears, as the tooth contacts and rolls, the entire load is placed near the top of a single tooth, putting it under great strain. In angled helical teeth, more than one tooth is in contact, and the load is spread over the entire surface of the tooth, not concentrated at the top.

A parallel-shaft helical gear drive must have the same pitch and the same pressure angle, with one gear having a *right-hand* helix angle and the other gear a *left-hand* helix angle. The *hand* (or direction of the twist of the tooth) can be determined by placing the gear vertically and noting whether the left side or right side of a single tooth is lower. If the tooth is lower on the left side, it is a left-hand gear. If the tooth is lower on the right side, it is a right-hand gear.

If helical gears with the same hand are used together, the shafts must cross each other at an angle determined by the sum of the mating gear's helix angles. For example, if two right-hand 45° helix angle gears make up a gear drive, the shafts would cross each other at 90° (45° + 45° = 90°) (see Figure 9.18). Helical gears are quieter and stronger than spur gears, but they are more difficult to make and, therefore, more expensive. The helix angle teeth that give helical gears all the advantages over spur gears also cause helical gears the problem of shaft *end,* or *axial thrust.* **Axial thrust** is a force, or push, along the axis of the shaft. Bearings must be used that are designed to support the shafts and to operate with the thrust load that the helical gear develops.

To overcome the axial thrust problem with helical gears, you can use a *double helical* gear that creates an equal but opposing force by adding another helical gear of opposite hand on the same shaft as the original gear (see Figure 9.19). A space separates the two opposed gears. This design produces no axial thrust because the helix tooth angle on one side counteracts the other. The gap between

FIGURE 9.18
Right-hand helical gears.

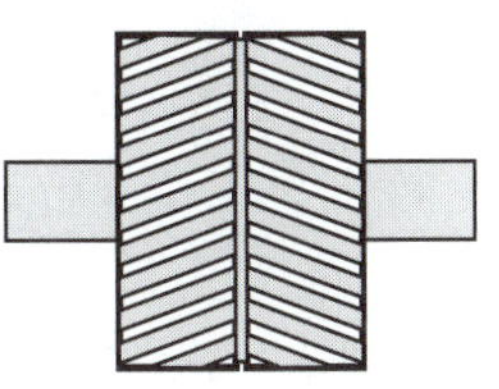

FIGURE 9.19
Double helical gear.

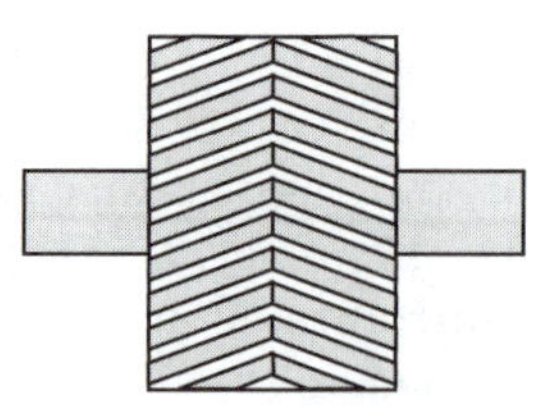

FIGURE 9.20
Herringbone gear.

opposing teeth sets makes this gear easier to manufacture, although still much more difficult and expensive than standard helical gears. The gap also allows for lubrication to escape, preventing lubrication entrapment. Gear drives using double helical gears often handle very high loads and speeds.

The *herringbone* gear is another type of double helical gear, but it has no gap (see Figure 9.20). Two sets of opposing helical teeth are cut on a common blank and meet at their apexes. Herringbone gears are more expensive and difficult to make than double helical gears. They require special machine cutters for manufacture, but they are stronger and operate well under very heavy loads and high speeds.

9-5 BEVEL GEARS

There are four types of bevel gears commonly used in industry: *straight* bevel, *spiral* bevel, *zerol* bevel, and *hypoid* bevel. Bevel gears are typically used where a gear drive is needed for intersecting shafts, usually at 90°. The larger gear is commonly referred to as the *ring* gear (or sometimes as the *crown* gear). The smaller gear is referred to as the *pinion* gear. When mating bevel gears have an equal number of teeth (1:1 ratio) and 90° shafts, they are referred to as *miter* gears. Miter gear drives do not reduce or increase speed. They only change direction.

Straight bevel gears are manufactured using cone-shaped blanks, with tooth forms that are like spur gears (see Figure 9.21). Bevel gears have the same advantages and disadvantages of spur gears. They are economical and easy to

FIGURE 9.21
Straight bevel gear set.

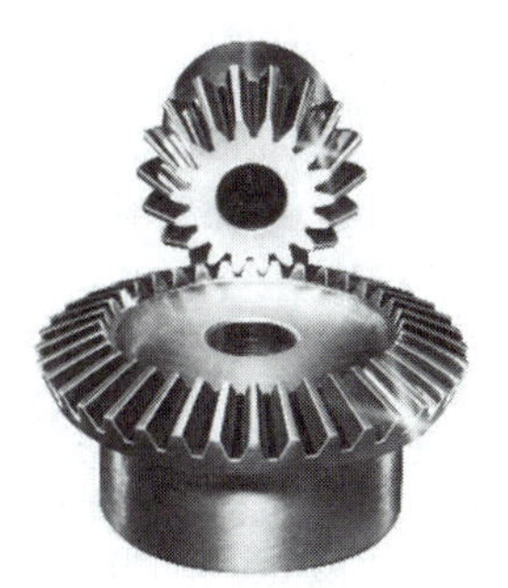

FIGURE 9.22
Spiral bevel gear set.

manufacture, but meshing is rough and noisy. Speeds must be kept low or moderate, and ratios are usually limited to 4:1 or less.

Spiral bevel gears are to straight bevel gears what helical gears are to spur gears (see Figure 9.22). The spiral of the teeth provides gradual engagement on multiple teeth, resulting in smoother running, higher speeds, less noise, and increased load capacity. Spiral bevel gear drive ratios can be commonly found up to 6:1.

Zerol bevel gears combine the best of both the straight and spiral gear designs. Zerol bevel gear teeth have a 0° angle. In other words, they have no inherent tooth angle; the teeth are curved. They are smooth and quiet for the same reason as spiral bevel gears, but because the helix angle is 0°, this type of bevel gear drive has no axial thrust, as spiral gears do.

Hypoid bevel gears look similar to spiral bevel gears, except the centerlines of the input and output shafts are offset. Because the shafts are offset, each shaft can have a bearing at both ends for better support. Hypoid gear design also allows the pinion gear to be larger and stronger. They have a higher tooth-surface contact ratio. This permits higher gear ratios and more rugged construction.

9-6 WORM GEARS

A worm gear set consists of two very different gears (see Figure 9.23). A *worm gear* looks very much like a standard helical gear, and in some applications, it is a standard helical gear. The *worm* looks like a hollow cylinder with *ACME*

FIGURE 9.23
Worm gear set.

(a square-type thread) threads on the outside. Worm gear sets are used to transmit mechanical power between two nonintersecting shafts that are commonly at right angles to each other. Worm gear sets can have high ratios in a compact area because they use the principle of the screw. Unlike the other gears we have studied, worm gear sets operate strictly with a sliding contact between gears. This makes them quieter with less vibration. They handle shock loads well and can have large ratios. Ratios of 60:1 are not uncommon, and ratios of up to 100:1 are possible, but rare. The sliding action of worm gear sets creates a lot of friction. This causes worm gear sets to run hotter and less efficiently than other types of gears. Despite these disadvantages, the economy and versatility of the worm gear set make it a popular drive.

In this unique gear set, worms are always the input gears, and worm gears are always the output gears. Worm gear sets are always used as speed reducers, never as speed increasers. Worm gear sets also tend to be *self-locking.* A worm gear drive is said to be self-locking when motion cannot be started by applying torque to the output shaft (worm gear). This is an important feature in applications, where the load may try to reverse the drive when the motor is turned off. An example would be a loaded, inclined conveyor, trying to run backward and dump the load when power is turned off.

In general, worm gear drives are *not* self-locking with ratios less than 30:1 under static conditions. Manufacturers generally do not guarantee that a worm gear drive will be self-locking under any field conditions. In any application where self-locking is a necessity, you should never rely solely on the self-locking *tendency* of worm gear drives; instead, you should use a *backstop.* A backstop is a bearinglike device that allows the shaft to rotate in only one direction. (See Chapter 11, "Clutches and Brakes.")

The gear ratio of a worm gear set is determined by the number of teeth on the worm gear divided by the number of *starts* (or threads) on the worm. For example, a worm gear set with a 40-tooth worm gear and a single-start worm has a ratio of 40 divided by 1, or 40 to 1 (40:1). A 40-tooth worm gear with a two-start worm has a ratio of 40 divided by 2, or 20 to 1 (20:1). The number of starts on the worm can be determined by looking at one end of the worm and then counting them. A start is a single thread running continuously down the worm. One-, two-, three-, or four-start worms are typically manufactured for most worm gears sets (see Figure 9.24). Just as with helical gears, worms can be right-handed or left-handed. A right-hand worm has a right-hand thread, just like a standard right-hand bolt. You can determine whether a worm has a right-hand or left hand thread by placing the worm shaft horizontally and choosing a single thread. If the bottom of that thread is closer to the right than the top of the same thread, it is a right-hand thread. If the bottom of the thread is closer to

FIGURE 9.24
Different starts on worms.

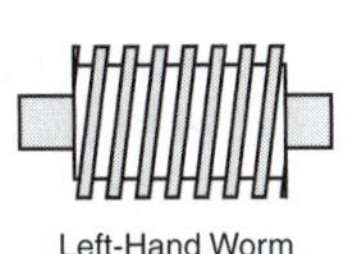

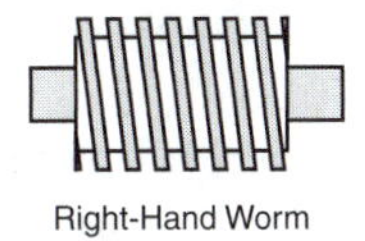

FIGURE 9.25
Worms can be left- or right-handed.

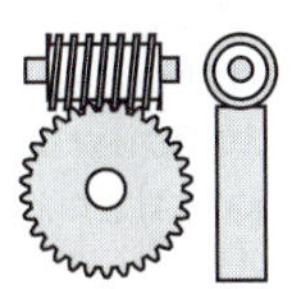

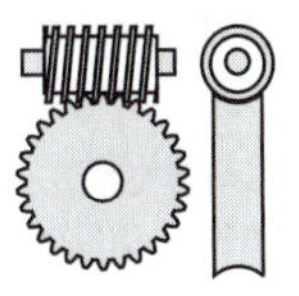

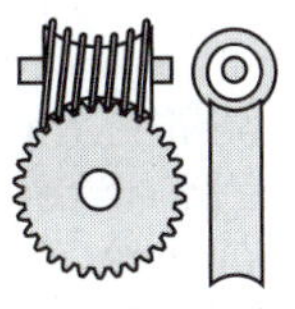

FIGURE 9.26
Worm gear set configurations.

the left than the top of the same thread, it is a left-hand thread (see Figure 9.25). The amount of tooth-contact area plays a large part in determining the load capabilities of worm gear sets. There are three types of worm gear tooth contact (see Figure 9.26):

1. A *nonthroated* worm gear set (also called a *nonenveloping* worm gear set), the simplest type, is a straight worm and a standard straight helical gear. This configuration has a small tooth-contact area and is designed for light loads and higher speeds.
2. A *single-enveloping* worm gear set is also called a single-throated worm gear set. In this type, the worm gear teeth are machined in a concave shape that wraps around the worm, producing a higher tooth-contact surface area for heavier loads and better shock resistance. Even though it is more costly to produce, the single-enveloping worm gear set is the most common type used because of its more rugged construction.
3. A *double-enveloping,* or *double-throated,* worm gear set has a worm gear with curved teeth that also curves to wrap around the worm gear. The special hourglass worm increases the number of teeth engaged and the surface contact area. The double-enveloping worm gear set provides the highest load capacity and greatest shock resistance of all three types. It is not commonly used because it is costly to produce and can be used only at slow speeds due to the excessive friction and heat generated.

Worms and worm gears are usually made of different materials. Worms are usually made of unhardened steel or hardened steel with ground or polished threads. Worm gears are typically constructed of bronze or cast iron. Cast iron is abrasion-resistant and is somewhat stronger than bronze. Bronze worm gears make up the majority of all worm gears, because bronze works well with the high-sliding-friction loads found in worm gear drives. A worm gear made of a softer material does not necessarily have a shorter life than the worm because the worm rotates many times more than the worm gear depending on the gear set ratio.

9-7 LUBRICATING OPEN GEARS

Gear drives must be kept clean and lubricated in order to have a long, useful life. With enclosed gear drives this is much easier, because seals and gaskets keep contaminants out and lubrication in. Contaminants can shorten gear drive life due to abrasion over a long period, but it generally causes more problems in a short time by filling up the clearances between gear teeth. This interferes with the lubrication film and can overload or even lock down drives if it is severe enough. Although keeping the gear drive clean is a serious problem, it is not usually the cause of most gear drive failures. Misalignment is probably the greatest cause of open gear drive failures, and it can be avoided by training technicians properly on alignment and installation. Lubrication problems are likely the second greatest cause of open gear failure. Lubricant contamination, overheating, low oil levels, and incorrect lubrication are problems that can be eliminated with good maintenance procedures. Proper lubrication is critical to the reliability and life of any gear drive.

9-8 TROUBLESHOOTING OPEN GEARS

A good examination of the gear teeth can usually tell you a great deal about any open-gear problems. All gears will eventually show some signs of wear. Normal gear-tooth wear will generally be below the middle of the gear tooth if alignment and lubrication are correct. Small, shallow pits will be evident in the area of normal wear.

The following is a list of common gear problems you may encounter:

Abrasive wear — The tooth surface shows scratch marks or grooves, caused by foreign material in the lubricating system.

Corrosive wear — The tooth surface shows etched, corroded areas over the entire tooth surface, caused by outside chemical contaminants or oil additive breakdown.

Broken teeth	Broken teeth are usually caused by overload conditions or excessive loading.
Destructive scoring	The tooth surface shows radical scratch and tear marks in the direction of tooth sliding, caused by excessive heat due to high operating temperature, surface loading, surface speed, or lubrication failure.
Rolling and peening	The tooth surface has been worked over the tips and ends of the gear teeth, giving a finned appearance. It is caused by excessively heavy load rolling and peening the load surface of the tooth.
Spalling	The tooth surface has large chunks of metal ripped out. It is caused by excessively high contact stresses.

9-9 ENCLOSED GEARS

An enclosed gear drive is a set of gears supported by bearings, all mounted in a sealed housing (see Figure 9.27). The gearbox case provides rigid support for the gears and bearings with sealed positive lubrication, including gaskets, seals, and air breathers. Enclosed gear drives are also commonly called *gearboxes,* or *speed reducers,* because the majority of enclosed gear drives are used to reduce standard-motor speeds rather than to increase them.

The advantages of gearboxes over other types of drives include the following:

Simplicity	Open gears, V-belt drives, and chain drives need many components to complete the drive (which requires a lot of space). Shafts, bearings, and possibly lubrication are all required. Drive components must be accurately aligned and installed. In gear boxes, shafts, gears, and bearings are already mounted and accurately aligned, saving many hours of installation.

FIGURE 9.27
Speed reducer (gearbox).

They also require only one point of lubrication, compared with several other types of drives.

Size — Compared with other types of drives, gearboxes are very compact. This is a definite advantage in today's manufacturing plants, where space is always at a premium.

High ratios — The ratios of V-belt and chain drives are usually limited to 7:1 or less. Gearboxes can provide extremely high ratios in a single compact housing. Single-reduction worm gear units are commonly manufactured in ratios of up to 60:1. Double-reduction units have ratios of up to 3,600:1. Triple-reduction units, although uncommon, can have even higher ratios.

Versatility — Gearboxes come in a wide range of shaft configurations. Input and output shafts can be right angled, parallel, or in line. Mounting arrangements include the following: close-coupled to a motor (called a gearmotor), mounted on a baseplate with couplings, V-belts or chain drives connected to the input or output shafts, foot-mounted, or shaft-mounted. Shaft arrangements can be horizontal or vertical or somewhere in between. Gearboxes can usually be modified for mounting in unusual positions with only minor changes in the location of lubricant fill, vent, level, and drain plugs.

Low maintenance and long life — Gearboxes are designed for long service life with no required maintenance except changing the oil or greasing bearings. It is not uncommon for gearboxes to operate under severe conditions for many years if they are properly selected for the application and normal maintenance procedures are followed.

9-10 ENCLOSED GEAR DRIVE TERMINOLOGY

Most of the information covered in Chapter 3, "Basic Principles of Mechanical Systems," also applies to working with gearboxes. The following are some additional terms and principles with which you should be familiar:

AGMA — The American Gear Manufacturers Association (AGMA) establishes standards to help standardize the design and application of gear products. The AGMA is made up of manufacturers and technical members who

FIGURE 9.28
Longer center distance means larger gears.

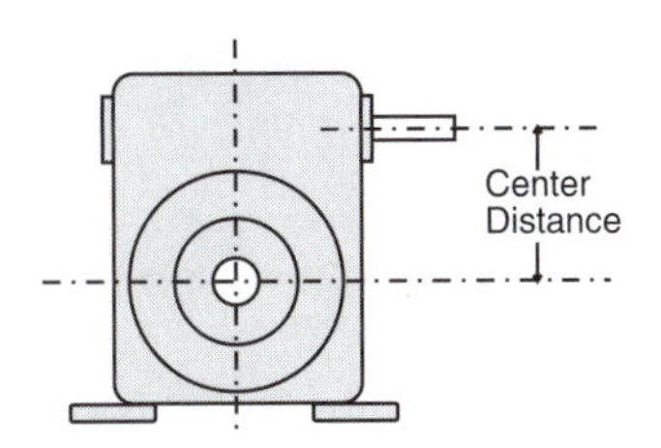

assure gear product users that products built, selected, and applied in accordance with AGMA standards will perform well.

Center Distance Center distance refers to the distance between the centerline of the input and output shafts of a single-reduction gearbox. It doesn't matter if the shafts are parallel or at right angles. In multiple-stage gearboxes, you should use the dimension of the center distance of the slowest-speed gear set. Typically, the bigger the center distance, the more torque the gearbox is capable of producing (see Figure 9.28).

Efficiency Efficiency in any power system can be expressed by the following formula:

$$\text{Efficiency in percent} = (\text{power output divided by power input}) \times 100\%$$

Because the mechanical unit for power is horsepower, the formula can also be written as

$$\text{Efficiency in percent} = (\text{horsepower output divided by horsepower input}) \times 100\%$$

Using the manufacturer's listed input horsepower and output torque, you can calculate output horsepower by

$$\text{Output hp} = \frac{(\text{Output torque, in.-lb})(\text{output speed, rev/min})}{63{,}025}$$

The amount of gearbox loading affects efficiency. A gearbox operating at its rated capacity is more efficient than when operating at light loads because some internal losses are relatively constant. This makes them proportionally higher with light loads. It is impossible for any gearbox output to be equal to its input (that would be 100% efficient). There will always be some losses, but some single-reduction parallel-shaft gearboxes approach 99% efficiency.

Mechanical Rating The mechanical rating of a gearbox is the maximum torque or horsepower that it can transmit. The gearbox can be no stronger than its weakest mechanical component. Gearboxes are designed with a safety factor of two to three times their mechanical ratings. This means a gearbox can withstand short overloads of 200 to 300% of its manufacturer's mechanical rating.

Overhung load The **overhung load** is the force applied at right angles at the end of the input or output shaft. An example is the force applied to a shaft because of the tension of a V-belt drive. Any side pull on a shaft makes up its overhung load. Overhung load puts stress not only on the bearings but also on the shaft. Bearings and shaft must be sufficient to sustain overhung load without damage to the gearbox. The overhung load rating of a gearbox is determined by gearbox design characteristics such as bearing size and location, shaft strength, and case strength. You must consider both the amount and the location of the overhung load when selecting a gearbox. The total overhung load is made up of three different types of force (see Figure 9.29). The first force comes from dead weight on the shafts. This force is the weight of anything mounted or attached to the shaft. The second force is created by the torque generated by the gearbox. Using the formula

$$\text{Torque} = \text{force} \times \text{radius}$$

the second force can be calculated by dividing the torque (ft-lb) by the radius (ft) of the sprocket, sheave, or pulley. The third force is that on the slack side of a V-belt drive and is combined with the force on the tight side (the second force) to create the tension needed to keep the V-belt from slipping. All three of these forces are added, if they exist, to calculate overhung load. (A chain drive may have no tension on the slack side of the chain.)

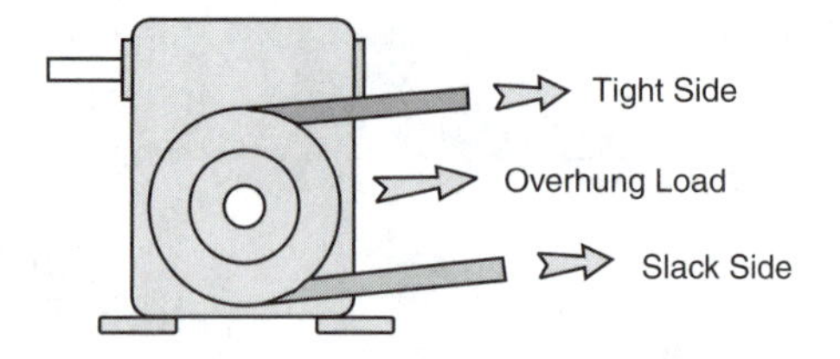

FIGURE 9.29
All forces are added to calculate overhung load.

Prime mover The prime mover is the machine that provides power to a drive. Examples of prime movers are electric motors, internal combustion engines, hydraulic motors, and pneumatic motors. Different types of prime movers affect the selection of the gearbox because of vibration, shock, and torque characteristics of each type power source.

Service factor A service factor is a constant used to multiply the horsepower by a multiplier in the selection of a gearbox. This multiplier varies according to the type of prime mover, the duty cycle (how many hours per day in operation), and the type of service in which the gearbox is to be used (uniform load, moderate shock load, heavy shock load, and extreme shock load). The values for service factors are set by AGMA. Table 9.2 is a typical AGMA service factor table.

For example, a 10-hp electric motor will be used to drive a worm gear reducer. The reducer is coupled to a conveyor. The conveyor will operate 8 h a day with moderate shock loading. From Table 9.2 the correct service factor is 1.25. The service factor is multiplied by the horsepower (1.25×10), requiring the gearbox to have a mechanical horsepower of 12.5 hp. (Table 9.2 is not the ultimate authority. You should always use manufacturers' tables when possible.)

TABLE 9.2
Service factors for worm gear-type reducers.

		Driven-Machine-Load Classifications			
Prime Mover	**Duration of Service per Day**	**Uniform**	**Moderate Shock**	**Heavy Shock**	**Extreme Shock**
Electric motor	Occasional 2 h	—	—	1.00	1.25
Hydraulic motor	Less than 3 h	1.00	1.00	1.25	1.50
Pneumatic motor	3–10 h	1.00	1.25	1.50	1.75
	More than 10 h	1.25	1.50	1.75	2.00
Multicylinder	Occasional 2 h	—	—	1.25	1.50
internal-	Less than 3 h	1.25	1.25	1.50	1.75
combustion	3–10 h	1.25	1.50	1.75	2.00
engine	More than 10 h	1.50	1.75	2.00	2.25
Single-cylinder	Occasional 2 h	—	—	1.50	1.75
internal-	Less than 3 h	1.50	1.50	1.75	2.00
combustion	3–10 h	1.50	1.75	2.00	2.25
engine	More than 10 h	1.75	2.00	2.25	2.50

Thermal rating — The **thermal rating** of a gearbox is the maximum torque or power that it can transmit continuously based on its ability to dissipate heat generated by friction. If the thermal rating of a gearbox is lower than its required output, then you must select a larger gearbox or provide a means of keeping the gearbox cool. Fans or heat exchangers may be used to keep the oil cool. Exceeding the thermal rating of a gearbox will result in shortened gearbox life.

Thrust Load — The thrust load is the force applied to the shaft parallel to the shaft's axis (also called *axial* thrust) (see Figure 9.30). Some gearboxes have an internal thrust load due to the type of gear used (helical gears, for example). At this point we are concerned only with the thrust load applied externally. Thrust load is commonly found in applications such as fans, blowers, mixers, and agitator drives. Use manufacturers' tables to determine the thrust capabilities of any selected gearbox. Do not exceed manufacturers' recommendations. In some cases, you can add an external thrust bearing to help support the load and reduce gearbox thrust loading.

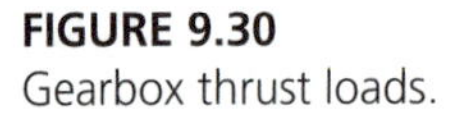

FIGURE 9.30
Gearbox thrust loads.

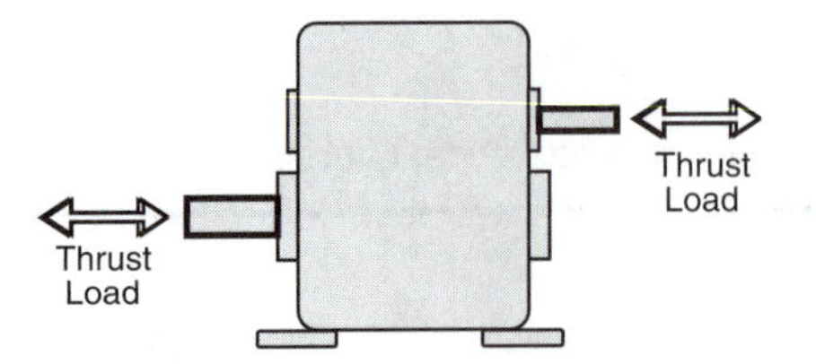

9-11 COMMON ENCLOSED GEAR DRIVES

There are many different types and configurations of enclosed gear drives, or gearboxes. The three most common types of gearboxes used in industry are the *worm, helical,* and *bevel* gearboxes.

Worm gearboxes are compact, economical, and capable of higher ratios. Single-reduction units are always offset-shaft, right-angle drives. Because of the continuous sliding contact to transmit power, the worm gearbox is very quiet. This type of gear has low vibration and can tolerate substantial shock loads.

FIGURE 9.31
Single-reduction worm.

The sliding contact that gives the worm gearbox all these advantages also causes its greatest disadvantage: high frictional losses. The sliding contact generates heat and causes the worm gearbox to be the lowest-efficiency gearbox discussed in this section. The heat generated must be absorbed by the lubricant and dissipated through the case. Some gearboxes have cooling fins and possibly a fan attached to the high-speed shaft to blow air across the surface of the case to help keep it cool.

Most worm gearboxes are single-reduction (see Figure 9.31). Single-reduction worm gearboxes can have ratios as high as 100:1, but most manufacturers offer a maximum reduction of 60:1.

Higher ratios are usually obtained by using a double-reduction gearbox (see Figure 9.32). In a double-reduction gearbox, the output worm gear drives the input worm of the second set of gears. Double-reduction gearboxes have offset parallel input and output shafts (see Figure 9.33). Extremely high ratios

FIGURE 9.32
Double-reduction worm gearbox (parallel shafts).

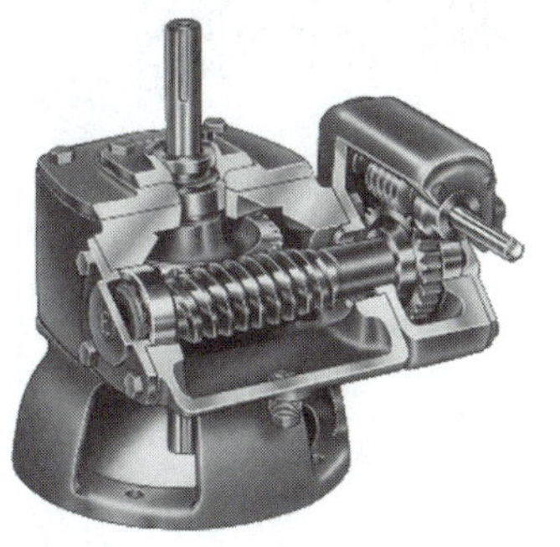

FIGURE 9.33
Double-reduction worm gearbox (vertical output shaft).

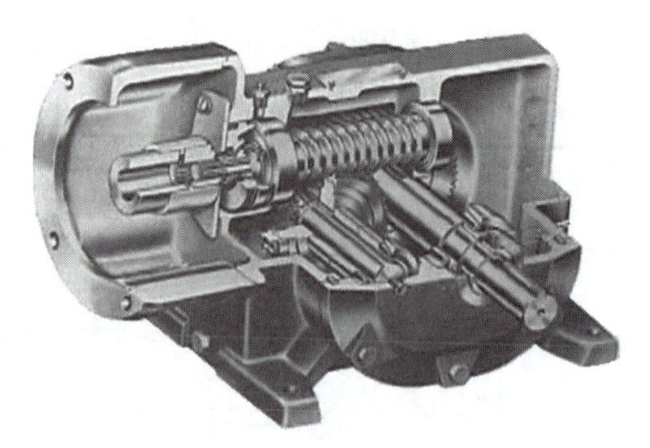

FIGURE 9.34
Combination worm and helical gearbox.

can be obtained with a double-reduction gearbox. For example, if both worm gear sets were 60:1, then the total reduction would be 60×60, or 3,600:1! With a standard 1,800-rev/min electric motor and a 3,600:1 reducer, the output revolutions per minute would be 1,800 divided by 3,600, or 0.5 rev/min. Very few applications need a slower output than this.

Combination worm gear and helical gear reducers are not uncommon in industry (see Figure 9.34). Usually, the worm gear set is the primary reduction, and the helical gear set is the secondary reduction. This unique combination allows high ratios with better efficiency than a standard double-reduction worm gearbox. Combination worm and helical gearboxes also maintain the right-angle output of a single-reduction worm gearbox.

Modern helical gearboxes are durable, cool-running, and efficient. Because of the helical tooth configuration, typical efficiency is 95 to 99%. That's a great advantage compared to worm gearboxes with an efficiency of 60% or lower. A single-reduction helical gear set does not usually exceed an 8:1 ratio, but double-, triple-, and even quadruple-reduction gearboxes are not uncommon (see Figure 9.35). Single-reduction helical gearboxes always have parallel offset shafts. Multiple-reduction helical gearbox shafts must also be parallel, but they usually have concentric configurations. Input and output shafts are parallel and also in line (along the same centerline) with each other. The exception to this type of configuration is *hollow-shaft*

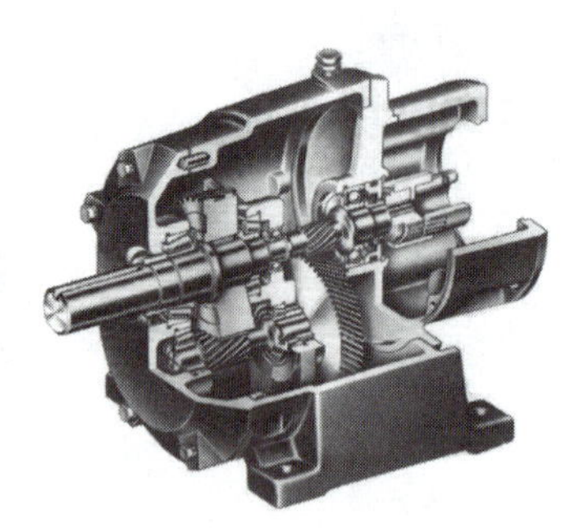

FIGURE 9.35
Triple-reduction gearbox.

FIGURE 9.36
Hollow-shaft helical gearbox.

helical gearboxes, which for obvious reasons always have offset parallel shafts (see Figure 9.36).

As with worm gearboxes, the total gear ratio of multiple-reduction helical gearboxes is obtained by multiplying the individual ratios of each stage by each other. For example, a triple-reduction gearbox with a 5:1 primary ratio, a 6:1 second reduction, and a third ratio of 3:1 has a total gear ratio of $5 \times 6 \times 3 = 90{:}1$.

Bevel gearboxes are all right-angle-type gearboxes (see Figure 9.37). They are used primarily to change direction rather than to change speed, but low-ratio bevel gearboxes are available. In industry, common bevel gearbox ratios are 1:1, 1.5:1, and 2:1. Bevel gearboxes with a 1:1 ratio, using identical gears for input and output, are called *miter boxes*. Miter boxes with spiral gears make up the majority of bevel gearboxes used in industry, but straight spur gears and zerol bevel gears are not uncommon.

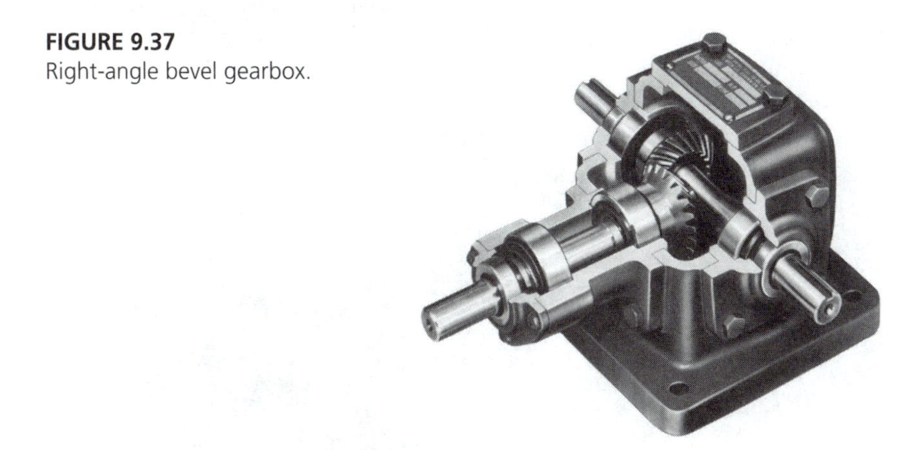

FIGURE 9.37
Right-angle bevel gearbox.

9-12 COMMON GEARBOX CONFIGURATIONS

One of the most common and most versatile gearbox configurations is the *foot-mounted* reducer (see Figure 9.38). This gearbox may be direct-coupled to a motor, or it may use a chain or V-belt drive. Many of these gearboxes have a double foot, so the input shaft may be configured high or low.

Vertical-output-shaft gearboxes have the same internal parts as standard worm gearboxes but have a special mount suited to vertical mounting (see Figure 9.39). Common applications for this type of gearbox include conveyor drives, agitators, and mixers.

Hollow-shaft gearboxes use a hollow output shaft for applications in which it may be hard to provide a base due to the lack of space or simply for convenience (see Figure 9.40). A hollow-shaft reducer mounts directly on the driven machine shaft, making a base and a coupling between the gearbox and the driven shaft unnecessary. This type of gearbox may use worm or helical gears.

A special type of hollow-shaft reducer is the *torque-arm* configuration. Torque-arm reducers have one end of an arm attached to the gearbox and the

FIGURE 9.38
Foot-mounted worm gearbox.

FIGURE 9.39
Vertical-output-shaft gearbox.

FIGURE 9.40
Hollow-shaft worm gearbox.

FIGURE 9.41
Torque-arm gearbox.

FIGURE 9.42
Worm-type gearmotor.

other end attached to a suitable mounting surface (see Figure 9.41). The arm keeps the gearbox from trying to rotate around the driven shaft.

A gearmotor is a gearbox with an integrally attached electric motor (see Figure 9.42). Typically, the motor shaft has teeth or worm threads machined as part of the shaft. Gearmotors provide a very economical drive package but do not have the versatility or interchangeability of other types of drives.

A variation of the gearmotor is the *C-face* reducer (see Figure 9.43). C-face reducers have a special motor mount that accepts standard footless C-face motors. Because they use a NEMA (National Electric Manufacturers Association) standard motor, motor replacement is fast and easy. NEMA sets standards for

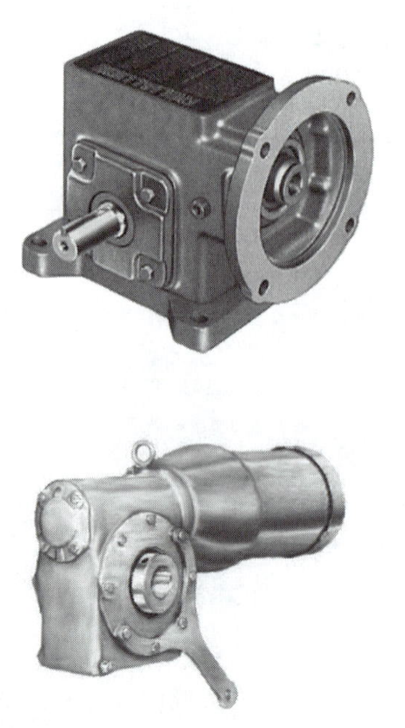

FIGURE 9.43
C-face worm gearbox.

FIGURE 9.44
Hollow-shaft gearmotor.

electric motor terminology and dimensions, thus permitting interchangeability among motors of different manufacturers. Gearmotors and C-face reducers are available with worm gear, helical gear, and combination worm/helical gear sets. They can have horizontal, vertical, or hollow output shafts (see Figure 9.44).

9-13 GEARBOX INSTALLATION

Many different types and configurations of enclosed gear drives require different types of installation procedures. Some of the more common areas that apply to most gearboxes are discussed here. Always follow manufacturers' recommendations for specific applications.

A solid foundation is basic to all gearboxes. The type of foundation may be concrete, structural steel, or a fabricated baseplate. Concrete is used primarily for large, permanent gearbox installations. The concrete should be completely cured before the gearbox is secured. It is better to install steel base pads on which to mount the gearbox rather than to grout them in the concrete.

Steel pads allow for more accurate positioning and aligning (see Figure 9.45). Structural steel and fabricated base plates are used for small- to medium-size gearboxes. The steel should be thick enough to prevent warping caused by weight and torque stresses. As a rule of thumb, the baseplate thickness should be at least equal to the diameter of the mounting bolts. After 40 to 80 h of operation, you should retorque the mounting bolts.

FIGURE 9.45
Steel pads allow for more accurate positioning and aligning of gearboxes.

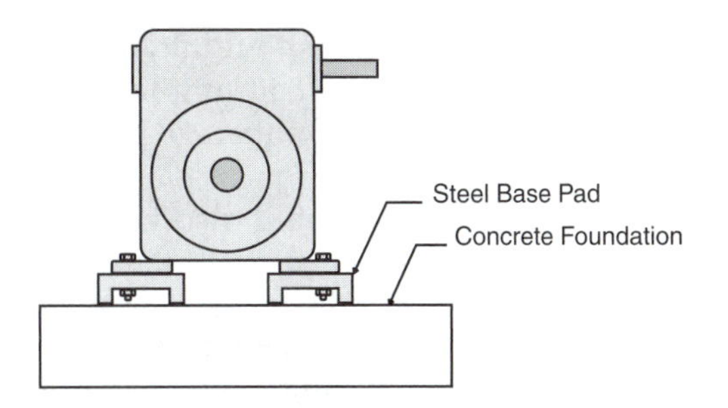

Leveling and checking for **soft foot** are important steps to ensure correct coupling alignment and long gearbox life. Alignment of the gearbox, motor, and machinery is critical to the life of the installation. Misalignment can cause serious stresses and vibration. (See Chapter 10 for more information on couplings and alignment.) When V-belt drives, chain drives, or open gears are used with a gearbox, they must be installed correctly for long life of the installation. Shafts must be paralleled and squared. V-belt and chain drives must be tensioned according to manufacturers' recommendations. (Appropriate sections in this book give you general guidelines for installation and tensioning of V-belt and chain drives.)

You should check the oil level of the gearbox. Some gearboxes are shipped dry, although most already have oil filled to the correct level. Check enclosed manufacturers' recommendations for break-in and regular oil changes. Note that some manufacturers now ship gearboxes with a lifetime oil, usually a high-grade synthetic, that never needs to be changed under normal operating conditions.

After installing and aligning all the drive components, run the gearbox without the load to ensure that all components run freely with no unusual noise or binding. Make any necessary adjustments and check the torque on all bolts. Replace all guards and safety devices before returning the drive back to regular service.

9-14 MAINTAINING AND REBUILDING A GEARBOX

Maintaining gearboxes requires proper installation, proper loading, proper lubrication, and periodic inspection. Gearboxes should be given a routine inspection periodically, consisting of the following:

- Look for loose or damaged parts, loose guards, and oil leaks. Repair or replace them as necessary.
- Listen for unusual noises: If an unusual noise occurs, shut down the drive and determine the cause. Repair or replace parts as necessary.
- Perform a hands-on inspection: Shut down the drive and check the oil level and the oil temperature. For most applications, oil temperature

should not exceed 180°F (82°C). Monitoring the load is also helpful. The simplest way to do this is by using an ammeter. The motor amperage relates directly to the load on the drive.

You should also keep the unit clean for better heat dissipation and maintain records of scheduled maintenance, repair, and dates when parts are replaced.

If you need to overhaul a gearbox, obtain a copy of the manufacturer's service manual. For most gearboxes, a specific sequence of disassembly and reassembly must be followed for correct gearbox operation.

Overhauling a Gearbox

The following gives a general sequence of steps for overhauling most gearboxes:

- Lock out the machinery involved. Block or lock all sources for potential mechanical energy to obtain *zero energy*. Ensure a safe working environment.
- Clean the gearbox thoroughly before starting disassembly. This makes it easier to keep dirt and trash out of the gearbox during disassembly.
- If the gearbox is to be moved to the shop for overhaul, remove the guards, couplings, and mounting bolts. Transport the gearbox to the shop area.
- Some gearboxes have backstops that prevent the gears from rotating backward. These may be mounted externally and have their own source of lubricant. Drain the lubricant and remove the backstop.
- Drain all oil from the gearbox and consult the service manual for the disassembly procedure.
- Handle gear assemblies carefully. Gear teeth can be damaged if they bump against each other. Inspect the gear teeth for excessive wear and damage. Replace them as necessary. Note that some shafts and gears are machined from a single piece of metal and must be replaced as a unit. Also note that many manufacturers recommend replacing mating gears as a set. In general, this is a good practice for most gear sets.
- Examine the shaft for wear or damage. The area of the shaft where the seal runs should always be checked carefully. Grooves may be worn into the shaft in this area. For a new seal to work correctly, it must run on a smooth, uniform surface. Sometimes, shafts or seals can be moved slightly to enable the new seal to ride on a new, undamaged section of the old shaft.
- Clean and inspect all bearings. Replace any that are questionable. Always use the correct tools when working with bearings. Never use a hard-face hammer or chisel because these can damage the shaft. Never spin bearings with an air gun to clean or dry them. Never use a torch or any spot-heating source to expand a new bearing. Use an oil bath, oven, or induction heating for expanding new bearings to prevent damage. Make sure that pressure is applied evenly around the inner

bearing ring when pressing it onto a shaft. If you are pressing it into a housing, apply pressure evenly to the outer ring of the bearing.

- Thoroughly clean the case and all parts of the gearbox to be reused. Always use new seals and gaskets.
- Replace the gear assembly and make any adjustments. Some common adjustments include shaft-end float and preloading bearings. Sometimes, special gaskets called shim-gaskets are used. Shim-gaskets are used to adjust running clearances and are obtained from the manufacturer. Always consult the manufacturer's service manual for adjustment procedures.
- Reinstall any backstops removed during overhaul and make adjustments as required.
- Add the correct lubricant to the proper fill level for the mounting position of the gearbox.
- Hand-turn the input shaft to check for binding or rough spots. Make any adjustments or repairs as necessary.
- Run the gearbox without a load for an extended period of time, checking temperature and oil levels.
- Replace all guards, remove lockouts and blocks, and return the drive to production.

QUESTIONS

1. In most gear sets the smaller gear is called the ___________.
2. What is the difference between a speed reducer and a speed increaser gear drive?
3. What is an idler gear?
4. Most new gear drive applications use a ___________° pressure angle.
5. What is the difference between gear tooth clearance and backlash?
6. What are the three major configurations of spur gear drives?
7. What are the four types of bevel gears commonly used in industry?
8. What are miter gear drives?
9. What is meant by the statement, Worm gear sets tend to be self-locking?
10. To ensure that a gear drive cannot be reversed, what should be used?
11. What is the AGMA?
12. Which type of gearbox approaches 99% efficiency?
13. What is the service factor of a gearbox?
14. What is the thermal rating of a gearbox?
15. What is another term for thrust load?
16. What is a torque-arm reducer?
17. What is a gearmotor?
18. What are the main advantages of modern helical gearboxes?

CHAPTER 10

Couplings

10-1 INTRODUCTION

A *coupling* is a mechanical device used to connect two rotating shafts for the purpose of transmitting torque from the driver to the driven piece of rotating equipment. Couplings have been around for a long time. It has been documented that the ancient Greeks used crude couplings to transmit mechanical power around 300 B.C. In the seventeenth century, Robert Hooke used a simple universal joint design in a clock drive. The first modern industrial couplings can be traced to the early part of the twentieth century, when mass production of the automobile necessitated longer-life, better-designed couplings.

There are two basic classes of industrial couplings: *rigid* and *flexible.* Rigid couplings are not designed to allow for any misalignment. Flexible couplings can compensate for some misalignment. Flexible couplings also allow for shaft **end-float,** or *end-play*, which is the inclination of shafts to move back and forth across their bearings (see Figure 10.1). All couplings should be selected according to the type of application in which they are to be used. Couplings must also be properly aligned to ensure long life and efficient transmission of power. Without proper selection and alignment, couplings, bearings, and shafts may be damaged or worn out prematurely. Equipment failures that may occur result in possible damage to machinery, lost production, and expensive downtime.

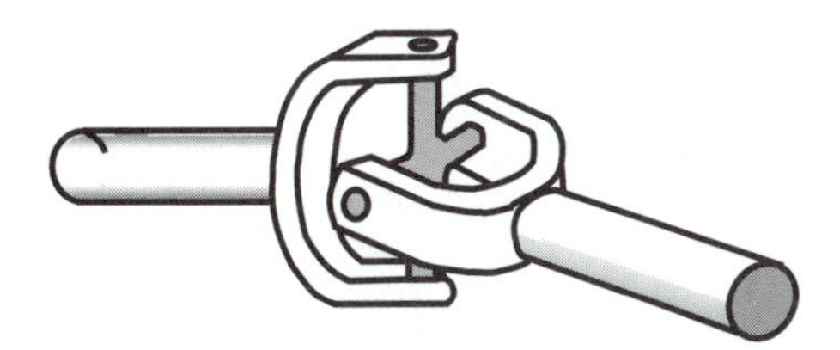

FIGURE 10.1
Early coupling, simple universal.

10-2 RIGID COUPLINGS

Rigid couplings, as the name implies, are used to connect shafts rigidly. They are not designed to compensate for any misalignment or end-float. They join shafts together in the same fixed manner as if they were welded, effectively making the two shafts into one. Near-perfect alignment and proper support for the shafts are required. Without these, there will be excessive bearing wear and, eventually, equipment failure. Because of its rigidity, this type of coupling does not dampen or cushion any of the shock loads that may be transmitted. Historically, rigid couplings were widely used to couple lengths of **line shafting.** During the Industrial Revolution, a single steam engine was typically used to power a shaft running the length of the factory. This line shaft had pulleys mounted at intervals using a flat belt at each machine to power it. Although this method of power transmission is no longer used, the term *line shafting* still refers to rigidly coupled shafts used to transmit mechanical power. Rigid couplings have two major advantages: (1) They require no lubrication, and (2) they require no maintenance if properly installed and aligned. There are three common types of rigid couplings: *solid-sleeve, ribbed*, and *flanged.*

Sleeve Couplings

Industrial sleeve couplings are manufactured in two basic configurations: *finished-bore* and *tapered bushing* mounting. Finished-bore sleeve couplings are usually machined from a single piece of bar steel with two set screws to lock shafts in place (see Figure 10.2). Sleeve couplings may have a keyway cut the full length of the bore. Sleeve couplings are mounted so that the shafts to be coupled meet in the middle; the set screws or tapered bushings lock them in place. Tapered bushings are industry-standard bushings that are also used for sprockets, sheaves, and other industrial mechanical power-transmission products. Sleeve couplings require that shafts be perfectly aligned. All misalignment is absorbed by shaft flexing and bearing tolerance. This type of rigid coupling is used primarily with light to moderate loads. Sleeve couplings are selected by their bore size and are suitable for rated load capacities of standard 1018 shafting. This means that the mechanic doesn't have to go through a complicated procedure to size this type of coupling: If the shaft size is 1 in., the coupling selected should be 1 in. The coupling will carry the rated load of a 1-in. shaft with no problems.

FIGURE 10.2
Rigid-sleeve coupling.

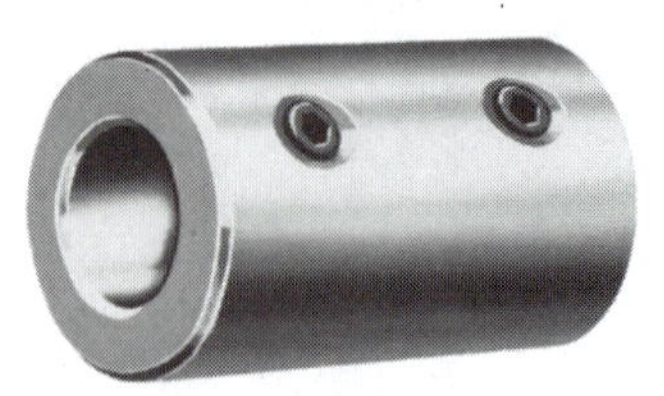

FIGURE 10.3
Rigid ribbed coupling.

Ribbed Couplings

Ribbed couplings can be thought of as split-sleeve couplings. The two halves of this type of coupling are bolted together, joining only the same-size shafts (see Figure 10.3). A single key is used to help transmit torque and facilitate alignment. Ribbed couplings are suitable for rated torque capacities of standard 1018 steel shafting, the same as the sleeve couplings are. Like sleeve couplings, ribbed couplings have no tolerance for misalignment and must be perfectly aligned to ensure long, trouble-free life. An advantage of the split design is that couplings can be easily replaced without moving either shaft. The ribbed design also protects bolt heads for greater safety. Ribbed couplings are typically made from gray iron and are available for shaft sizes up to 7 in. They are typically used in moderate to heavy loads and torque services. They are not available in a tapered bushing construction because of the way they are designed.

Flange Couplings

The rigid flange coupling is also a split-type coupling, but unlike the ribbed coupling, the flange coupling is split vertically (see Figure 10.4). The two flanges are mounted on the shafts and then bolted together. The two halves can be bored to different diameters, or the flange coupling can be purchased in a tapered bushing construction, allowing much more diversity. Rigid flange couplings are available for shaft sizes up to 6 in., and even larger sizes can be

FIGURE 10.4
Rigid flange coupling.

special ordered. This type of coupling is the most common rigid coupling used by industry. These couplings, like all rigid couplings, tolerate no misalignment and have no shock-absorbing capabilities. Their advantages are easy installation, no required maintenance, no lubrication, and no wear if properly installed and aligned. Rigid flange couplings are designed for heavy-duty applications and can handle high torques and speeds.

10-3 FLEXIBLE COUPLINGS

Flexible couplings are designed to allow for some misalignment and end-float. There are many types of flexible couplings for use in many different applications. There is no such thing as the perfect coupling that works in all situations with any amount of misalignment. You should always try to choose the best type of coupling available for your application. Many times, more than one type of coupling may work well. The ultimate goal is to have a coupling that provides long life and trouble-free service. To achieve this goal, you must have accurate alignment and proper installation along with the correct coupling suited for your application. Selecting a coupling that can accept gross amounts of misalignment to compensate for bad practices in alignment is not acceptable. You should always pay attention to the term *allowable misalignment* when selecting a flexible coupling. The coupling manufacturer's specification should be considered to be the *maximum* allowable misalignment under adverse conditions. It should not be the standard to use under normal conditions. Good flexible-coupling alignment will ensure a long, trouble-free life for your coupling. You will have to decide how accurate the alignment should be, given any set of circumstances. Laser alignment is very accurate but, in most cases, may not be economically feasible. Later, we discuss factors to consider when setting a standard for flexible coupling alignment.

A flexible coupling has several roles, including the following:

- Compensating for some misalignment.
- Transmitting mechanical power.
- Dampening shock loading without damage.
- Allowing for shaft end-float.
- Withstanding environmental and frictional temperatures.
- Withstanding exposure to adverse environments (chemicals, petroleum products, outside conditions, and so on).

Most industrial flexible couplings are available in an industry-standard tapered-bushing construction, as well as finished bore and **minimum plain bores** (an unfinished small hole with no keyway) that can be finish-bored at any machine shop.

Flexible couplings usually provide for misalignment by flexing in two different ways:

1. Mechanical motion
2. Resilient compression or extension

Industrial flexible couplings can be divided into two categories according to their flexible member's construction:

1. Metallic couplings use metal elements for their flexible member.
2. Elastomeric couplings generally use a nonmetal element—typically rubber, nylon, or plastic. In the next section, we discuss some of the more common metallic-element couplings.

10-4 METALLIC-ELEMENT COUPLINGS

Metallic-element couplings allow sliding or rolling of their components to compensate for coupling misalignment. Because of their metal-to-metal movement, most of these couplings require some type of lubrication to minimize wear, noise, and heat generated by friction. This can be a disadvantage because they require additional maintenance and the lubricant attracts contaminants that can accelerate coupling wear. Metallic-element couplings typically are better suited for higher temperatures and are not damaged by the petroleum products found in some environments. This type of coupling does not have as good shock-absorbing capabilities as elastomeric couplings but is better than rigid couplings in absorbing shock.

There are many different types of metallic-element couplings available, but we discuss only three of the most popular types found in industry: *chain couplings, gear couplings*, and *grid couplings.*

10-5 CHAIN COUPLINGS

Chain couplings have been used by industry for many decades (see Figure 10.5). Their availability and low cost make them a good choice for many applications. There are three types of chain couplings that are widely used in industry: *roller chain* couplings, *silent chain* couplings, and *nylon* (or *synthetic*) *chain* couplings.

Chain couplings have little shock- or vibration-dampening capabilities and cannot tolerate as much misalignment as some other flexible couplings can. If a chain coupling is used on an application that has shock and vibration, it will allow most of the shock to be transmitted to the machinery; another type of flexible coupling should be used for this type of application.

FIGURE 10.5
Roller chain coupling.

FIGURE 10.6
Roller chain coupling with cover.

The chain part of the coupling can be replaced without unbolting the drive or driven machinery; however, you should always check alignment when you replace a coupling even if the equipment has not been disturbed. Because of their construction (metal-to-metal moving parts), chain couplings require lubrication. The exception is the nylon chain coupling, which does not need any lubrication. A chain cover is also required to prevent the lubricant from being thrown off by the centrifugal force of the rotating coupling. Covers usually have a grease fitting to permit lubrication without disassembly of the cover (see Figure 10.6).

Roller Chain Couplings

Roller chain couplings use standard industry sprockets of the same pitch and number of teeth. The two sprockets are usually connected by double-strand roller chain connected by a master link. This coupling accommodates shaft misalignment because of the clearances between the chain and sprocket teeth and within the chain. A major disadvantage is the noise created by this coupling. Roller chain couplings are best suited for lower-speed, higher-torque applications.

Silent Chain Couplings

Silent chain couplings are used in applications that require more torque and higher speeds than roller chain couplings (see Figure 10.7). Although silent chain

FIGURE 10.7
Silent chain coupling.

FIGURE 10.8
Synthetic chain coupling.

couplings are quieter and smoother than their equivalent roller chain counterparts, they are still noisy compared to elastomeric flexible couplings. Like roller chain, silent chain requires lubrication and covers to contain lubrication and prevent contamination.

Nylon (or Synthetic) Chain Couplings

Nylon chain couplings can be found in light-duty applications where lubrication and noise are not desirable (see Figure 10.8). Because of its nylon chain, it also works well in some corrosive environments. Nylon chain couplings are generally found in food processing, textiles, printing, and other industries where contamination with lubricants cannot be tolerated.

10-6 GEAR COUPLINGS

Gear couplings use two hubs with external teeth. A sleeve or hub cover with internal teeth connects the two shaft-mounted gears to provide power transfer (see Figure 10.9). The sleeve is typically a two-piece construction that bolts together, but single-piece sleeves are also used. Gear-tooth clearances provide for a small amount of shaft misalignment. This coupling must be more

FIGURE 10.9
Gear coupling.

precisely aligned than most other types of flexible couplings. Older gear-coupling designs used straight-sided gear teeth. This design has been improved and now most gear couplings have curved-face gear teeth with straight pitch lines, called an *involute* tooth design. Involute-tooth design permits more angular misalignment than the original straight-tooth design (see Figure 10.10). Gear couplings usually resemble rigid couplings and, like rigid couplings, tolerate little misalignment. Similar to rigid couplings (see Figure 10.4), gear couplings have almost no shock- or vibration-dampening abilities. The main advantage of gear couplings is their ability to tolerate end movement of shafts without transmitting large amounts of axial thrust to the coupled equipment. Typically, gear couplings are made with all-metal construction and require lubrication. There are some exceptions that use nylon or synthetic materials that do not require any lubrication. Gear couplings are relatively quiet, are capable of high speeds and horsepower, and are a good choice for larger applications if they are precisely aligned.

FIGURE 10.10
Different couplings have different tooth designs.

Straight Tooth

Involute Tooth

FIGURE 10.11
Metallic grid coupling.

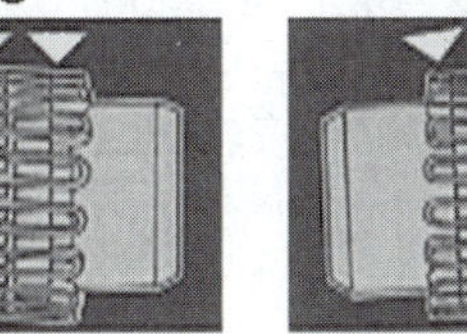
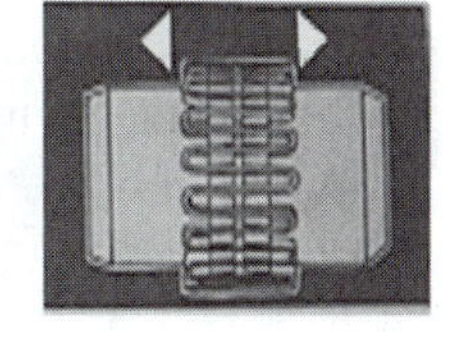

FIGURE 10.12
Advantages of metallic-grid coupling.

10-7 METALLIC-GRID COUPLINGS

Metallic-grid couplings have two hubs resembling gears (see Figure 10.11). The gear teeth are replaced with slots, and a continuous S-shaped spring grid fits into the slots, connecting the two halves together. The spring grid allows end-float, and angular and parallel misalignment and has good shock-absorbing capabilities (see Figure 10.12). Because of the all-metal construction, lubrication is a requirement. A cover is also needed to contain the lubrication and keep the spring grid in place. Covers may be split vertically and bolt together like flanges, or they may be split horizontally.

10-8 ELASTOMERIC FLEXIBLE COUPLINGS

All the flexible couplings discussed so far typically are all-metal construction. In this section, we discuss elastomeric couplings that generally use non-metallic inserts (see Figure 10.13). The elastomeric insert will be loaded either in *compression* or *shear*. Compression inserts may be pinned, bolted, or molded to the coupling hubs. Shear-type inserts are usually bolted or

FIGURE 10.13
Elastomeric coupling (nonmetallic insert).

clamped to hubs, can withstand more misalignment, and provide better shock-absorbing abilities. Compression-type inserts tend to work better for higher loads. In general, all elastomeric couplings work better than other types in applications where shock and vibration are problems. Elastomeric couplings are typically more sensitive to high temperatures than other couplings are. Common types of elastomeric couplings that are compression-loaded are *jaw* and *rubber bushing.* Some common shear-loaded elastomeric couplings are *rubber tire* and *donut.*

Jaw Couplings

The jaw coupling is one of the most widely used couplings for light industrial applications. It offers a simple design that is easy to install and maintain. Two hubs are separated by a spider (a compression-type elastomeric insert) that requires no lubrication (see Figure 10.14). The jaw-coupling spider can be replaced without unbolting machinery by loosening the hubs and sliding them away from the spider. Spiders are available in a variety of materials for chemical and heat resistance (see Figure 10.15). Jaw couplings can accommodate moderate misalignment and end-float. It is also a good coupling in applications where shock and vibration are a problem.

FIGURE 10.14
Jaw-type coupling.

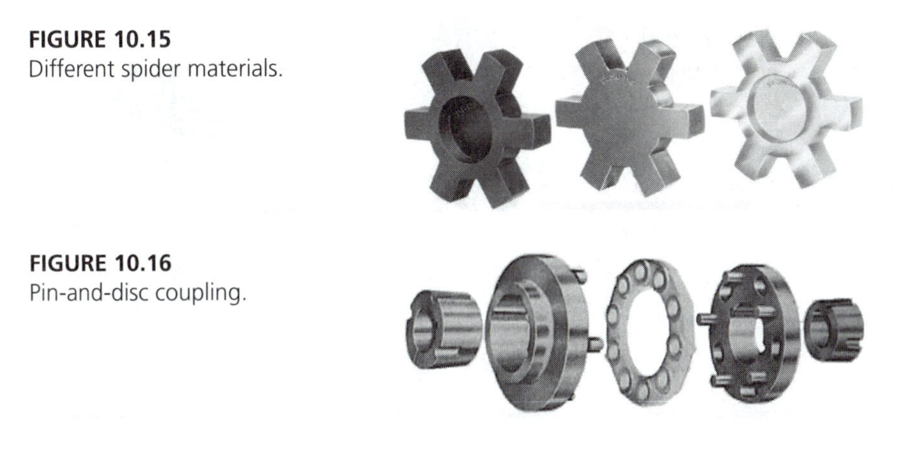

FIGURE 10.15
Different spider materials.

FIGURE 10.16
Pin-and-disc coupling.

Pin-and-Disc Couplings

The pin-and-disc coupling is a newer design of coupling consisting of two flanges with pins pressed into them (see Figure 10.16). The disc is typically a polyurethane compound that is resistant to most chemicals and petroleum products. It can withstand temperatures of up to 170°F (77°C). The disc is assembled with a light-press fit over the pins that prevents the accumulation of abrasive particles between the disc and pins and reduces sliding friction. Misalignment is accommodated by the flexing of the disc rather than the usual clearance between the disc and pins. Like the jaw coupling, the pin-and-disc couplings work well in applications with shock and vibration.

Rubber-Bushing Couplings

There are several types of rubber-bushing couplings. One of the most popular used has neoprene rubber bushings mounted under compression in a metal housing (see Figure 10.17). This type of coupling is used where high misalignment may result or is expected. It also cushions shock loads and absorbs vibration. Rubber-bushing couplings can be used in ambient temperatures to about 200°F (93°C) (see Figure 10.18).

FIGURE 10.17
Rubber-bushing coupling.

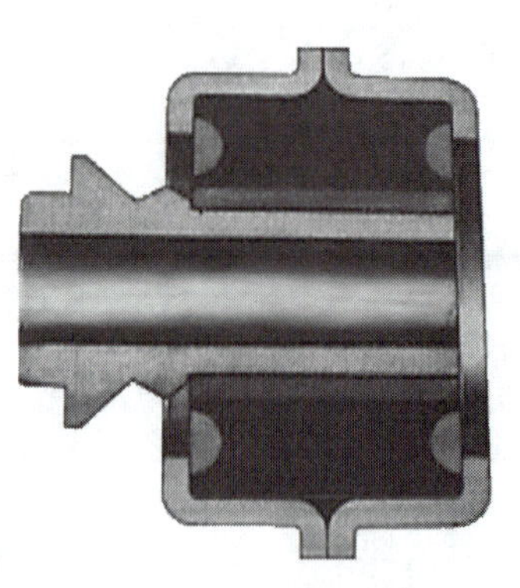

FIGURE 10.18
Rubber-bushing insert.

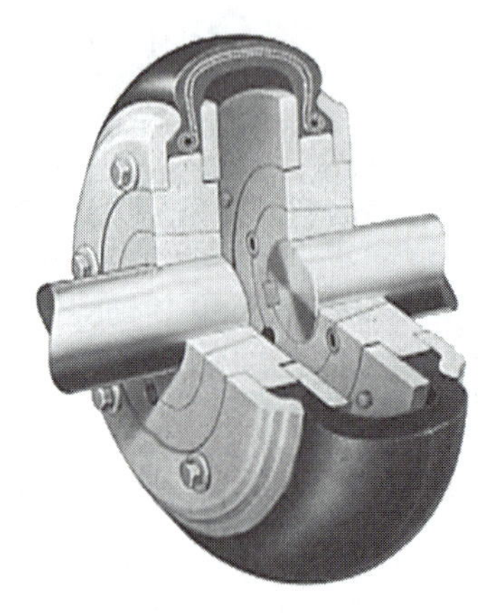

FIGURE 10.19
Rubber-tire coupling.

Rubber-Tire Couplings

The rubber-tire coupling is one of the most popular medium- to heavy-duty elastomeric couplings used in industry (see Figure 10.19). This coupling's long life and ability to handle large amounts of misalignment is due to the same technological advances used in automobile tires today. The elastomeric element of this coupling is constructed much in the same way as an automobile tire, giving it the ability to carry large loads and withstand high-shock loads (see Figure 10.20). Twin flanges clamp the rubber tire in place, allowing it to flex as needed. This mounting system also allows for the removal of the rubber tire without moving the driver or driven coupling flange or machine (coupling alignment should always be checked anyway). No lubrication is required, and rubber-tire couplings can be used in ambient temperature to about 210°F (99°C).

FIGURE 10.20
Elastomeric element.

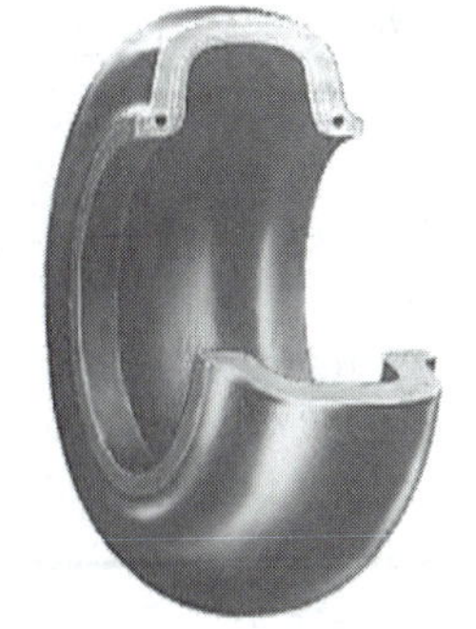

10-9 COUPLING INSTALLATION

Coupling manufacturers generally supply information on the installation, maintenance, and lubrication of their couplings. You should always consult this information before installing a new coupling. If for any reason you do not have this information, consult your coupling vendor for another copy. In general, you should follow these steps:

- Inspect the coupling components to ensure that all needed parts are on hand.
- Check for damage on the shafts where the coupling is to be mounted. Remove any burrs or nicks.
- Check the shafts with a dial indicator for runout and end-play. Follow the manufacturer's recommendations for tolerances. Typically, runout should never exceed 0.002 in. Repair or replace any shaft that is out of tolerance.

Most coupling hubs should be mounted before starting the alignment procedures. Nontapered bushed couplings may either be a *clearance* fit or an *interference* fit. With a **clearance fit,** the hub will slide onto the shaft without forcing or heating. With an **interference fit,** the hub must be heated so it expands enough to slide onto the shaft. The coupling manufacturer usually supplies information on how to heat the hub and to what temperature it should be heated. Oil bath or oven heating should be used to heat hubs. Never use an open flame or torch; uneven heating could cause distortion, which can damage or crack the hub. Oil-bath heating can be used to heat the hub to about 350°F (177°C), depending on the flash point of the oil being used. Ovens can be used for higher temperatures, but hubs usually shouldn't be heated to more than 600°F (316°C). Heated hubs should be handled very carefully. Safety is of utmost importance. Procedures for mounting tapered bushed hubs are the same as discussed in Chapters 7 and 8. Refer to those chapters if you need to review the procedure.

10-10 COUPLING ALIGNMENT

There are three basic methods for coupling alignment: *straightedge, dial indicator*, and *laser.* The straightedge and dial indicator methods are more common. Laser equipment is very expensive and is usually used only when high precision is needed. Good coupling alignment is important for prolonging coupling life and machinery life. Ideally, if all couplings were perfectly aligned with zero misalignment, flexible couplings wouldn't be needed. Tests have proven that precision alignment can increase bearing and seal life of machinery an average of six years. You must decide how carefully and accurately the coupling should be aligned. Some say that laser alignment tools should always be used and alignment should be less than 0.001 in. However, if the machine on which the coupling is mounted has a sloppy shaft runout (say, 0.002 to 0.003 in.), it cannot be aligned to the target precision of 0.001 in. If the machine and coupling can be accurately aligned to only 0.002 to 0.003 in., it is better to use a dial indicator rather than a laser alignment. You should select the alignment procedure and tools to fit the situation—in other words, choose the proper tool for the application. As in most other applications, use common sense. To be a good troubleshooter and a safe worker, always think through the problem first.

10-11 RUNOUT

Runout is the amount of wobble caused by loose machinery bearings, bent shafts, or poorly machined coupling hubs. Shaft or coupling runout controls how good the coupling alignment is. Use a dial indicator to check runout. Mount the dial indicator on a stationary object or use a magnetic base (as shown in Figure 10.21). The dial indicator plunger should be placed on the shaft where the coupling hub will be mounted (see Figure 10.22). Rotate the shaft to the edge of the keyseat. Do *not* allow the plunger to drop into the keyseat. Set the dial indicator to zero and rotate the shaft to the other side of the keyseat, noting any change in the indicator reading. Note that the indicator plunger should always be perpendicular to the object being measured to ensure accuracy (see Figure 10.23).

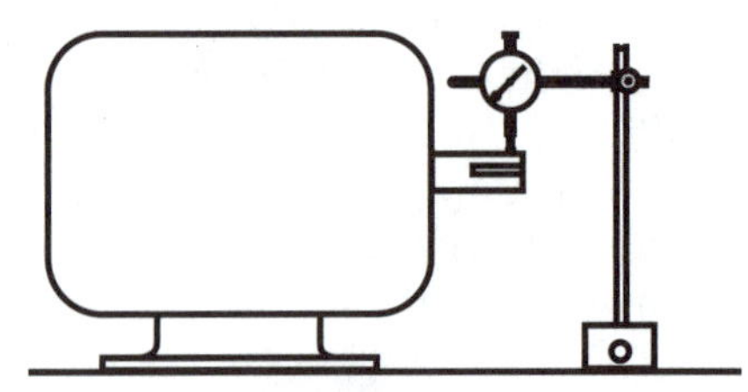

FIGURE 10.21
Use a dial indicator to check shaft runout.

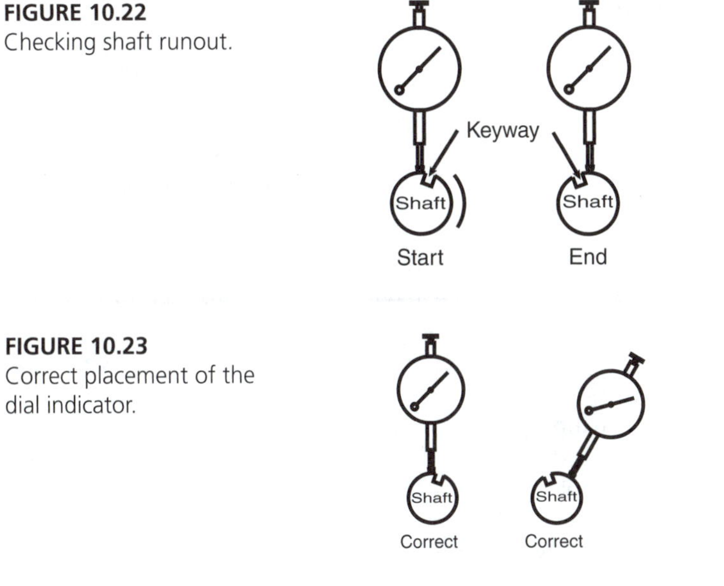

FIGURE 10.22
Checking shaft runout.

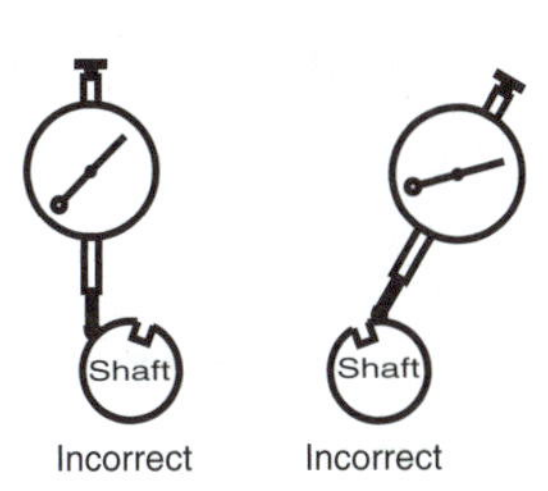

FIGURE 10.23
Correct placement of the dial indicator.

Check runout several times to ensure consistency. If the shaft is not within tolerance, it should be repaired or replaced and then rechecked. After the coupling hubs are mounted, repeat the runout procedure on the coupling hubs before attempting alignment.

10-12 SOFT FOOT

Soft foot refers to the condition that exists when the feet of a machine do not equally support its weight. It can be compared to a chair with one short leg, the consequence being that the chair rocks. If a machine has this condition, the same thing happens. Soft foot can make alignment difficult or even impossible. The dial indicator readings will be different every time the bolts are tightened. Uncorrected soft foot can also cause the foot of the machine to crack or break off when it is pulled tight to the baseplate. This may happen immediately or over a period of time as vibration weakens the foot. Tightening down a soft foot can also cause warping of the machine frame and misalignment of its internal

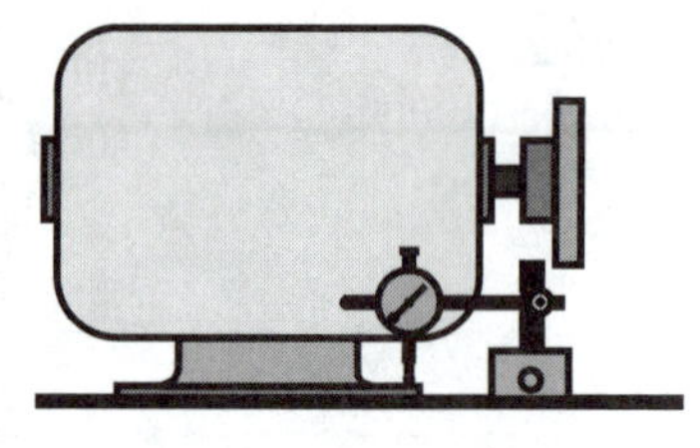

FIGURE 10.24
Checking for soft foot with a dial indicator.

rotating parts, leading to early bearing failure. Before checking for soft foot, you should always make sure the baseplate and feet of both the driving and driven machines are clean and free of debris, burrs, and rust.

You can initially check for soft foot by trying to rock the machine just as you would do a chair with a short leg. If it can be rocked, it has a soft foot that must be corrected. A feeler gauge or dial indicator can be used to determine the amount of shim needed under the soft foot to correct its condition (see Figure 10.24). Make sure that any shims used are flat and free of burrs. Precut stainless steel shims are the best choice if they are available. For more accuracy, you can use a dial indicator on each foot. Set the indicator to zero and tighten the mounting bolt, noting any change in the indicator reading. Shim the soft foot and recheck with the dial indicator. Repeat this procedure until soft foot is corrected.

10-13 SHIMS

Although precut stainless steel shims are ideal for alignment, most mechanics, unfortunately, have to fabricate their own shims. If you must do this, make sure that the fabricated shim is large enough to support the entire foot being shimmed. Shims should not be bent or have any burrs. Cut off all sharp corners at either a 45° angle or a radius, as shown in Figure 10.25. Shims should be U-shaped to fit around the bolt and have a tab for easier installation and removal. You should mark each shim with its thickness. This will prevent any confusion later when several different thicknesses of shims are being used. Thick shims should be drilled and then slotted with a fine-tooth hacksaw. Never use a chisel to slot a shim because it produces a ragged edge and deforms the shim. Use as few shims as possible to achieve the desired thickness, because more shims in a pack tend to compress when tightened, thus changing the alignment. You should never stack more than five shims in

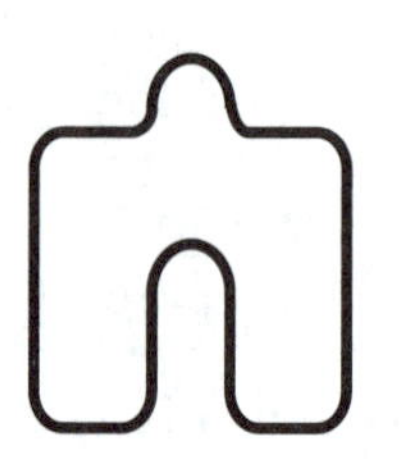

FIGURE 10.25
Alignment shim.

one pack. Some mechanics use a machine shop to fabricate a single shim to replace a pack. A surface grinder can be used to obtain a precision thickness with a good finish.

10-14 TYPES OF MISALIGNMENT

Two shafts are in perfect alignment when their axes are in the same line in all planes. Unfortunately, perfect alignment is not easy to obtain. Figure 10.26 illustrates the two types of misalignment:

1. Angular misalignment
2. Parallel misalignment

Angular misalignment occurs when the axes of the two shafts cross each other. With parallel misalignment, the axes do not cross. They are parallel but not in alignment because they are offset from each other. Both angular and parallel misalignment can occur in the horizontal and vertical planes, further complicating correct alignment. Use the systematic approaches outlined in the next sections to make coupling alignment simpler. Several methods are used to align a coupling: *straightedge alignment, single-dial indicator alignment, dual-dial indicator alignment, reverse-dial indicator alignment,* and *laser coupling alignment.*

FIGURE 10.26
Types of coupling misalignment.

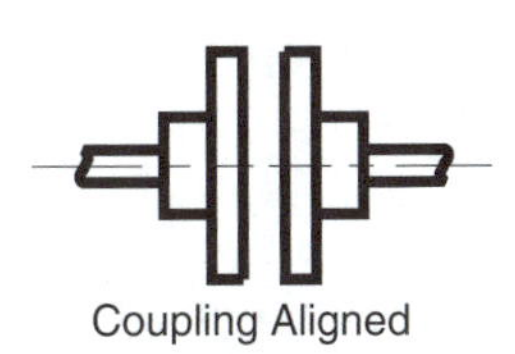

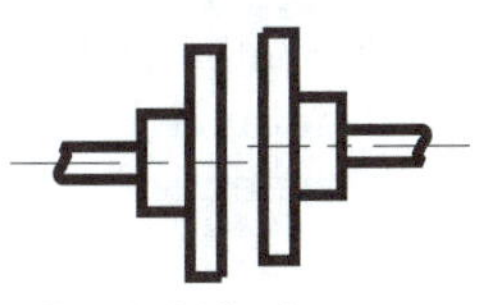

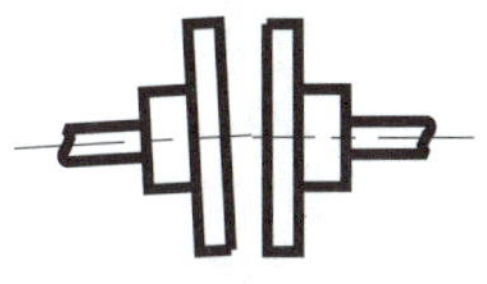

10-15 TYPES OF ALIGNMENT

Straightedge Alignment

Straightedge alignment is the least accurate method of alignment. It should be used to obtain a rough alignment before using the dial indicator or laser methods of alignment (see Figure 10.27). It is often used as the only means of alignment with some success when using a flexible coupling to compensate for some misalignment. The following steps detail the procedure for the straightedge alignment method:

- *Check the baseplate.* Make sure that it is clean and has no rust. File any burrs. The baseplate should rest solidly with no flexing.
- *Check runout and end-float.* Check runout on both driver and driven shafts. Maximum allowable runout depends on the application but, as a general rule, should be no more than 0.002 in.
- *Check for soft foot.* Check both driver and driven units and correct them as necessary.
- *Bolt one unit in place.* The driven unit is usually bolted, and all alignment adjustment is done on the other unit. For easier alignment, the shaft centerline of the first mounted unit must be higher than the shaft centerline of the second unit. It is easier to raise the second unit than to readjust both units.
- *Mount the coupling halves and check runout on them.* (Note: On some applications, this will be the last step, depending on the type of coupling.)
- *Check and correct vertical angular misalignment.* You can use a feeler gauge to compare the air gap at the top and bottom of the coupling halves (see Figure 10.28). You can also use a micrometer for more accuracy. The thickness of the required shims may be determined by trial and error or by calculation. To calculate the shim size, you need to measure three things:

  a. The center distance lengthwise of the mounting bolts of the driver
  b. The diameter of the flange face of the coupling
  c. The difference between the top and bottom flange-gap measurements

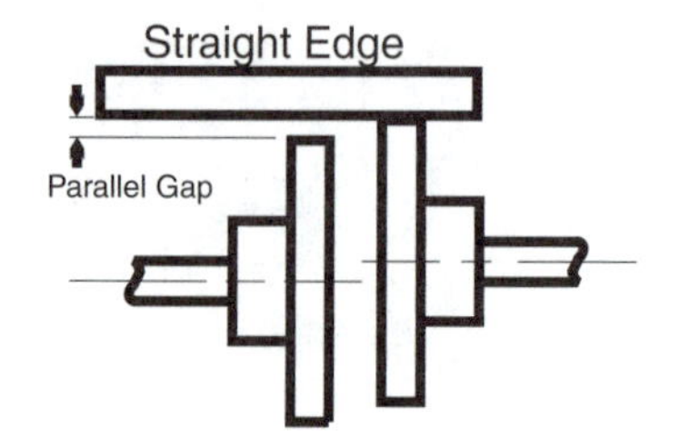

FIGURE 10.27
Using a straightedge for coupling alignment.

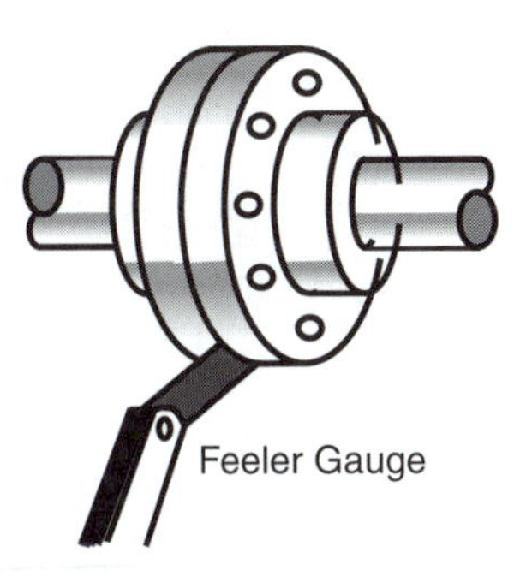

FIGURE 10.28
Using a feeler gauge to check angular misalignment.

Shim size is calculated by multiplying the center distance of the mounting bolts (CD) by the gap difference (G) between the flange faces, and then dividing by the diameter of the flange face (D).

$$\text{Shim size} = \frac{\text{CD} \times G}{D}$$

Insert the shims under the front or rear feet as needed. Tighten the mounting bolts. Measure the gap and repeat the procedure if necessary to obtain the required tolerance.

- *Check and correct vertical parallel misalignment.* Vertical parallel misalignment is the difference in height between the centerline of the driver and the centerline of the driven shafts. Put the straightedge

VERTICAL ANGULAR SHIM COMPUTATION

In this example, the motor bolt center distance is 12 in. The flange face diameter is 6 in. The air gap at top measures 0.011 in. and at the bottom measures 0.006 in.

Find the shim size required:

$$\text{Shim size} = \frac{\text{CD} \times G}{D}$$

$$\text{Shim size} = \frac{12\text{ in.} \times (0.011\text{ in.} - 0.006\text{ in.})}{6\text{ in.}}$$

$$\text{Shim size} = \frac{12\text{ in.} \times 0.005\text{ in.}}{6\text{ in.}}$$

$$\text{Shim size} = \frac{0.060\text{ in.}}{6\text{ in.}}$$

$$\text{Shim size} = 0.010\text{ in.}$$

Because the gap is larger at the *top*, the shims should be placed under the *rear* feet for correct alignment.

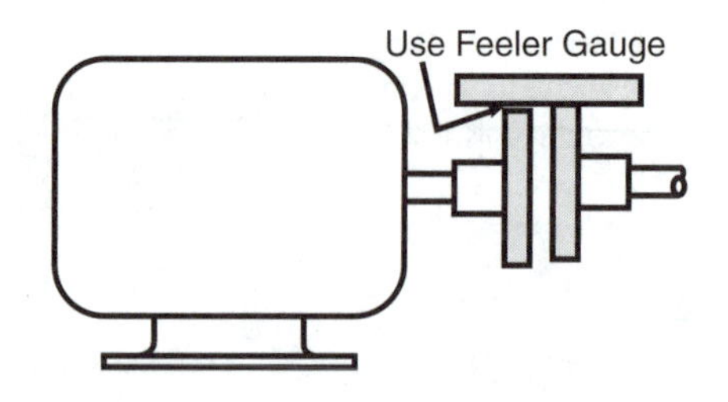

FIGURE 10.29
Use a feeler gauge to check parallel misalignment.

across the rim of the higher shaft. You can then measure the gap between the straightedge and the other rim using a feeler gauge. An alternative method is to use precut shims inserted in the gap to determine the total thickness of shims needed to correct vertical parallel misalignment. Place shims equal to the gap under all four feet. Do not remove any shims already being used to correct soft foot or vertical angular misalignment. Tighten the mounting bolts. Measure the gap, and repeat the procedure if necessary to obtain required tolerance. You are now finished with the shims. All horizontal adjustment is made by moving the driver, not shimming (see Figure 10.29).

- *Check and correct horizontal angular misalignment.* Use a feeler gauge or micrometer to compare the air gap on the left and right sides of the coupling halves. Move the driver until the air gap is the same on both sides. Use a soft-face hammer with light taps near the base. *Never* strike the coupling or shaft because this can damage bearings and warp the shaft. On larger units, jack bolts are the best way to make adjustments (see Figure 10.30).
- *Check and correct horizontal parallel misalignment.* This is done the same way as with vertical parallel misalignment except you line up the driver and driven shafts from left to right instead of up and down. Use a soft-face hammer or jack bolts to move driver into place. Once alignment is achieved, you should recheck for horizontal angular misalignment (step 8). Changing either horizontal angular or parallel adjustments can cause the other to become misaligned. If the driver must be moved, you must recheck *both* horizontal angular and parallel alignment until your target alignment is achieved (see Figure 10.31).
- *Perform a final preoperation check.* Make sure all mounting bolts are tight. Recheck the vertical angular and parallel alignment. Recheck horizontal angular and parallel alignment. Realign if misalignment is

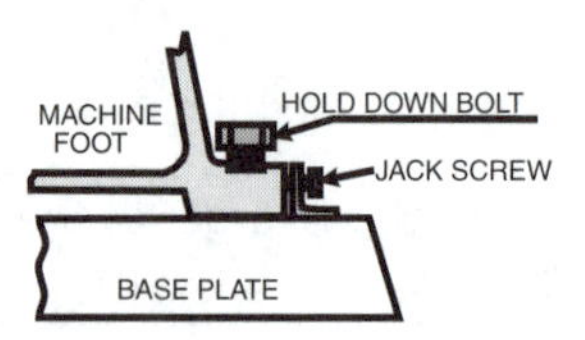

FIGURE 10.30
Use jack screws to make adjustments.

FIGURE 10.31
Check horizontal misalignment.

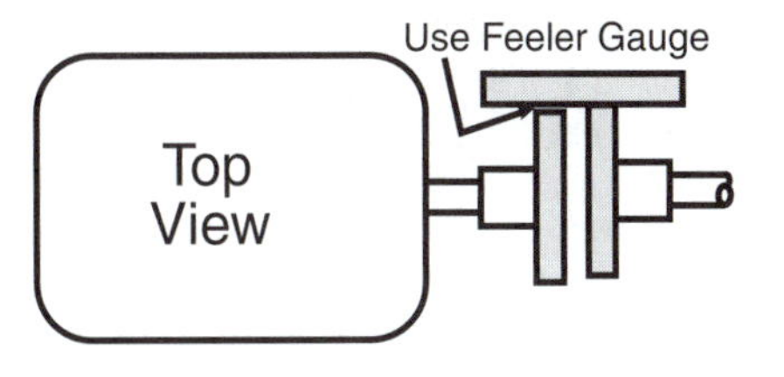

not within tolerance. If alignment is good, finish assembling the coupling. Replace all guards.

- *Perform a postoperation check.* After the unit has been run to operating temperature, remove the guards. Recheck all alignments and correct any misalignment. Be sure to replace all guards before returning to operation.

Single-Dial Indicator Alignment

Dial indicator alignment is the best choice for most industrial applications. The single-dial indicator alignment method follows the same general procedure as the straightedge method, but dial indicators increase the precision of all measurements. Because dial indicators have limited measurement travel, the straightedge method can be used to get a rough alignment before using the dial indicator. Dial indicators are precision-measuring tools, and they must be handled carefully because they can be damaged by rough usage. A faulty dial indicator cannot provide accurate alignment measurements. Make sure that the measurement required does not exceed the maximum plunger travel of the dial indicator. How the dial indicator is mounted varies according to the type of coupling. A magnetic-base mount can be used for most runout and end-float measurements. It can sometimes also be used to mount dial indicators for angular and parallel readings. In most cases, standard or special-purpose clamps are used for mounting indicators. With any mount, the indicator must be mounted tightly enough to eliminate any movement. Even the smallest movement can cause a false reading on the dial indicator. You should also ensure that the plunger movement is always perpendicular to the surface measured. If it is not mounted at right angles to the surface, you cannot get a true measurement. Use the following steps to align a coupling with a single-dial indicator:

- *Check the baseplate.* Make sure it is clean and has no rust. File any burrs. The baseplate should rest solidly with no flexing.
- *Check runout and end-float.* Run-out should be checked on both the driver and driven shafts. Maximum allowable runout depends on the application but, as a general rule, should be no more than 0.002 in.
- *Check for soft foot.* Both the driver and driven units should be checked and corrected as necessary.

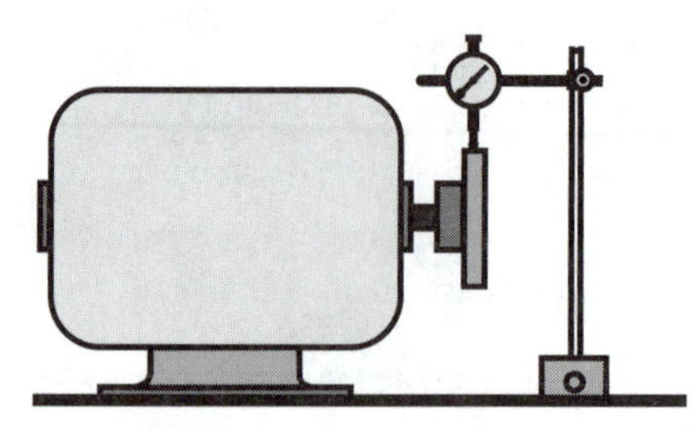

FIGURE 10.32
Using a dial indicator to check runout.

- *Bolt one unit in place.* The driven unit is usually bolted down, and all alignment adjustment is done on the other unit. (Note: For easier alignment, the shaft centerline of the first mounted unit must be higher than the shaft centerline of the second unit. It is easier to raise the second unit than to readjust both units.)
- *Mount the coupling halves and check runout.* You can use a magnetic-base mount on the baseplate or mount the dial indicator on one coupling half, holding it stationary and rotating the other coupling half to get runout readings (see Figure 10.32).
- *Check and correct the vertical angular misalignment.* The dial indicator should be securely attached to one coupling flange and its plunger placed against the face of the other flange (see Figure 10.33). The plunger should be preloaded to allow movement in both directions. Rotate the dial indicator to the 6 o'clock position and set the dial to zero. Rotate the flanges to the 12 o'clock position and take the reading.

CAUTION: *A common error is to rotate only one flange, but this measures only surface deformities. To get a true reading, both flanges must be rotated so that the plunger stays on the same physical spot when rotated. This measures angular misalignment only, not surface irregularities.*

Always take multiple readings for consistency. The shim thickness can be calculated using the same method described in

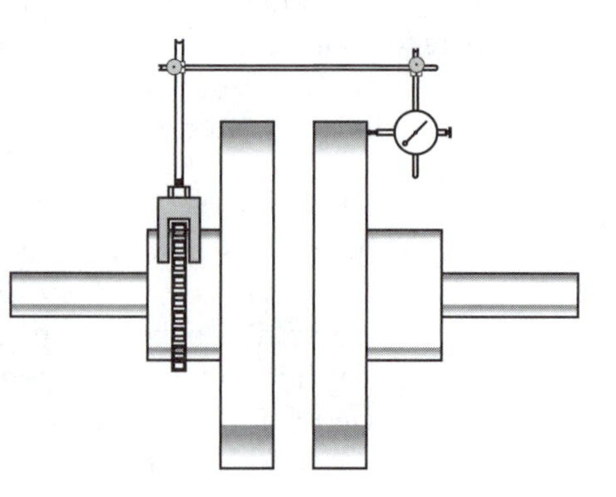

FIGURE 10.33
Check angular misalignment using a dial indicator.

the straightedge method. For this calculation, you must know the following:

a. The center distance lengthwise of the mounting bolts of the driver.
b. The diameter of the flange face (where the plunger rests).
c. The dial indicator reading of the difference of the top and bottom flange gaps.

Shim thickness is calculated by multiplying the center distance of the mounting bolts (CD) by the gap difference (G) between the flange faces, and then dividing by the diameter of the flange face (D).

$$\text{Shim size} = \frac{\text{CD} \times G}{D}$$

Insert the shims under the front or rear feet as needed. Tighten the mounting bolts. Remeasure the gap, and repeat the procedure if necessary to obtain the required tolerance.

- *Check and correct vertical parallel misalignment.* Vertical misalignment is the difference in the height between the centerline of the driver and the driven shafts. The dial indicator should be mounted with its plunger contacting the *outside diameter,* or *rim,* of the coupling flange (see Figure 10.34). Rotate the dial indicator to the 6 o'clock position and set the dial to zero. Rotate the flanges *together*, as described in step 6, to the 12 o'clock position, and take the reading. The shim thickness needed to correct parallel misalignment will be *one-half* of the dial indicator reading. This shim should be placed under all feet. (Be careful not to remove any shims already in place to correct angular misalignment and soft foot.) Tighten the mounting bolts. Take the dial indicator readings again, and repeat the procedure if necessary to obtain required tolerance. You are now finished with the shims. All horizontal adjustment should now be made by moving the driver, not shimming.
- *Check and correct horizontal angular misalignment.* Mount the dial indicator to read from the face of the coupling flange exactly as described in step 6. In step 6, we found the vertical angular misalignment by reading the dial indicator at 6 o'clock and 12 o'clock

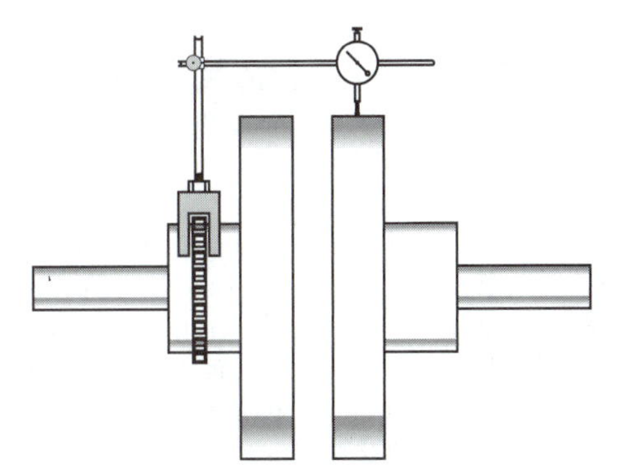

FIGURE 10.34
Check parallel misalignment using a dial indicator.

(bottom to top). For horizontal angular misalignment, read the dial indicator at 3 o'clock and 9 o'clock (side to side). Set the indicator to zero on the side away from you (the one hardest to read), rotate the coupling, and take the reading on the opposite side. As before, make sure both coupling halves rotate together and the plunger touches the coupling flange at the same spot on both sides. Move the driver until the reading on the dial indicator is in tolerance. Use a soft-face hammer with light taps near the base. *Never* strike the coupling or shaft. This can damage bearings and warp the shaft. On larger units, jack bolts are the best way to make adjustments.

- *Check and correct horizontal parallel misalignment.* Mount the dial indicator to read from the outside diameter or rim of the coupling flange, as described in step 7. In step 7, we found the vertical parallel misalignment by reading the dial indicator at 6 o'clock and 12 o'clock (bottom and top). For horizontal parallel misalignment, read the dial indicator at 3 o'clock and 9 o'clock (side to side). Set the indicator to zero on the side away from you (the one hardest to read), rotate the coupling, and take the reading on the opposite side. Always rotate both couplings. Use a soft-face hammer or jack bolts to move driver into place. Move the front and rear of the driver unit equally until the reading on the dial indicator is in tolerance. Tighten the mounting bolts and recheck the horizontal angular and parallel misalignment. Remember that any change in horizontal angular or parallel alignment can cause the other to become misaligned. If the driver must be moved, you must recheck *both* horizontal angular and parallel alignment until both are in tolerance.
- *Perform a preoperation check.* Make sure all mounting bolts are tight. Recheck vertical angular and parallel alignment. Recheck the horizontal angular and parallel alignment. Realign if misalignment is not within tolerance. If the alignment is good, finish assembling the coupling. Replace all guards.
- *Perform a postoperation check.* After the unit has been run to operating temperature, remove guards and recheck all alignments. Correct for any misalignment. Be sure to replace all the guards before returning to operation.

Dual-Dial Indicator Alignment

The dual-dial indicator method can more accurately be called the *face-peripheral dial indicator method.* This alignment procedure uses exactly the same steps as the single-dial indicator method, except that two indicators are used to read misalignment (see Figure 10.35). One indicator is placed to read angular misalignment, and the other simultaneously reads parallel misalignment. An experienced alignment technician can align a coupling quickly using this method. If you use dial indicators to align couplings only occasionally, you will probably find the single-dial indicator method to be easier and faster.

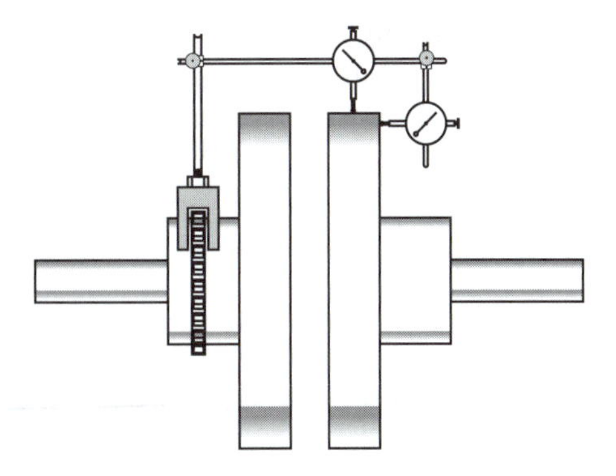

FIGURE 10.35
Face-peripheral dial indicator method.

Reverse-Dial Indicator Alignment

The reverse-dial indicator alignment method is a sophisticated procedure for alignment using two dial indicators to increase the accuracy of the indicator readings (see Figure 10.36). Indicators are mounted opposite each other, with one reading the top of the driven-unit coupling and the other reading the bottom of the driver-unit coupling. This alignment method is more difficult than any alignment discussed previously. This method should not be used if the distance between the shafts is less than the coupling flange diameter.

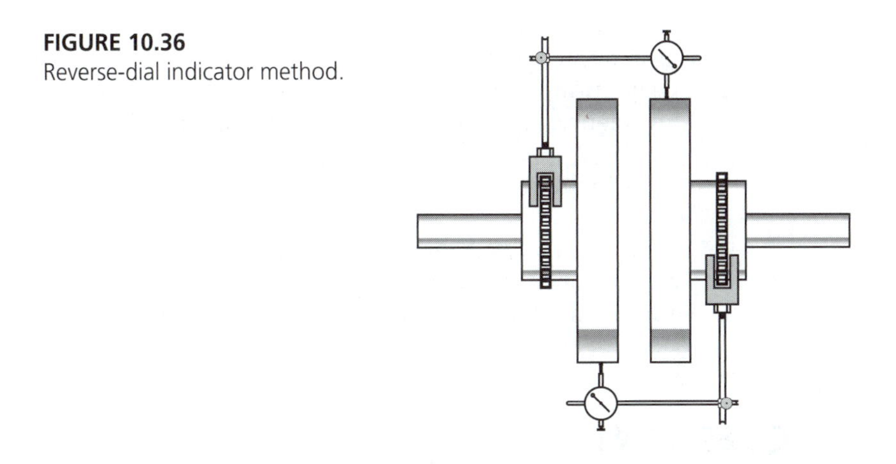

FIGURE 10.36
Reverse-dial indicator method.

Laser Coupling Alignment

Laser coupling alignment is the most accurate method for coupling alignment, but it is not necessarily the best choice of alignment methods. Laser systems can be accurate to 0.0005 in. or more. If a flexible coupled piece of machinery has a target alignment of only 0.010 in., the dial indicator method of alignment would probably be the best choice. Laser alignment is best used on larger or rigid-type couplings that require more accurate alignment. The most common

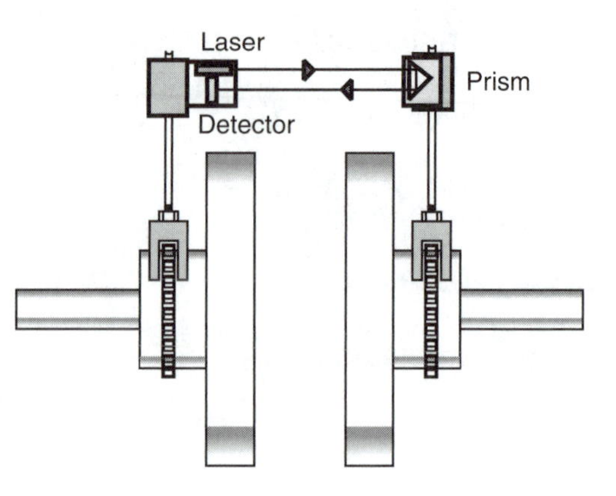

FIGURE 10.37
Laser coupling alignment.

laser system uses a combination laser send-and-receive head that attaches to one side of the coupling. An adjustable prism head acts as the target and attaches to the other side (see Figure 10.37). A small microprocessor accepts inputs from the laser head, and the technician uses a keypad to enter the distances between the laser and prism and the distances of the machinery feet. The direction and the amount of movement needed for alignment are then computed and displayed on an LCD screen. Advantages of laser alignment include the following:

- It can handle spans up to 6 ft.
- It is extremely accurate.
- The coupling can be kept in place during alignment.
- Movement needed for alignment is automatically calculated.

Disadvantages include the following:

- The equipment is expensive.
- The equipment must be handled with care to prevent damage.
- Periodic manufacturer's recalibration of the equipment is required to ensure accuracy.

QUESTIONS

1. What is a coupling?
2. What are the three common types of rigid couplings?
3. Why must rigid couplings be perfectly aligned?
4. Which type of rigid coupling is designed for heavy-duty applications and high torques and speeds?
5. What is the ultimate goal when selecting a coupling?
6. Industrial flexible couplings can be divided into two categories by their flexible member's construction: ______________ and ______________.

7. Name the three most common types of metallic-element couplings found in industry.
8. Why do most chain couplings require lubrication?
9. What is the main disadvantage of roller chain couplings?
10. Where are nylon chain couplings generally found?
11. The elastomeric coupling insert is loaded either in compression or shear. What is the difference?
12. The two common types of elastomeric couplings that are loaded in compression are ______________ and ______________.
13. The two common types of elastomeric couplings that are loaded in shear are ______________ and ______________.
14. What is the difference between a clearance and an interference fit in installing a coupling?
15. What are the three basic methods for coupling alignment?
16. Tests have proven that precision alignment can increase bearing and seal life of machinery to an average of ______________ years.
17. What is runout?
18. What is soft foot?
19. What are the two types of misalignment?
20. What is the most accurate method of coupling alignment?

CHAPTER 11

Clutches and Brakes

11-1 INTRODUCTION

As the age of the steam engine blossomed in the 1800s, the ability to engage and disengage machinery from a motor or engine became much more important (see Figure 11.1). In many instances, a single engine was used to drive several different pieces of equipment. Sometimes, this was done by changing a flat belt from the engine to each machine. Inside factories, it was more convenient to have the engine drive a *line shaft.* A line shaft was a rotating shaft driven by an engine that ran the length of a room. A pulley was located on the line shaft at each machine. The engine and line shaft ran constantly. To disengage each

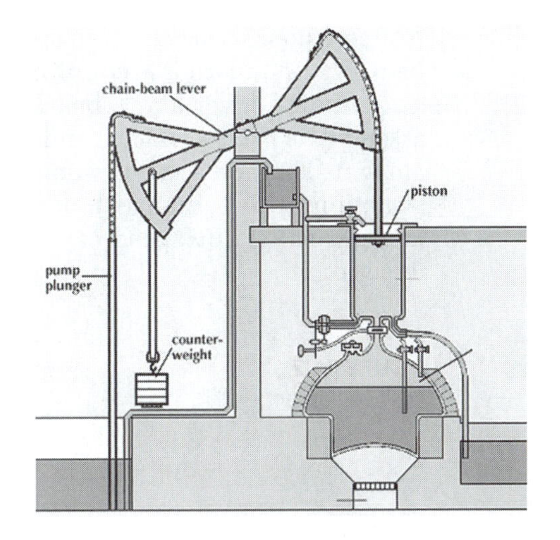

FIGURE 11.1
Early steam engine.

machine, the flat belt running it would be "run off" the pulley by an operator using a wooden stick. It was a very tricky procedure; many operators were hurt or maimed trying to run a belt on or off a rotating line shaft. Clutches began to be used as an improvement over flipping a belt on or off a line shaft.

Modern clutches have many different forms and variations. This chapter covers only the more common basic types. The function of a clutch is to engage or disengage a machine (or machine component) without starting or stopping the driver. Different types of clutches can also provide the following:

- Slower, smooth engagement and disengagement under full speed.
- Quick engagement and disengagement.
- Overload protection by limiting the maximum torque loads.
- Prevention of accidental machine reversal.

Clutches for industrial use are classified by the type of clutch and the method of clutch actuation.

You may have noticed that up to this point brakes have not been mentioned. In physical characteristics, most brakes are actually clutches with one side locked down so when the clutch/brake engages the rotating shaft stops. As we discuss different types of clutches, keep in mind that the same principles and configurations will also apply to most brakes.

11-2 POSITIVE-CONTACT CLUTCHES

Positive-contact clutches allow no slippage when engaging or disengaging. They will allow no slippage when running under extreme load up until they physically break. The earliest basic positive-contact clutch is the *square-jaw* (sometimes called a "dog") clutch. Figure 11.2 illustrates this type of clutch. Notice it looks very similar to a standard jaw coupling. The only difference between a standard jaw coupling and a square-jaw clutch is that the clutch has an actuation lever and a bearing assembly called a *throw-out* bearing. This assembly is used to engage and disengage the clutch by moving one of the jaws along its shaft on a key or sometimes a spline. Because of the square-jaw construction, this clutch should always be engaged *only* when stopped. Engagement at any speed can cause heavy shock loading and probable clutch damage.

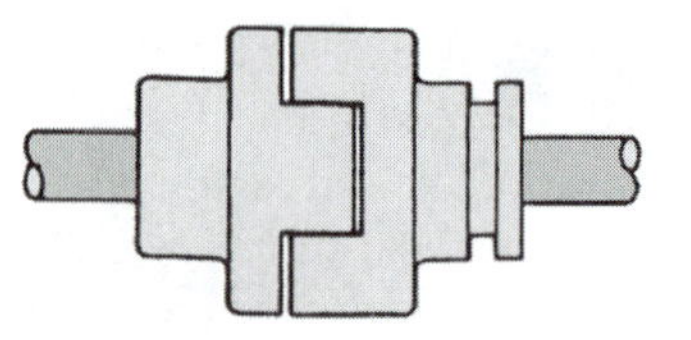

FIGURE 11.2
Positive-contact clutch.

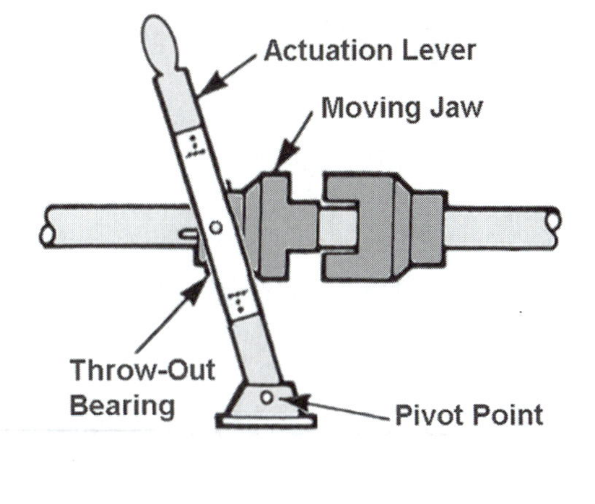

FIGURE 11.3
Square-jaw clutch operation.

FIGURE 11.4
Spiral-jaw clutch.

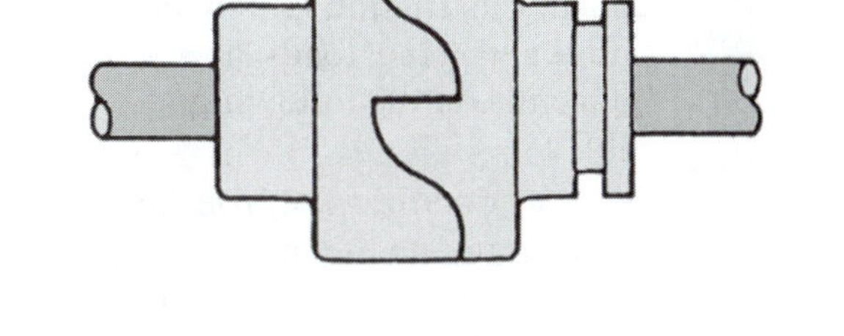

Figure 11.3 shows a simple square-jaw clutch application. Notice that the actuation lever has a pivot attached to a solid surface; the other end is moved to engage or disengage the clutch. The moving jaw is attached to the actuating lever through the throw-out bearing assembly and is keyed to the shaft. The key is used to transmit the rotary motion of the clutch to the shaft. The movable jaw is designed to slide freely on the shaft while its opposite end is locked in place, just as a coupling half would be.

Figure 11.4 shows a modification of the square-jaw clutch called the *spiral-jaw* clutch. Because of its taper-tooth design, this clutch is better suited to be engaged at low speeds. It requires some sort of locking device to keep it engaged. In some cases, this locking device may be spring-loaded, which allows a spiral-jaw clutch to become an overload mechanism. Depending on the amount of force applied by the spring, the spiral-jaw clutch automatically disengages when its threshold torque is reached. This overload function should be used for intermittent operation only because of the excessive wear created when operating under overload conditions.

A taper-toothed clutch is shown in Figure 11.5. With many more teeth, these clutches can be engaged at slow speeds. The taper-tooth design can also function as an overload device, no matter which direction the clutch is turning. Some modern electromagnetic clutches use a taper-toothed clutch mechanism. In these clutches, there are many more teeth of a smaller design than on those found on the older mechanical-engagement taper-tooth clutches.

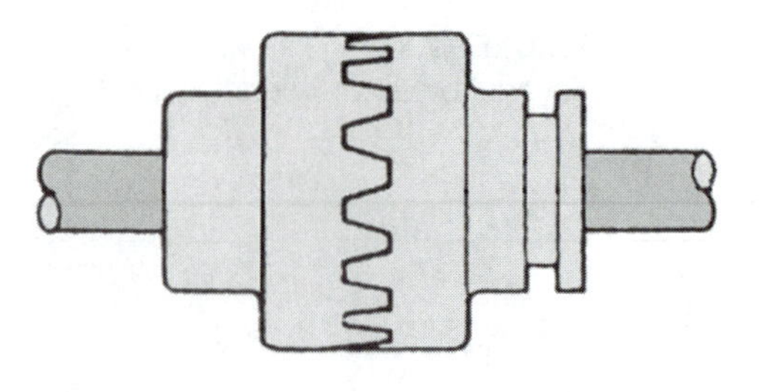

FIGURE 11.5
Taper-toothed clutch.

11-3 FRICTION CLUTCHES

Because of the limited ability of early positive-contact clutches to engage under extreme loads and higher speeds, friction-type clutches and brakes were developed. The friction clutch and brake have the ability to transfer power smoothly under varying loads and speeds. The effectiveness of this type of clutch depends on the following:

- *Coefficient of friction of the mating materials:* Friction is the force that resists the movement of one surface against another. This resistance to sliding can be expressed as a ratio of the force holding the surfaces together divided by the force that resists sliding. This ratio is called the **coefficient of friction**, or μ.

$$\text{Coefficient of friction} = \frac{\text{force pressing object together}}{\text{force that resists sliding}}$$

$$\mu = \frac{N}{F}$$

  A larger value for the coefficient of friction translates to more resistance to movement and better braking. Unfortunately, this sliding motion generates heat, sometimes lots of it. Clutch and brake liner material must be heat-resistant. Originally, asbestos was the most widely used liner material because of its excellent heat resistance. Asbestos has since been found to be a hazardous material and has been discontinued as a clutch or brake liner material. Organic materials are now typically used in modern clutches and brakes. Sometimes soft metal particles are added to liners to help dissipate heat and increase liner life.
- *Surface area:* Increasing the surface area of the mating surfaces of a brake or clutch increases the effectiveness of it proportionally. Physical space requirements usually limit the maximum size of a clutch or brake in any given application.

- *Amount of force used:* The greater the force pressing the clutch or brake surface together, the greater the effectiveness. Power-assisted clutches and brakes are common today.

Friction clutches and brakes can be one of two basic types: *radial* or *axial.* In **radial-type friction clutches,** contact pressure is applied to the peripheral of a drum or rim. An example of a radial friction brake is the drum brake found on most older automobiles. Before the automotive drum brake, horse-drawn wagons also used a radial-type friction brake. A block of wood pressing on the outside of the wheel provided the braking (see Figure 11.6).

The *band* or *strap* brake shown in Figure 11.7 is one of the simplest and oldest radial friction brakes used by industry. A flexible friction strap is wrapped around a smooth drum. Braking is accomplished when the linkage tightens the strap. This type of brake is easy to manufacture, is easy to use, and has very predictable performance. The disadvantages of this type of brake are its inability to dissipate heat efficiently, its open construction, and its uneven friction material wear.

Centrifugal clutches, like the one shown in Figure 11.8, are a good example of radial friction clutches. This type of clutch has a steel drum and friction shoes. The shoes are normally held away from the drum by a heavy spring. The motor or engine is attached to the rotating shoe. As the motor speed increases, centrifugal force overcomes the spring pressure, and the shoes are forced against the drum. This type of clutch provides a soft start and some slippage

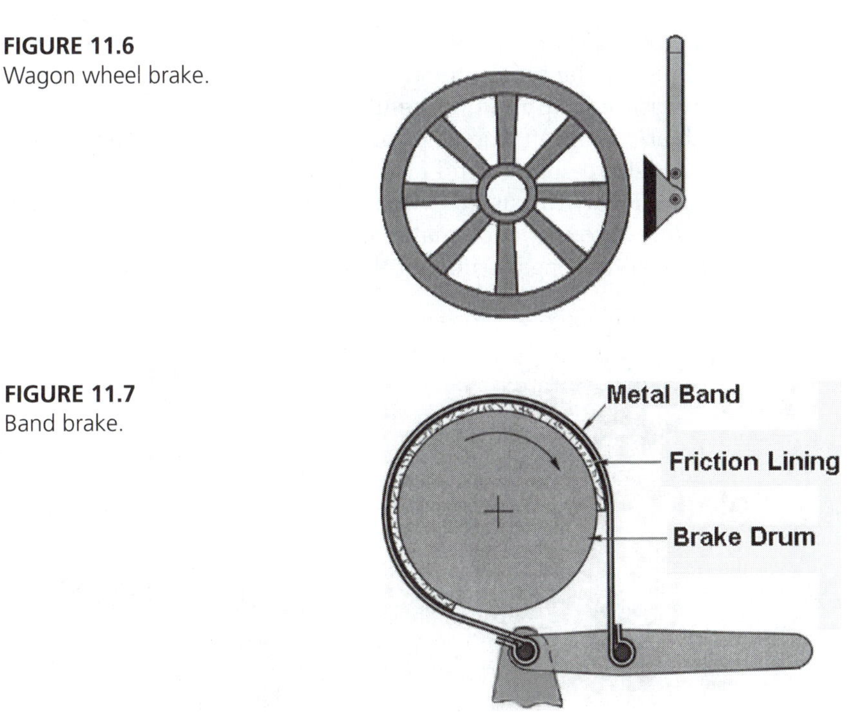

FIGURE 11.6
Wagon wheel brake.

FIGURE 11.7
Band brake.

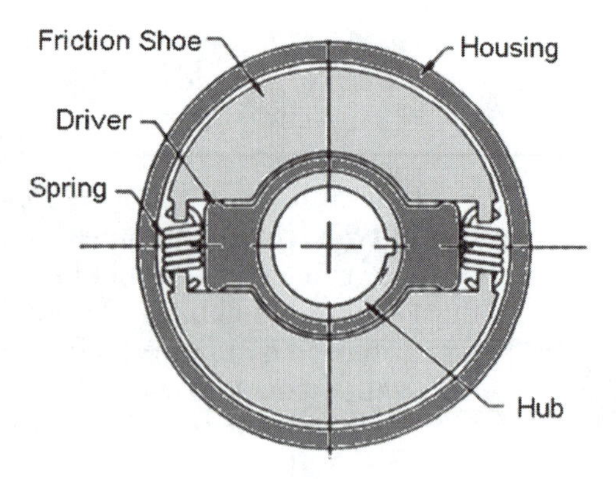

FIGURE 11.8
Centrifugal clutch.

under overload conditions. Under normal conditions and speeds, these clutches lock with the drum, allowing no slippage. This type of clutch will overheat and can be damaged if subjected to constant slippage. Always make sure this type of centrifugal clutch has enough capacity to lock up under normal operating conditions. The capacity of the centrifugal clutch is determined by the shoe-contact area, the weight of the shoe, and the type of friction material used. The spring also contributes somewhat to the capacity, but its real purpose is to determine the speed at which the clutch engages. A heavy spring gives a later (higher-speed) engagement; a light spring allows an earlier (lower-speed) engagement.

So far, we have discussed radial friction brakes and clutches in which contact pressure is applied to a drum or rim. The other type of friction brake and clutch is the **axial-type friction clutch** shown in Figure 11.9, in which contact pressure is applied perpendicular to the rotating shaft. The *disc* brake and clutch are common examples of axial-type friction brakes and clutches.

Axial-type brakes and clutches provide a larger surface area than the radial type previously discussed. More surface area means higher horsepower capabilities. With various friction materials and constructions, these clutches may be used to engage or disengage only, or they may act as torque limiters by slipping under

FIGURE 11.9
Axial-type clutch.

FIGURE 11.10
Torque limiter.

excessive torque (see Figure 11.10). These clutches may be designed to run *dry* or *wet*. A wet clutch is one that is immersed in oil or coolant, which helps dissipate heat when the clutch slips. Wet clutches are especially designed to squeegee and wipe away oil so that friction surfaces do not ride on a film of lubricant.

Axial-type clutches consist of one or more metal pressure plates and friction discs (see Figure 11.9). Axial clutches are usually designed so the friction disc is free to move or float on the driven shaft by using keyways or splines. A spring-loaded pressure plate is used to engage and disengage friction disc. A throw-out bearing is also used for support when the clutch is disengaged. Dry clutches are typically air-cooled. Sometimes an external fan must be used to keep the clutch from overheating. Oil or contaminants in a dry clutch can cause slippage and shorten clutch life.

Multiple-disc clutches may be used to increase torque while keeping space requirements minimized (see Figure 11.11). Additional friction discs are sandwiched between metal plates, adding more surface area and thus increasing torque capacities. Automatic transmissions in modern automobiles may use three or more multiple-disc clutch packs. Motorcycle clutches are also multiple-disc clutches to keep space and weight requirements down. Multiple-disc clutches are usually wet clutches for better heat-dissipation purposes.

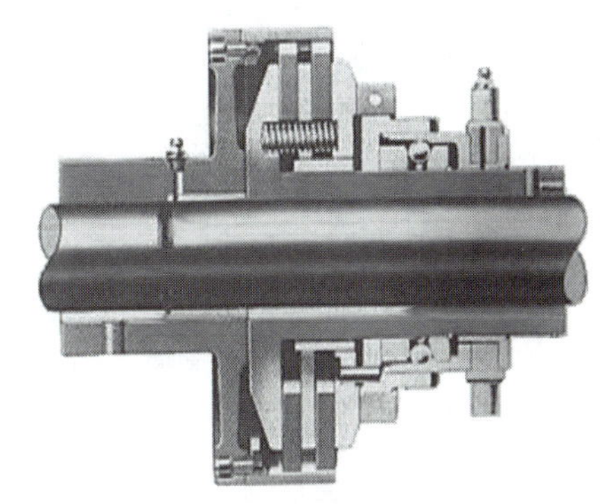

FIGURE 11.11
Multiple-disc axial clutch.

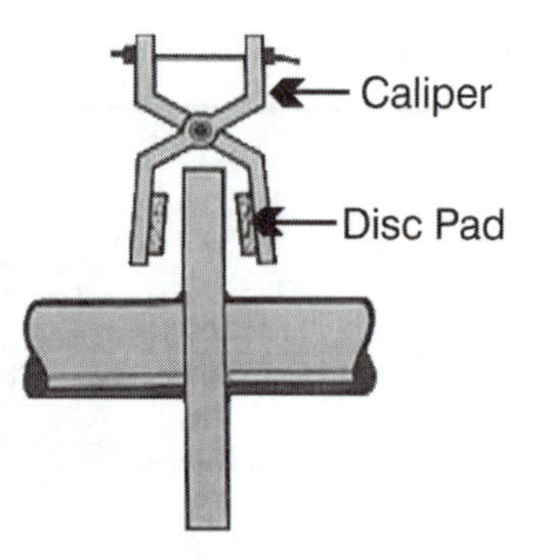

FIGURE 11.12
Axial (disc) brake.

Figure 11.12 shows a simple mechanical axial (disc) brake. A typical disc brake uses a floating caliper with two friction discs, one on each side of the disc. The floating caliper moves so that equal pressure is applied to both sides of the metal disc. With the disc brake, the caliper is always stationary, and the disc rotates. This type of brake generates a large amount of heat. To help dissipate excessive heat, hollow discs (also called *rotors*) that have ribs or slots for air to flow through between the braking surfaces may be used (see Figure 11.13). Disc brakes are relatively low maintenance, dependable brakes, but they require more force to brake than radial or drum brakes. Automobiles generally use power-assisted hydraulics to actuate disc brakes. For industrial applications, disc brakes may be actuated by mechanical, pneumatic, hydraulic, or electric actuators.

Figure 11.14 shows a *cone* clutch. Cone clutches are a combination of radial- and axial-type clutches. Because of the wedging action of the cone inside a matching drum, the force required for activation is less than purely radial or axial clutches. This type of clutch is usually found on high-speed applications with lower horsepower requirements.

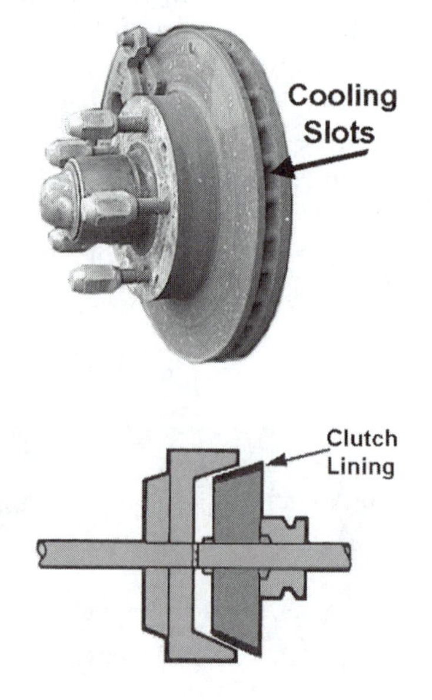

FIGURE 11.13
Automotive disc brake.

FIGURE 11.14
Cone clutch.

11-4 FLUID CLUTCHES

Probably the most common fluid clutch is the *torque converter* used in automotive automatic transmissions. A rotating impeller imparts energy by centrifugal force to a viscous fluid. The liquid, in turn, imparts its motion to the runner attached to the output shaft. The impeller and the runner vanes have a small amount of clearance between them, allowing little slippage (see Figure 11.15). Varying the amount of fluid in the viscous fluid clutch can vary the amount of slippage. More fluid means less slippage. Efficiency for this type of clutch will always be less than 100% but can typically be between 97% and 99% in a well-designed system. The advantages of the viscous fluid clutch are low maintenance, soft starts, and overload and shock protection. Some of the disadvantages are possible fluid leakage, no positive lockup, and the need for an external cooling system for heat dissipation.

Another common industrial fluid clutch uses dry pellets called *shot* instead of oil or liquids as the fluid media (see Figure 11.16). The motor or driver turns the clutch housing. A rotor attached to the driven shafts is inside the housing. Loose pellets, generally steel shot, are also used in the housing.

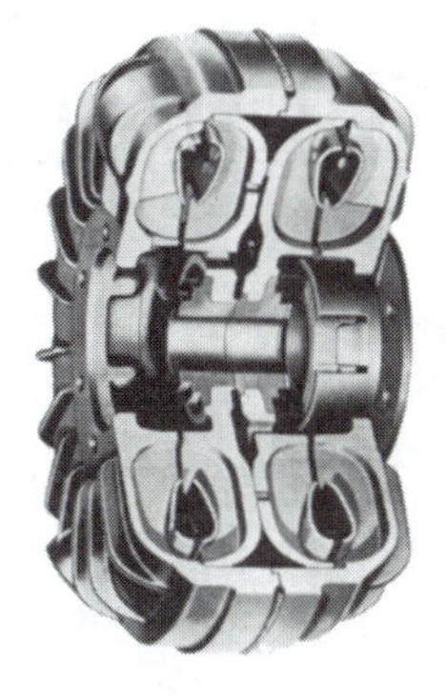

FIGURE 11.15
Fluid-clutch torque converter.

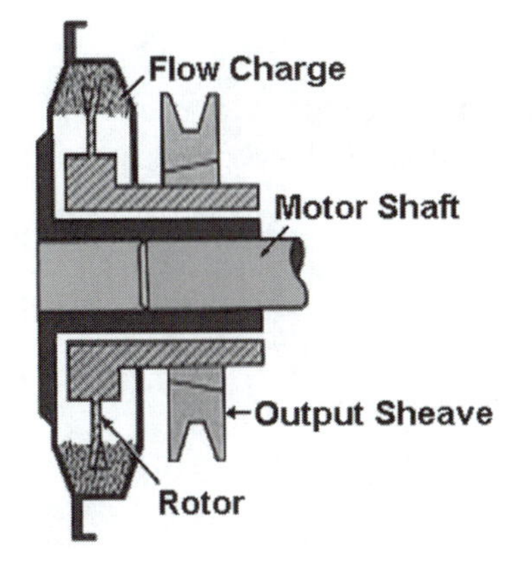

FIGURE 11.16
Dry fluid clutch.

The amount of shot used in this dry clutch is called the *flow charge.* When there is no rotation, the shot lies at the bottom of the housing. When the clutch housing starts to rotate, the shot becomes evenly distributed and starts to turn the rotor because of the friction between the shot and housing. As the speed increases, centrifugal force locks the shot into place, and the housing and rotor become locked together at the same speed. At this point, this type of clutch becomes 100% efficient because there is no slippage.

Proper selection of dry fluid clutches is very important. The clutch should be sized so that it will slip under high overload conditions to prevent damage to equipment. Correct selection also allows for soft starts with less initial amperage draw, which can be critical with large electric-motor applications. If the dry fluid clutch is sized so that too much slip occurs, clutch life can be greatly shortened. The shot used in these clutches is highly abrasive, and excessive slip will wear out the rotor and housing quickly. Good selection ensures that the clutch locks up under normal operating conditions and still slips if high overloads occur. The dry fluid clutch is an economical solution for industrial applications needing soft start and overload protection. It is used quite extensively in industry because of its simplicity and efficiency.

11-5 MAGNETIC CLUTCHES AND BRAKES

Some of the most common magnetic clutches and brakes used in industry are friction clutches and brakes. Electricity applied to an electromagnet causes the activation of the clutch or brake (see Figure 11.17). Electromagnetic clutches and brakes are used mainly for starting or stopping and are not suited for continuous-slip operation. They are manufactured in a wide range of configurations and sizes. They may be integral to a motor or another piece of equipment or be mounted separately. Horsepower capabilities range from fractional horsepower to several hundred horsepower.

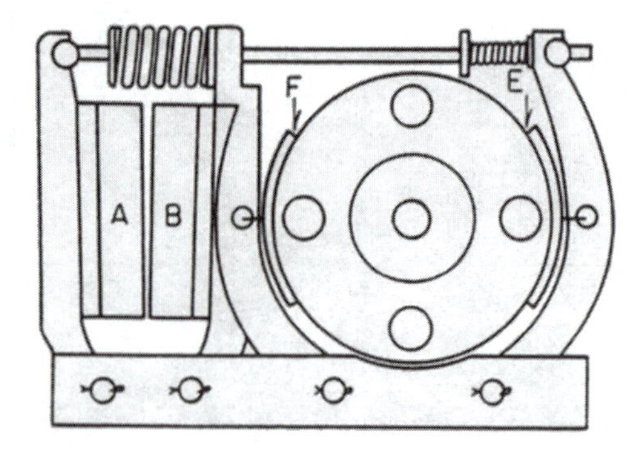

FIGURE 11.17
Electromagnetic brake.

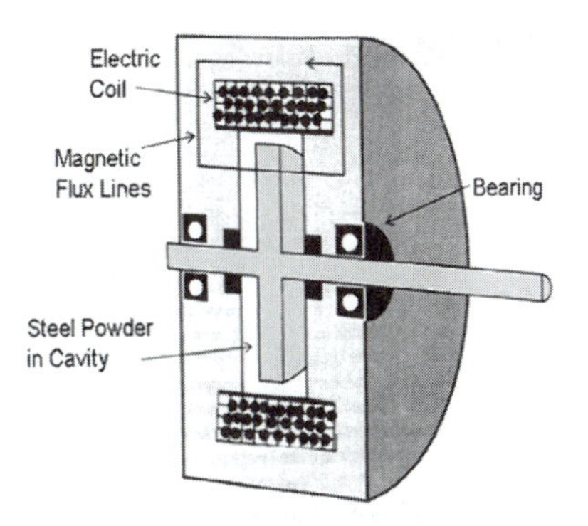

FIGURE 11.18
Magnetic-particle clutch.

Magnetic-particle clutches are similar to the dry fluid clutch previously discussed. A rotor is mounted in a sealed housing, as shown in Figure 11.18. Dry, fine, stainless steel powder is added inside the housing. Around the perimeter of the housing, an electromagnetic coil is made as an integral part of the housing. When the electromagnet is energized, the metal particles lock into patterns formed by the magnetic flux lines and immobilize the rotor inside the housing. When the electromagnet is deenergized, there is no magnetic field, and the rotor turns inside the housing. Relatively little heat is generated by the friction of the rotor and metal powder. A magnetic-particle clutch or brake is frequently used in applications when rapid cycling is required. Lock-and-unlock torque characteristics are very consistent and predictable, making this type of clutch a good choice for automated production processes. If properly selected and sized, the lifespan of magnetic-particle clutches is excellent, and almost no maintenance is required.

When selecting or sizing a magnetic-particle clutch or brake, always make sure that the heat generated by cycling and slip does not exceed a value that would cause damage to the metal powder. Always consult the clutch or brake manufacturer for temperature limitations of specific models of clutches or brakes. Specifications will vary from manufacturer to manufacturer and even from model to model.

Another limitation of this type of clutch or brake is rotational speed. Rotational speeds that exceed the manufacturer's recommendations will generate excessive centrifugal force. This excessive centrifugal force will cause the slip and torque to become erratic. In extreme cases, the clutch may even engage because of centrifugal force locking the metal powder and the rotor.

Another variety of magnetic clutches and brakes is the *eddy-current* clutch (see Figure 11.19). This type of clutch uses a noncontacting, nonferrous rotor (usually copper or aluminum) sandwiched between two magnetic discs mounted in a housing. The electric current induced in the rotor by the magnetic field causes the rotor to rotate. The eddy-current clutch will never lock completely and always has some slippage. The rotor speed never reaches the housing speed. The torque can be adjusted by misaligning the magnetic poles

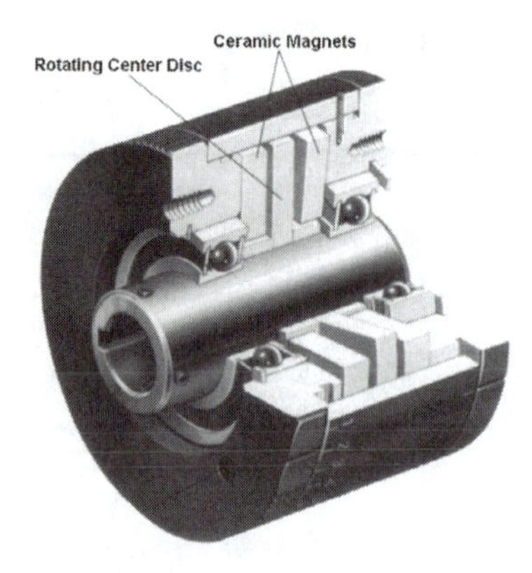

FIGURE 11.19
Eddy-current clutch.

of the two magnetic disks on either side of the rotor. Maximum torque is obtained by a north–south magnetic-pole arrangement and minimum torque is obtained by a north–north and south–south magnetic-pole arrangement.

The rotation speed required to produce the eddy currents that make these clutches operate also causes them to be poor choices for brakes. They are most widely used in industry as clutches in applications where overloads frequently occur. Frequent overloads do not harm the clutch or equipment. The eddy-current clutch works well in applications that require constant torque, long life, and no maintenance. They should never be used in applications where synchronization between the input and output is required.

Hysteresis clutches are very similar to eddy-current clutches. Physically, the main difference between them is that the rotors in hysteresis clutches are made of a ferrous (iron or steel) material. Like eddy-current clutches, hysteresis clutches have no contact surface to wear, so clutch characteristics are stable over a long life. Unlike the eddy-current clutches, hysteresis clutch input and output synchronize as long as the rated torque value is not exceeded. The torque can also be adjusted. Torque is proportional to the coil current. The performance and life of hysteresis clutches depend on the heat generation and dissipation. These clutches are typically used in tensioning applications, such as winding operations, conveyors, and rewinding. They also work well in any constant tension or torque applications.

11-6 OVERRUNNING CLUTCHES

An *overrunning* clutch is a mechanical device that allows rotation and torque to be transmitted only in one direction (see Figure 11.20). Physically, overrunning clutches look like bearings and are sometimes mistaken for them. These clutches are used for three basic types of applications: *overrunning, indexing,* and *backstopping.*

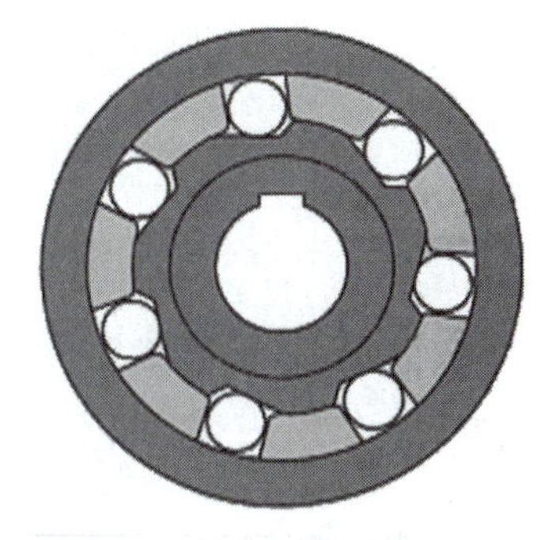

FIGURE 11.20
One-way roller-ramp clutch.

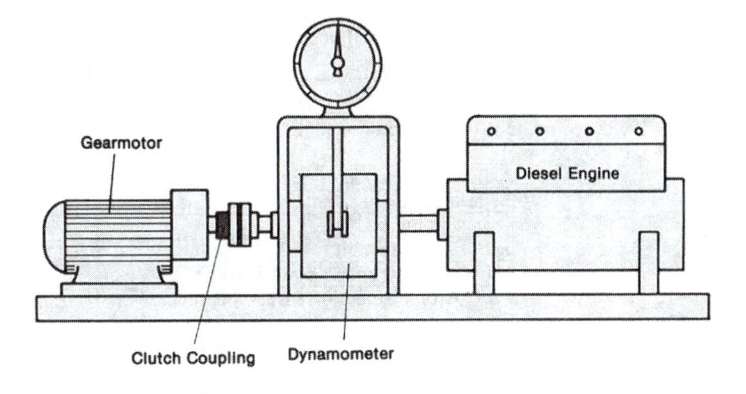

FIGURE 11.21
Overrunning clutch application.

Overrunning Applications

Overrunning (sometimes called *freewheeling*) applications are those that allow the driven device to run faster than the driver. An excellent example of this is a bicycle freewheel mechanism. The rider can pedal to make the rear wheel turn, but the wheel does not make the pedals turn, allowing the rider to coast downhill without pedaling. In industry, a typical application using an overrunning clutch is one in which two drivers (for example, an engine and an electric motor) run a machine (see Figure 11.21). Either driver could run the machine without the other. Another industrial application is one in which some machinery can speed up and damage the motor by forcing it to run too fast. An overrunning clutch allows the machine to run faster than the motor.

Indexing Applications

When an overrunning clutch is used for indexing applications, reciprocating motion is transformed to intermittent rotating motion in one direction only. An industrial example is a pneumatic cylinder used to index a disc with

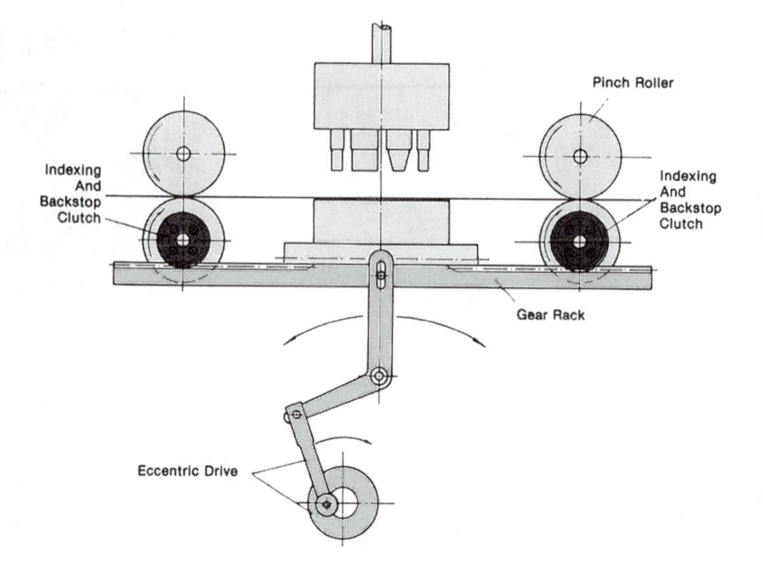

FIGURE 11.22
Indexing application.

blanks to be stamped in a press. Each stroke of the cylinder rotates the disc loaded with parts to be stamped one-half turn. While one part is being stamped, an operator can remove a finished part and insert another blank to be stamped (see Figure 11.22).

Backstop Applications

In backstop applications (sometimes called *holdbacks*), the function of the overrunning clutch is to prevent any rotation in the reverse direction. This may be considered to be a braking application. The housing of the overrunning clutch is permanently locked and mounted to the machine frame. With forward rotation, nothing happens. However, if reverse rotation is attempted, the clutch locks the shaft instantly and acts as a brake, preventing any reverse operation. Usually, there is no problem with stored energy because the clutch does not have to deal with any momentum from the shaft and machinery. However, because the braking action is so quick, even small dynamic loads can possibly cause machine damage. For this reason, always select the backstop for the absolute worst-case condition. Industrial centrifugal pumps sometimes use backstops to prevent reverse rotation when the motor is stopped and the pumpage may tend to run back through the pump. Figure 11.23 shows a backstop used with a screw pump system. Another common industrial application for backstops is with inclined conveyors. In this case, the function of the backstop is to keep the conveyor from running backward if a power failure occurs. Without a backstop, an inclined conveyor loaded with gravel can run backward during a power failure, damaging equipment and endangering people. Many industrial gearboxes have a built-in integral backstop ordered especially for this purpose.

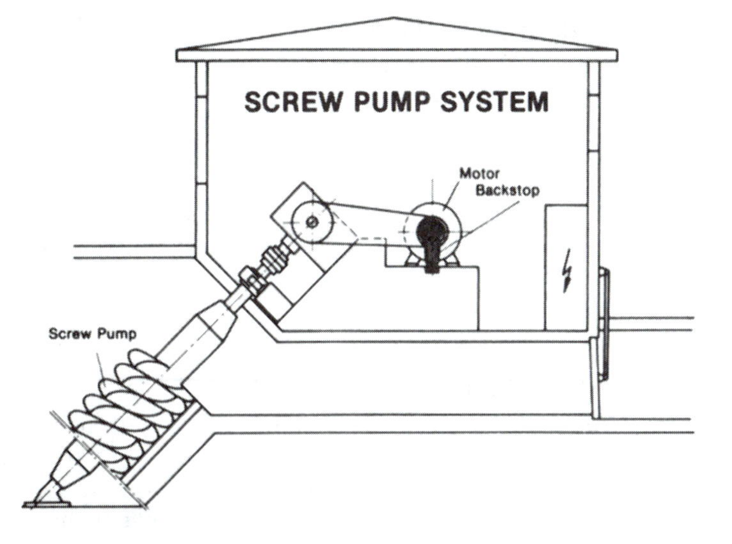

FIGURE 11.23
Backstop application.

Overrunning Clutch Configurations

Overrunning clutches are available in different types of configurations: *pawl-and-ratchet*, *roller*, *sprag* (or *cam*), and *wrap-spring*.

One of the first mechanical overrunning clutch designs is called the *pawl-and-ratchet* clutch (see Figure 11.24). The pawl is gravity-operated or spring-loaded. The ratchet is a specially cut gear that allows the pawl to slide over the teeth if it is rotated forward and locks the ratchet if it is reversed. A similar mechanism is used in older mechanical clocks; the pawl-and-ratchet creates the characteristic tick-tock sound of these clocks.

The *roller* type of overrunning clutch is shown in Figure 11.25. This spring-loaded ball-and-ramp assembly rotates freely, but the ball or roller runs up the ramp and wedges if reversed, thus effectively locking the clutch assembly. The spring keeps the rolling element engaged with both the housing and the hub. This causes very quick operation with less time for the machine to

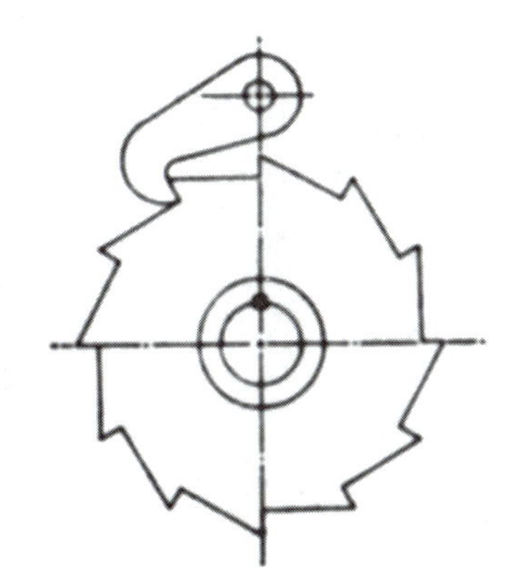

FIGURE 11.24
Pawl-and-ratchet clutch.

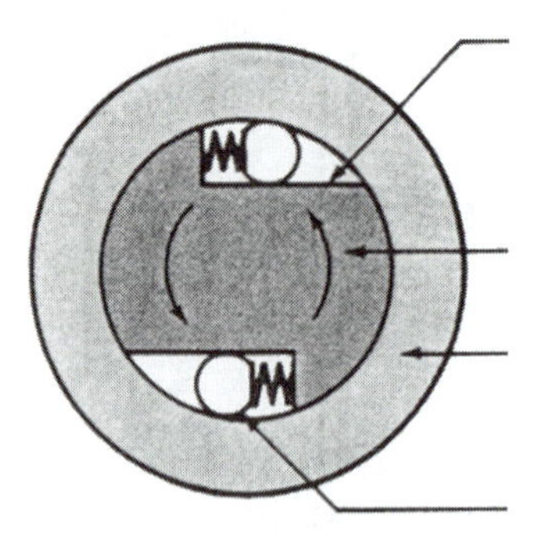

FIGURE 11.25
Ball-and-ramp clutch.

build up dynamic shock. When used for backstop applications, either the housing or the hub may be locked to the machine frame.

The *sprag,* or cam, type of overrunning clutch is shown in Figure 11.26. Sprags are special camlike pieces. A simple version of this type of clutch was used in wooden wagons to prevent them from rolling backward. A sharpened piece of wood or iron was attached to the rear axle and was dragged behind the wagon. When the wagon tried to roll backward, the sharp point dug into the dirt and wedged the wagon in place, effectively braking it. Figure 11.27 shows a modern sprag clutch. It has cylindrical hubs and housings with sprags filling the spaces between them. These sprags are shaped so that they slide when rotated but wedge and lock when reversed. Sprag clutches can be designed to slide from

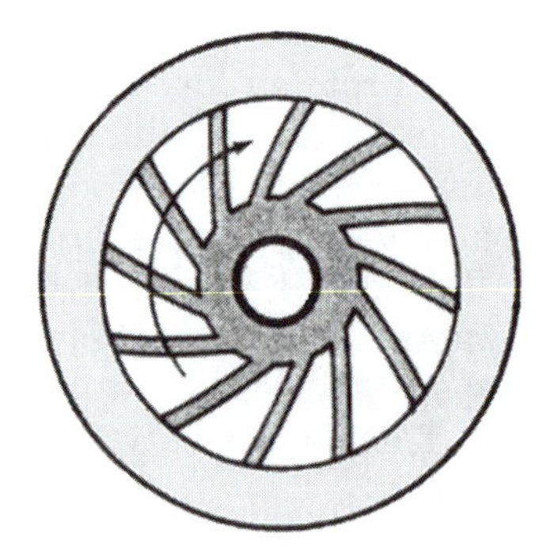

FIGURE 11.26
Sprag overrunning clutch.

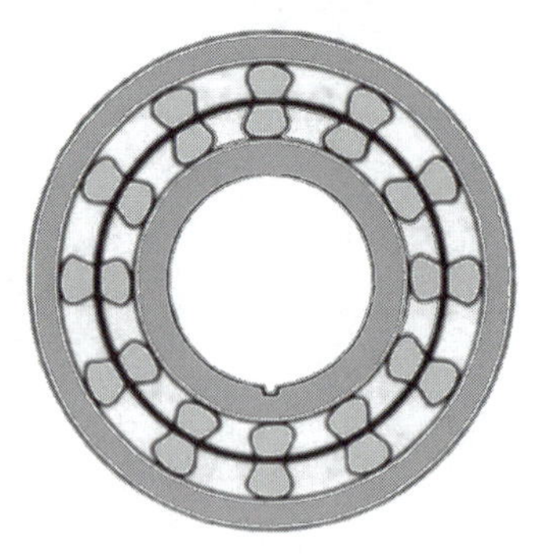

FIGURE 11.27
Modern sprag clutch.

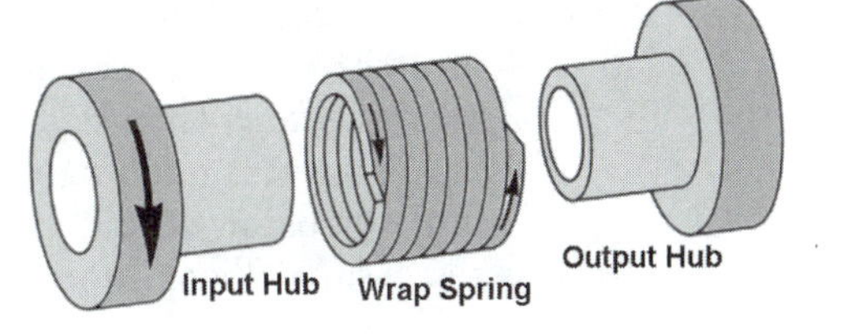

FIGURE 11.28
Wrap-spring overrunning clutch.

either the hub or the housing, depending on the type of application. They can be used in overrunning, indexing, or backstopping applications. However, because they operate by sliding instead of rolling, lubrication becomes a critical consideration. Sprag clutches have an advantage over roller clutches in that sprags can be packed more closely than rollers. More sprags mean that more torque can be transmitted in the same relative space than with a roller clutch.

Wrap-spring overrunning clutches consist of three basic parts: an input hub, an output hub, and a helical wound spring having an inside diameter slightly smaller than the outside diameter of the hubs with which it mates (see Figure 11.28). When the spring is rotated in one direction, it tightens on the hub and locks the clutch. Increasing torque causes the spring to get even tighter. When the spring is rotated in the opposite direction, it loosens and slips on the mating hub.

Wrap-spring clutches are simple, economical, and reliable. They are most often used in light-duty, slower-speed applications. Copy machines, packaging machines, and light conveyors are some of the many applications that depend on this type of clutch.

11-7 MAINTAINING AND TROUBLESHOOTING CLUTCHES AND BRAKES

We have discussed many different types of clutches and brakes, and each type has its own advantages and disadvantages. Each different type we have discussed requires somewhat different maintenance and troubleshooting procedures. The information in this section is rather general. You should always refer to the manufacturer's information for a specific type of clutch or brake.

If properly selected and installed, most clutches and brakes provide long life and trouble-free service. The excessive heat that they produce usually causes most problems with friction clutches. Excessive heat in the friction clutches and brakes is usually created by excessive slip. Excessive slip may be the result of the following:

- *Improper adjustment of clutch or brake:* The clutch or brake may not be fully engaging. Follow the manufacturer's adjustment procedures.
- *Oil or contaminant on friction surfaces:* Clean or replace the surfaces.

- *Worn-out friction components:* Check the components to see if they are within tolerances. Replace them if necessary.
- *Worn linkage or parts used in engaging clutch or brake:* Sometimes adjustment is adequate to compensate for wear. Check for obstruction and corrosion on moving parts. Clean or replace the parts as necessary. Check lubrication, and relubricate if required.
- *Too much torque:* This may be because an increased load exceeds design capacity or because of poor initial selection of a clutch or brake. Check the machine to determine if the increased load is temporary or permanent. Repairing or servicing a machine may reduce torque to acceptable levels. If not, the clutch or brake should be replaced with one designed for the increased torque loads required.
- *High-frequency cycling or high-inertia loads:* Generally, these cases of excessive heat can be solved only by changing to a clutch or brake with greater heat-dissipation ability. Sometimes a fan or blower may be used to increase air flow, thus cooling equipment. Shortening the slipping time during start-up can also reduce heat. Less slippage means less heat, but make sure that engagement is not so sudden that severe shock loads are created in the machine. Engaging clutches under the lightest possible start-up loads is always recommended.

With any type of clutch or brake, the following are generally recommended:

- The clutch or brake should always be the correct size for the application.
- Heat dissipation should always be adequate to ensure long life and low maintenance.
- Lubrication, if required, should be done on a periodic, regular schedule.
- Components should be checked regularly for adjustment and wear.
- Clutches and brakes should be kept clean and free from debris whenever possible.

QUESTIONS

1. What is the function of a clutch?
2. What are the two main types of early clutches?
3. What is the main advantage of the friction clutch over the positive-contact clutch?
4. The effectiveness of the friction clutch depends on several factors. Name three of them.
5. Friction clutches and brakes may be either a ____________ type or ____________ type.
6. What is the advantage of a multiple-disc clutch?
7. What is the most common type of fluid clutch?

8. The ______________ ______________ clutch is an economical solution for applications needing soft start and overload protection. It is used quite extensively in industry because of its simplicity and efficiency.
9. ______________ clutches and brakes are used mainly for starting or stopping and are not suited for continuous slip operation.
10. What is the flow charge of a dry fluid clutch?
11. This type of clutch uses a noncontacting, nonferrous rotor (usually copper or aluminum) sandwiched between two magnetic disks mounted in a housing. What is it called?
12. Unlike the ______________ ______________ clutch, the hysteresis clutch input and output synchronizes as long as its rated torque value is not exceeded.
13. What is an overrunning clutch?
14. How does a wrap-spring overrunning clutch work?
15. Most problems with friction-type clutches usually can be identified by the excessive ______________ that the clutch produces.

CHAPTER 12

Rigging

12-1 INTRODUCTION

Rigging has had a rich and widely varied history. Early people probably first used vines to tie things together. They discovered that weaving the vines together made them stronger and more flexible. Ropes were used to secure packs on the backs of camels in caravans. Early sailors used rope and timbers to build ships and to rig masts and sails. The Egyptian pyramids were probably built by using log rollers and ropes to move huge blocks of stone into place. Today, modern industry still has to move and install large machines (see Figure 12.1). In some cases, modern equipment makes the job easy;

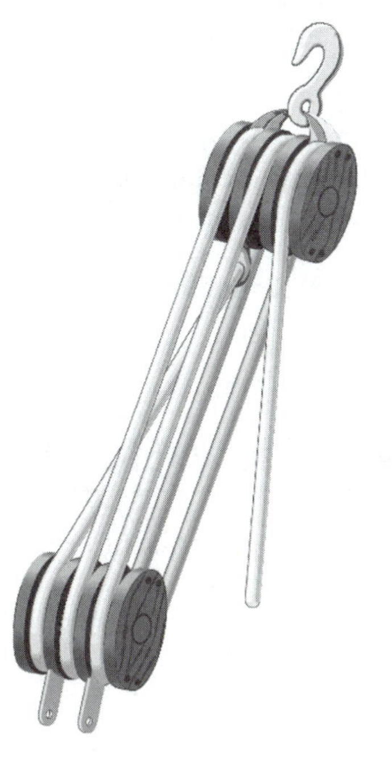

FIGURE 12.1
Block and tackle.

however, in other cases, ancient methods work as well or better than anything designed in the twentieth century. This chapter covers the different methods and techniques of rigging. The rigging practices presented are basic and generalized and may not fit all situations. Because this information is intended for industrial technicians, it focuses on basic theory, hardware identification, and usage, with an emphasis on safety.

Most people who are hurt while rigging or moving equipment are injured because of their own negligence. The majority have been trained and are aware that they are violating safety procedures but do so anyway. Some of the common reasons for accidents are the following:

- *In a hurry to finish*: It may be close to quitting time, or there is outside pressure to finish quickly. Always take time to be safe. Gambling for 10 minutes now against a lifetime of suffering is not worth it.
- *Too lazy to get the proper equipment*: Always take time to get proper personal protective equipment and the correct tools to accomplish the job.
- *Fatigue or inattention*: If fatigue becomes a safety concern, do not get into a situation that can cause injury to you or others. Inattention or carelessness can be fatal. Stay alert. Watch out for hazards and potential accidents. Always be aware of your surroundings for self-protection. Avoid potential injury areas when moving or lifting equipment.

Many accidents occur during routine maintenance because machinery has not been locked out properly. It is very dangerous to work around a machine that may start unexpectedly.

Many tons of material and machines are moved annually. To lift and move this material and machines, you should follow these precautions:

- Know how to do the job correctly and safely.
- Always put safety first. Be alert. Do not take unsafe shortcuts.
- Always use the proper personal safety gear and the correct tools and equipment to get the job done without injury to the equipment or personnel.

12-2 ESTIMATING WEIGHT AND CENTER OF GRAVITY

Before attempting to lift or move material or machinery, you should know its approximate weight and center of gravity. This ensures that the load matches the rigging equipment under the conditions given for lifting. Knowing the center of gravity determines where the load must be secured to keep it upright and stable. Sometimes the weight and lifting points are given, but they usually must be computed for moving the load safely.

To estimate the load weight, you should use the linear and cross-section measurements to compute the volume. If a solid object is the load, the weight can be computed by using the volume and weight per square foot.

Table 12.1 gives useful formulas for finding the area of different geometric objects. This should be a review for you. If it is not, you can find more detailed information for computing area and volume in most math books.

Table 12.2 gives some basic formulas for volume of common geometric shapes. Again, this text assumes you have worked with this type of calculation. For more detailed information, review a good math textbook.

Table 12.3 gives some examples of weights of a unit of volume for certain materials. Consult a good engineering handbook for a complete listing of many different materials. Using this information, we now do a sample problem.

EXAMPLE

Given: A steel plate 1/2 in. thick is 4 ft wide and 8 ft long (see Figure 12.2).

Find: The weight of the steel plate.

SOLUTION

From Table 12.2, the formula for volume of a rectangular solid is

$$\text{Volume} = \text{length} \times \text{width} \times \text{height}$$
$$= 8\text{ ft} \times 4\text{ ft} \times 0.5\text{ in.}$$

Notice that the length and width are in feet and the height is in inches. All the units must be the same. Because the values in Table 12.3 are in cubic feet, you should change inches to feet. Dividing 0.5 in. by 12 converts inches to feet.

$$0.5\text{ in.} \times \frac{1\text{ ft}}{12\text{ in.}} = \text{number of feet}$$

Therefore,

$$\frac{0.5\text{ ft}}{12} = 0.0417\text{ ft}$$

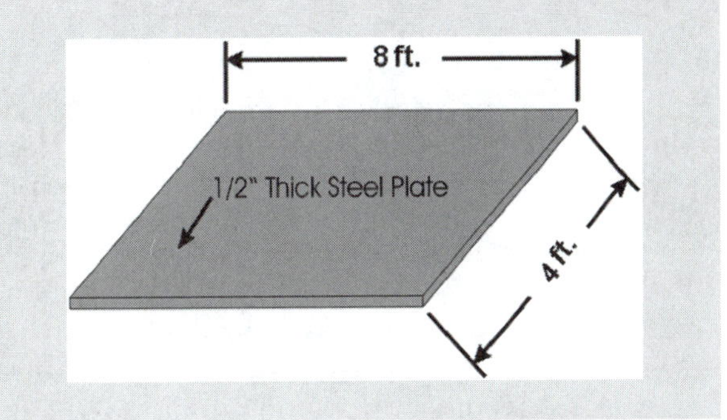

FIGURE 12.2
Find the weight of the plate.

(continued on page 274)

TABLE 12.1
Formulas for finding areas of different geometrical shapes.

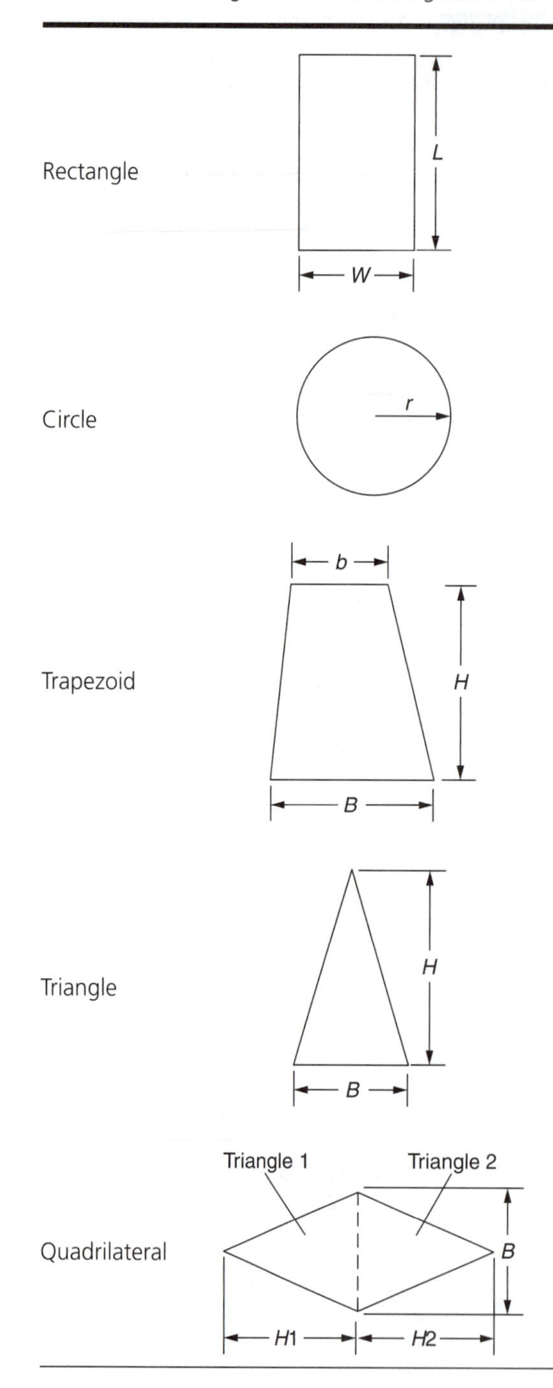

Area = length × width. The area of a rectangle is equal to its length times its width.

$$\text{Area} = \text{length} \times \text{width}$$
$$A = L \times W$$

$A = \pi r^2$ (Area = π (3.14) × radius squared.) The area of a circle is equal to pi times the radius of the circle, squared.

$$\text{Area} = \pi\ (3.14) \times \text{radius squared}$$
$$A = \pi \times r^2$$

The area of a trapezoid is equal to its height multiplied by the average of the bases. Find the average of the bases by adding the two widths and dividing by 2.

$$A = \frac{1}{2}(B + b)H$$

The area of a triangle is equal to its base times its height divided by two.

$$A = \frac{BH}{2}$$

To find the area of a quadrilateral, you use the formula for finding the area of a triangle. Find the area for each half of the quadrilateral (two triangles). Add them together.

TABLE 12.2
Finding volume.

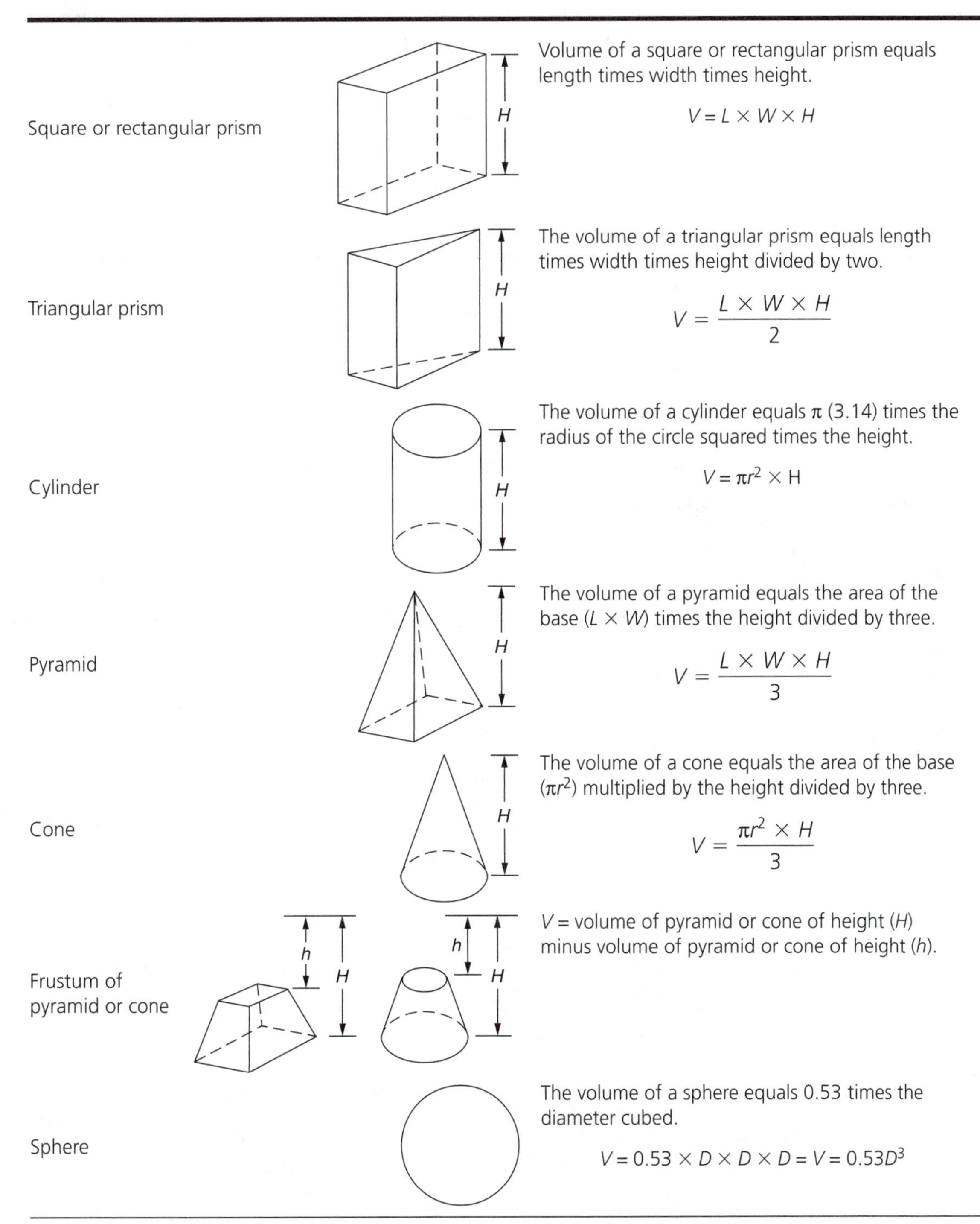

Shape	Description and formula
Square or rectangular prism	Volume of a square or rectangular prism equals length times width times height. $V = L \times W \times H$
Triangular prism	The volume of a triangular prism equals length times width times height divided by two. $V = \dfrac{L \times W \times H}{2}$
Cylinder	The volume of a cylinder equals π (3.14) times the radius of the circle squared times the height. $V = \pi r^2 \times \mathrm{H}$
Pyramid	The volume of a pyramid equals the area of the base ($L \times W$) times the height divided by three. $V = \dfrac{L \times W \times H}{3}$
Cone	The volume of a cone equals the area of the base (πr^2) multiplied by the height divided by three. $V = \dfrac{\pi r^2 \times H}{3}$
Frustum of pyramid or cone	V = volume of pyramid or cone of height (H) minus volume of pyramid or cone of height (h).
Sphere	The volume of a sphere equals 0.53 times the diameter cubed. $V = 0.53 \times D \times D \times D = V = 0.53D^3$

TABLE 12.3
Weights of common materials.

Metal	Weight (lb/ft³)	Material	Weight (lb/ft³)
Aluminum	166	Bluestone	160
Antimony	418	Brick, pressed	150
Bismuth	613	Brick, common	125
Brass, cast	504	Cement, Portland (packed)	100–120
Brass, rolled	523	Cement, Portland (loose)	70–90
Copper, cast	550	Cement, slag (packed)	80–100
Copper, rolled	555	Cement, slag (loose)	55–75
Gold, 24-carat	1,204	Chalk	156
Iron, cast	450	Charcoal	15–34
Iron, wrought	480	Cinder concrete	110
Lead, commercial	712	Clay, ordinary	120–150
Mercury, 60E F	846	Coal, hard, solid	93.5
Silver	655	Coal, hard, broken	54
Steel	490	Coal, soft, solid	84
Tin, cast	458	Coal, soft, broken	54
Zinc	437	Coke, loose	23–32
Wood	**Weight (lb/ft³)**	Concrete, or stone	140–155
Ash	35	Earth, rammed	90–100
Beech	37	Granite	165–170
Birch	40	Gravel	117–125
Cedar	22	Lime, quick (ground loose)	53
Cherry	30	Limestone	170
Chestnut	26	Marble	164
Cork	15	Plaster of Paris (cast)	80
Cypress	27	Sand	90–106
Ebony	71	Sandstone	151
Elm	30	Shale	162
Fir, balsam	22	Shale	160–180
Hemlock	31	Terra cotta	110
		Trap rock	170

Notice how the inches cancel out, leaving a simple division to convert to feet. To complete our example:

$$\begin{aligned}\text{Volume} &= \text{length} \times \text{width} \times \text{height}\\ &= 8\text{ ft} \times 4\text{ ft} \times 0.0417\text{ ft}\\ &= 1.33\text{ ft}^3\end{aligned}$$

From Table 12.3, we find that steel weighs 490 lb per cubic foot. Multiplying the number of cubic feet times the weight per cubic foot gives

$$\begin{aligned}\text{Weight} &= \text{number of cubic feet} \times \text{weight per cubic foot}\\ &= 1.33\ \text{ft}^3 \times 490\ \text{lb/ft}^3\\ &= 653.33\ \text{lb}\end{aligned}$$

The total weight of the steel plate is 653.33 lb.

Unfortunately, most objects are more complicated than a simple steel plate. The following is another example:

EXAMPLE

Given: A steel tank is 4 ft long, 3 ft wide, 2 ft high, and 1 in. thick (see Figure 12.3).

Find: Estimate the weight of the tank.

SOLUTION

Find the weight of each side and add all the weights together:

$$\begin{aligned}\text{Volume} &= \text{length} \times \text{width} \times \text{height}\\ \text{Volume of bottom} &= 4\ \text{ft} \times 3\ \text{ft} \times 1/12\ \text{ft}\\ &= 1.0\ \text{ft}^3\\ \text{Volume of short side} &= 3\ \text{ft} \times 2\ \text{ft} \times 1/12\ \text{ft}\\ &= 0.5\ \text{ft}^3\end{aligned}$$

The total volume is

$$\begin{aligned}\text{Total volume} &= 2(1.0 + 0.5 + 0.67)\\ &= 2(2.17)\\ &= 4.34\ \text{ft}^3\end{aligned}$$

From Table 12.3, steel weighs 490 lb per cubic foot. Therefore,

$$\begin{aligned}\text{Weight} &= 4.34\ \text{ft}^3 \times 490\ \text{lb/ft}^3\\ &= 2126.6\ \text{lb}\end{aligned}$$

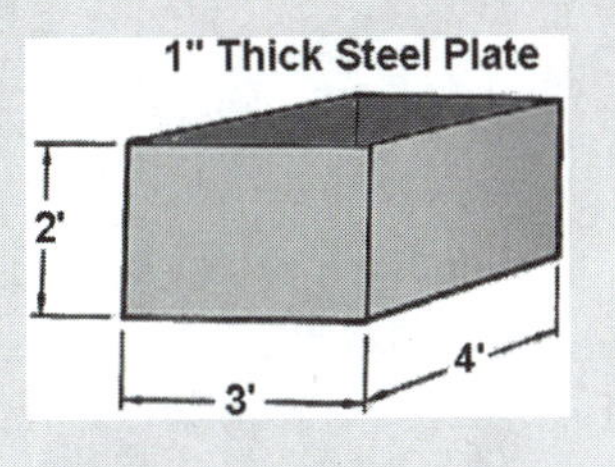

FIGURE 12.3
What is the weight of the tank?

Very complicated objects can be broken down into components so that the total weight can easily be calculated.

After you have estimated the weight of an object, you should add a *safety factor.* This provides a cushion, or margin of safety, for small miscalculations; 20% is a customary margin of safety. In our first example, we found the weight to be 653.33 lb. If we find 20% of 653.33, we get 130.67 lb. Adding this to our original estimate of 653.33 lb we get 783 lb. Rigging should be selected based on a total weight of 783 lb.

When lifting any object, the stability of the load is very important. To ensure that the weight of the load is equally distributed, you must know the center of gravity of the object being lifted. The **center of gravity** of an object is the point at which the suspended object will be balanced. The weight of the object is concentrated at that point. For a stable lift, the center of gravity of an object should be positioned directly below the lift point (see Figure 12.4). A suspended object will always tilt or move so that the center of gravity is below the lift point. An incorrectly attached load may shift radically and be dangerous if you do not attach it below the center of gravity. Figure 12.5 shows how an object shifts and tilts to the center of gravity.

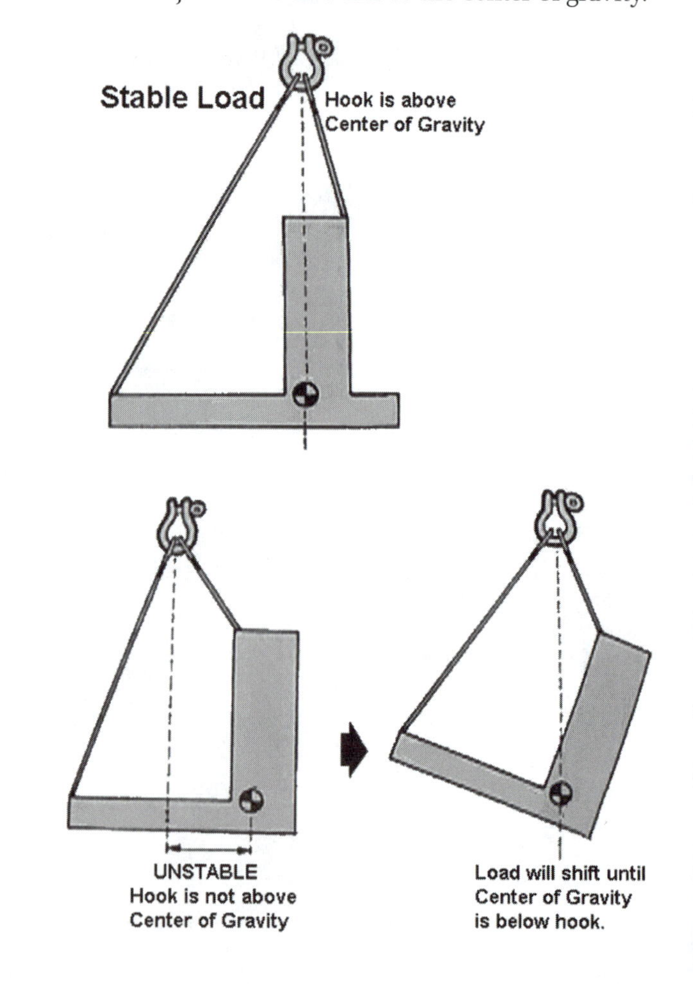

FIGURE 12.4
Stable load.

FIGURE 12.5
Load will shift due to center of gravity.

Several methods are used to determine the center of gravity of an object. If the object is of uniform size and weight, the center of gravity is located in the center of that object. For example, the center of gravity of a round ball is exactly at its physical center. The same is true of a cube. However, you sometimes must lift oddly shaped objects whose center of gravity cannot be accurately and easily found. To do this, you should first try to estimate the center of gravity and then attach the slings to the object accordingly. Slowly lift the object a few inches. If the object attempts to tilt, you should reevaluate your estimate of the center of gravity. The correct center of gravity will be toward the side that tilts *down*. Move the hoisting point toward the correct center of gravity. Carefully lift the object again and continue to make corrections until the load lifts evenly.

Sometimes it is necessary to rotate an object to mount it on a machine. Figure 12.6 shows the proper method for rotating an object using a single hoist. Notice that the hoisting point is placed on the corner of the object. As the hoist is moved, the object starts to tilt toward its center of gravity. Once the hoist is moved *past* the center of gravity, the object can be lowered gently into place. Using two hoists gives much more versatility and control. Figure 12.7 illustrates how a part can be picked up from a pallet and accurately positioned for

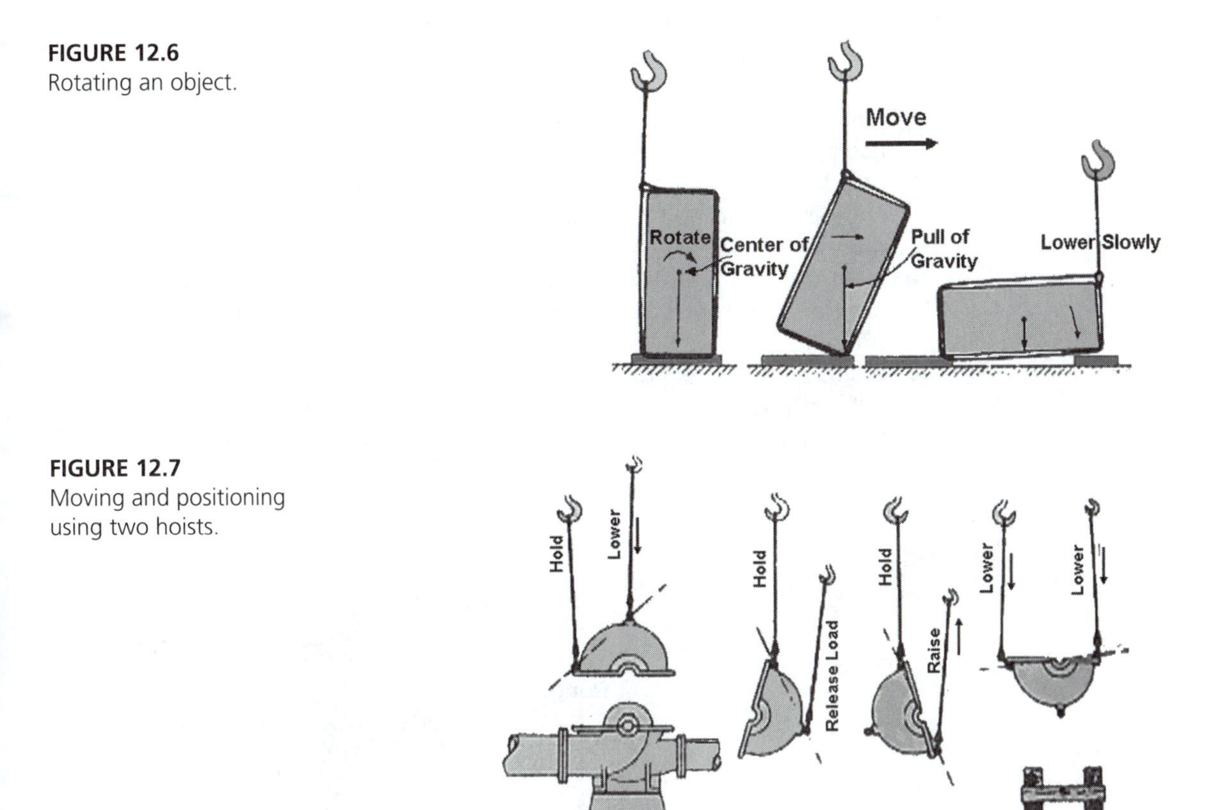

FIGURE 12.6
Rotating an object.

FIGURE 12.7
Moving and positioning using two hoists.

assembly. Notice that one hoist must support the entire load at the center point of the move. Be careful to select a hoist strong enough to support all the weight required. This operation could be accomplished with one hoist, but two make the job much easier for the rigger.

12-3 FIBER ROPE

Fiber ropes are made up of *fibers, yarns,* and *strands* (see Figure 12.8). A **fiber** is the smallest component of a rope. A bundle of fibers twisted together to the right forms **yarn.** Several yarns twisted together to the left form a **strand.** Three to six strands twisted together to the right will form a rope. The strands of a rope may be *hard-laid* (twisted tightly) or *soft-laid* (twisted loosely). **Hard-laid rope** is stiffer and has better resistance to abrasion and moisture penetration. **Soft-laid rope** is more flexible and has better tolerance to shock loading because it stretches and rebounds rather than breaking.

Fiber rope may be made from natural fibers or synthetic fibers. Natural-fiber ropes are usually no longer used in rigging because of their lower strength and rapid deterioration in adverse environments. The following is a list of some of the more common natural-fiber ropes:

- *Number one manila:* This is the *only* natural-fiber rope that should be used for rigging. Manila is made from fibers from a banana plant. It is pale yellow with a pearly luster and feels smooth and waxy. A special marker such as color-inlaid fibers and bands identifies number one manila.
- *Hemp:* Hemp fibers come from a hemp plant. It is an extremely strong natural-fiber rope but deteriorates very quickly when wet. It has a rough texture.
- *Sisal:* Sisal fibers come from a grasslike plant. It has only about 60% of the capacity of manila. It is lighter in color than manila, usually with a green tint. Like hemp, the surface is rough and scratchy.

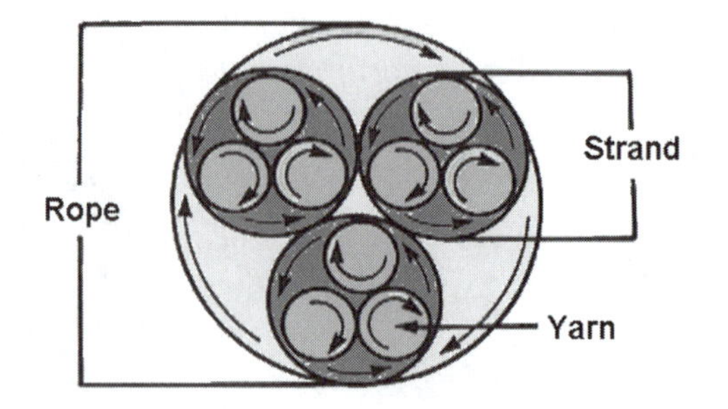

FIGURE 12.8
Cross section of rope.

- *Cotton:* Cotton fibers come from the cotton plant. It has only about 50% of the capacity of manila but is very flexible. It is white in color and has a smooth surface.

Synthetic-fiber ropes are much stronger than natural-fiber ropes because synthetic fibers are much longer than natural fibers. Synthetic fibers are resistant to rot and mildew and have a greater useful life than natural fibers. The following is a list of some of the more common synthetic-fiber ropes:

- *Nylon*: Nylon is one of the strongest ropes commonly used. It is 2 1/2 to 3 times stronger than number one manila. Nylon has good abrasion resistance, works well when wet or frozen, and resists most solvents and alkalis. Avoid using with paint products containing linseed oil and any acid-based chemicals. Ultraviolet light from sunlight weakens nylon fibers and over a long period of time destroys the rope. Nylon rope stretches when loaded. This can be an advantage because it absorbs shock loads without breaking, but it can be a hazard because of the snap-back quality. Nylon rope that snaps back under great tension can be extremely dangerous to people and equipment. Always ensure the weight capacity is adequate and the load is well secured.
- *Polyester* (*Terylene*): This synthetic rope is not as strong as nylon, but it still has about twice the strength of number one manila. It stretches far less than nylon but slightly more than manila. It has excellent resistance to chemicals, abrasion, and wet environments. Because it is similar to nylon, you should use it in applications where nylon would be used but excessive stretch is undesirable.
- *Polypropylene®*: Polypropylene is the lightest and most economical of all of the synthetic ropes discussed. It has 1 1/2 times the strength of number one manila. It floats on water and has good resistance to rot and to many chemicals. It stretches less than nylon or polyester.

All fiber rope should be inspected periodically for damage and wear. Any rope is only as strong as its weakest point. Make sure you check *all* the rope. The *outside* of the rope should be checked for the following:

- *Broken fibers and abrasion:* If 5% or more of the fibers are broken or worn through, the rope should be discarded.
- *Discoloration or stains:* This indicates exposure to paint, chemicals, solvents, or other contaminants that could weaken the rope.
- *Any indication of mold or mildew:* This is especially important for natural-fiber rope, because its strength drops greatly once it begins to deteriorate.
- *Change in flexibility:* If the rope becomes hard and brittle, it should be discarded.

The *inside* of the rope should be checked by untwisting the strands and yarns. Do not untwist the rope to the extent that the rope strand is damaged. Check inside the rope for the following:

- *Excessive broken fibers:* This indicates that the rope has been overstretched or overloaded and should be discarded.
- *Powder or rope dust:* If the interior of the rope shows evidence of breaking down and is not clean, it should be discarded. Rope dust particles are a sign of severe internal wear.
- *Internal discoloration:* This is evidence of dirt or chemical penetration. Discard the rope.

In general, when using rope you should keep the rope as clean and dry as possible. It should not be dragged over dirty or gritty surfaces, because this may possibly cause external or internal abrasion. Avoid sharp edges by protecting the rope with padding or protectors. Avoid kinking the rope and then putting it under load. Fiber rope should not be used in areas with moisture, chemicals, or excessive heat unless it is designed for the specific application. Store rope in a clean, well-ventilated area. Store ropes on wood-grate shelves or hang coiled ropes on large-diameter pegs.

Selecting and tying the proper knot for your application is an important part of rigging. Knots may be used to connect two ropes together, attach them to a load, secure a load, haul a load, and for other jobs. A good knot will not slip or come untied and can be untied easily when you're finished. There are many different types of knots for different applications. There are many good books dedicated to understanding different types of knots if you would like more information. Some basic knots are as follows:

- *Square knot:* This is typically a *binding knot,* one used for tying packages or bundles of materials together (see Figure 12.9). It should not be used for joining ropes.
- *Bowline knot:* The bowline is a *loop* knot (see Figure 12.10). It creates a loop of rope that will not slip or tighten around the object secured. It is usually used for mooring, hitching, and lifting loads.
- *Sheet bend knot:* This is a popular *bend* knot used to tie two ropes together (see Figure 12.11). This knot can be used to join two ropes of different diameters as well as two ropes of the same size.

FIGURE 12.9
Square knot.

FIGURE 12.10
Bowline knot.

FIGURE 12.11
Sheet bend knot.

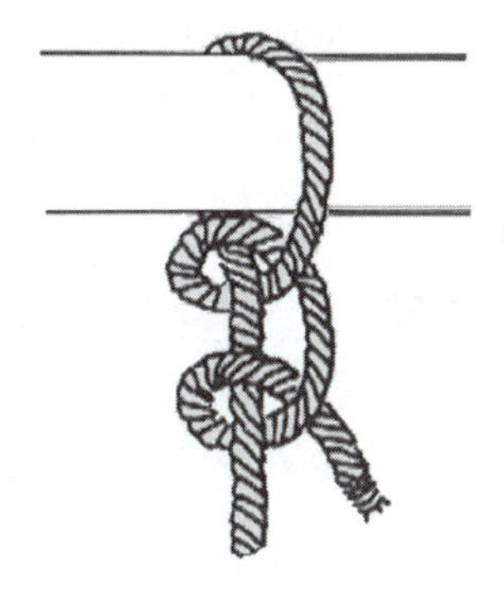

FIGURE 12.12
Half-hitch knot.

- *Half-hitch knot:* A *half hitch* is a temporary knot that is easy to untie. The half-hitch knot should not usually be used alone because it will untie too easily for most applications. Figure 12.12 shows *two* half hitches.

The knots just described are only a few of the many knots useful to the rigger. If you work with ropes often, you should learn more about the art of knot tying.

12-4 WIRE ROPE

Wire rope or cable uses metal wires wound into strands. The strands are then twisted around a fiber or metal core. The fiber may be a natural or synthetic fiber and is used to absorb shock and help cushion the strands. The fiber core is typically impregnated with a lubricant that is released when the wire rope is being used to help lubricate the wire strands. Wire ropes with wire cores are designated *IWRC,* which is the abbreviation for independent wire rope core. This wire core is usually just another strand. It has advantages over the fiber core in that it increases the overall wire rope capacity and helps resist crushing. It can also be used for high-heat applications where fiber core may be damaged.

Wire rope is specified by the number of strands and the type of wire and core material. Wires in ordinary wire rope are typically of the same size. Figure 12.13, however, shows a 6 × 7 wire rope, which means that it has six strands with seven wires in each strand. This wire rope is not very flexible. In general, the smaller the number of wires per strand and the more strands in the wire rope, the more flexible it is. Wire rope used with sheaves and drums should be flexible enough to wrap easily. Standard wire rope strands have 3 to 91 wires. More common wire ropes contain 7, 19, or 37 wires per strand. Figure 12.14 shows a 6 × 19 seale-type wire rope. Each strand has an outer ring of large wires, an inner ring of smaller wires, and a large center wire. This type of construction makes seale-type wire rope very flexible.

The direction in which the wires are twisted compared to the direction in which the strands are twisted is called the wire rope *lay*. Figure 12.15 shows a *right regular lay* wire rope. Notice that the wires in the strand are twisted

FIGURE 12.13
A 6 × 7 wire rope.

FIGURE 12.14
6 × 19 seale-type rope.

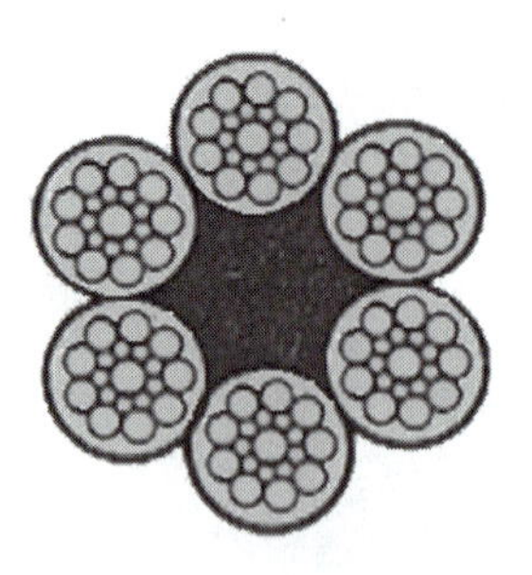

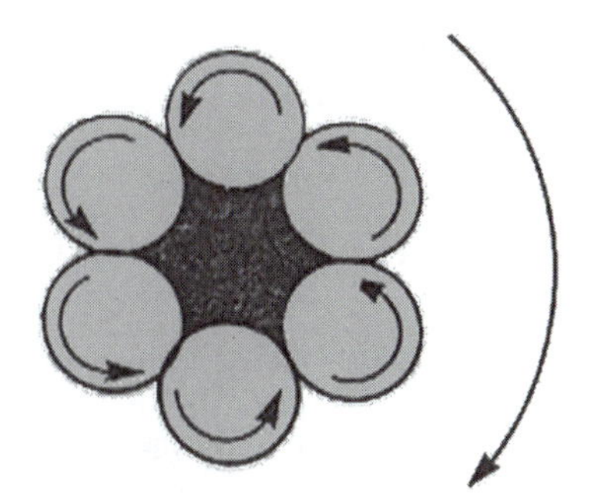

FIGURE 12.15
Right lay wire rope.

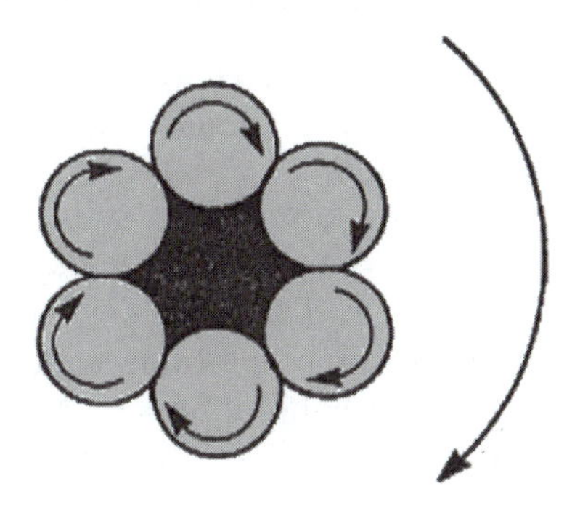

FIGURE 12.16
Right lang lay wire rope.

counterclockwise and the strand is twisted clockwise. This is the standard configuration of most wire ropes. The *lang* lay is an improved version of wire rope lay. Figure 12.16 shows a *right lang lay* wire rope. Both wires and strands are twisted clockwise. With lang lay ropes, more strand surface is exposed, making it more resistant to abrasion and more flexible. However, in some applications, lang lay ropes have a tendency to untwist more than regular lay wire ropes do.

The size of a wire rope is measured at its maximum outside diameter. The correct way to measure a wire rope is from the *top* of one strand to the *bottom* of the opposite strand. Do not measure the "valleys" between strands. Rotate your caliper around the diameter to obtain the maximum reading. New wire ropes will measure slightly higher in diameter.

Good safety procedures require that wire ropes be inspected periodically for wear and failure. The frequency and intensity of the inspection depend on the following:

- *Federal, state, and local regulations:* Depending on the application, laws may govern inspection of wire rope. An example is wire rope used in elevators.
- *Conditions of application:* The environment in which the wire rope operates may demand more frequent inspections.
- *Duty requirements:* Wire ropes that run 24 h a day require more frequent inspections than those running only 2 h a day.

- *Importance of operation and personnel safety:* A critical application or one involving possible hazards to people should be inspected more frequently.

Inspection of wire ropes should include the following:

- *Broken wires:* The number of broken wires that can be tolerated depends on application and any laws or regulations that may apply. You should check OSHA regulations and be familiar with ANSI standards that may apply to your application.
- *Wear or abrasion:* Wire rope should not be used if the outside individual wires show about 30% wear. If outside individual wires have only 70% or less of their original diameter, replace the wire rope.
- *Wire rope abuse:* Look for kinks, displaced strands, and crushed strands.
- *Corrosion or lack of lubrication.* If wire ropes are rusty, replace them. Rusty wires may break suddenly and without warning.

Wire rope comes prelubricated from the manufacturer. This will provide the wire rope protection for a reasonable period of time if it is stored under proper conditions. However, this lubrication is not designed to last for the life of the wire rope. Wire ropes should be relubricated periodically after they have been in operation. Always follow the manufacturer's instructions for the correct type of lubricant and method of application. Depending on the application and type of wire rope used, the lubricant ranges from a very thin oil to a thick grease. Before lubricating wire ropes, inspect and clean them thoroughly.

12-5 LINK CHAINS

Link chains have been used for rigging for hundreds of years. Chains withstand rough handling and do not kink. They are not affected by dirt and are much more resistant to abrasion and corrosion than wire ropes are. However, wire ropes work better with shock loads. Chains have almost no elasticity. If overloaded, they break with little warning.

Chains are available in many different materials and strengths. Only welded alloy chains should be used for lifting. This type of chain is identified by the letter *A* stamped on the links (see Figure 12.17). The size of the chain is determined by measuring the diameter or thickness of the link stock.

You should always make sure that any chain used for rigging has adequate capacity for the load. Don't just guess; make sure by checking the manufacturer's safe working-load limit given for that particular chain. Avoid jerking or shock-loading a chain. Apply loads smoothly and evenly. Straighten any

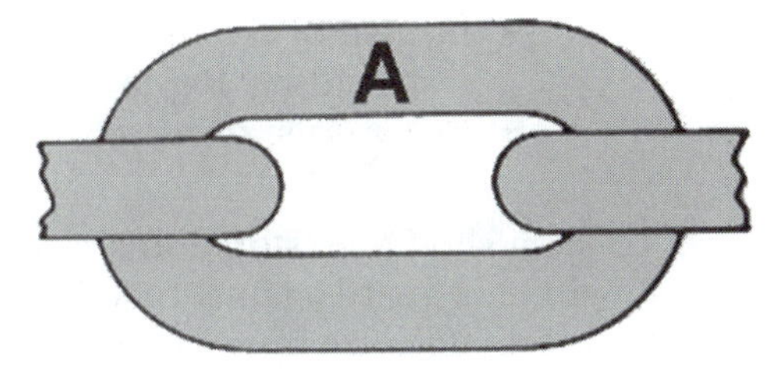

FIGURE 12.17
Alloy chain is marked with the letter A.

twists or kinks in the chain before applying a load; twists and kinks greatly reduce load capacity. Use wood blocks for softeners when lifting a load with sharp, hard corners. Ordinary bolts should not be used to attach or shorten a chain; always use proper chain attachments. Never use repaired links or weld an alloy chain. If the chain is used in environments with temperatures above 500°F (260°C), the safe working-load capacity of the chain should be reduced; always check the manufacturer's specifications.

Chains should be inspected regularly for defects, damage, or wear. You should always clean the chain thoroughly before an inspection. When inspecting the chain you should look for the following:

- *Deformed, bent, or twisted links:* A kinked chain stressed under load will be damaged.
- *Excessive wear:* Wear occurs at link ends where links rub against each other (see Table 12.4). Check the manufacturer's recommendations for maximum allowable wear. Replace any chain that exceeds specifications.
- *Excessive stretch:* Compare the current length of the chain with its new length. Stretch should not exceed 3% of its original length. Replace any excessively stretched chain.
- *Cracks or gouges:* If there is an excessive number, replace the chain.

TABLE 12.4
Maximum allowable wear at any point of chain link.

A Original Chain Material Size	B Minimum Chain Material Size	A Original Chain Material Size	B Minimum Chain Material Size
1/4	13/64	1	13/16
3/8	19/64	1 1/5	25/32
1/2	25/64	1 1/4	1
5/8	31/64	1 3/5	1 3/32
3/4	19/12	1 1/2	1 3/16
7/5	45/64	1 3/4	1 13/32

- *Stretch or separation of the weld at each link:* This is a sign of chain overloading. Replace the chain.
- *Severe rust or corrosion:* Replace the chain.

Chain should be stored in a clean, dry, well-ventilated room. A light coat of oil will help guard against rust. Never store chains on a concrete floor, as they will absorb moisture from the concrete.

12-6 SLINGS

A sling is a length of webbing, fiber rope, wire rope, or chain equipped at each end with a fitting. A sling is used to attach a load to a hook or hoist. The total configuration of a sling or slings and all the attachments connected to the load is called a **hitch.**

Most webbing and fiber-rope slings are made out of synthetic materials such as nylon or polypropylene. They are stronger and last longer than natural fiber. Because of relative softness and flexibility, fiber slings are typically used to handle loads with finished surfaces.

Wire ropes are probably the most common type of slings. They have excellent strength, wear well, and are easy to inspect. Avoid kinking wire rope slings, because kinking reduces the strength and possibly causes sling failure.

Chain slings are abrasion- and dirt-resistant but cannot handle the shock loading sometimes required of a sling. They are relatively heavy and have a smooth, hard surface that doesn't grip as well as wire rope or fiber slings do. Chain slings are typically used to hoist steel or metal beams and pipe. When hoisting heavy loads with hard, sharp corners, always use protective pads or cushions to protect the chain.

There are three basic types of slings:

1. *Single-part sling:* Also called a plain sling, this type uses a single web, rope, or chain, usually with fittings at both ends, to attach to the hoist and load. Figure 12.18 shows a single-part wire rope sling.
2. *Bridle sling:* A bridle sling is two or more single-part slings connected together by a common ring, as shown in Figure 12.19.
3. *Endless sling:* This sling is a continuous loop of webbing or rope. Wear is better distributed because the load contact can be varied, thus extending the sling life. Figure 12.20 shows an endless sling.

FIGURE 12.18
Single-part fiber sling.

FIGURE 12.19
Bridle sling.

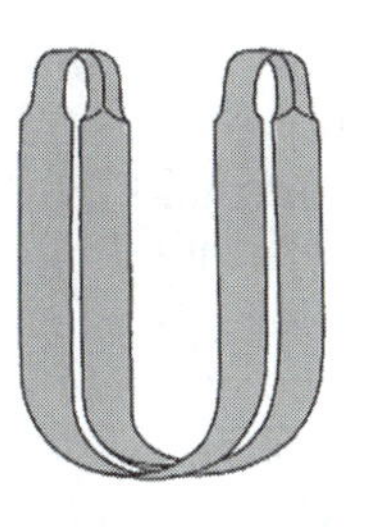

FIGURE 12.20
Endless fiber sling.

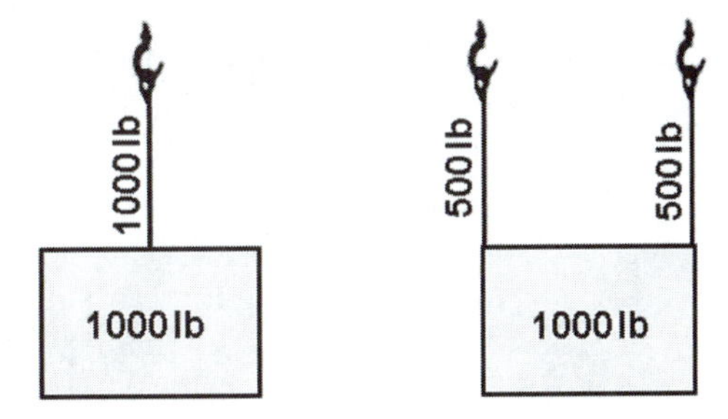

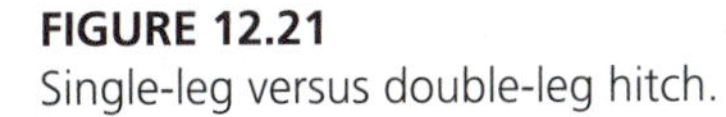

FIGURE 12.21
Single-leg versus double-leg hitch.

To determine the safe working load of a sling, you should always refer to manufacturers' tables. These tables usually give the safe working load for a single, vertical sling. If two vertical slings are used, the safe working load can be doubled (see Figure 12.21). If a three-leg vertical hitch is used, the safe working load of a single sling can be tripled, but *only* if the sling legs are evenly spaced and are the same length. This assures that the load distribution is equal with a uniform object. With a three-leg hitch, the worst-case scenario is when only two of the three legs are supporting the load and the third leg is only balancing but not actually supporting the load.

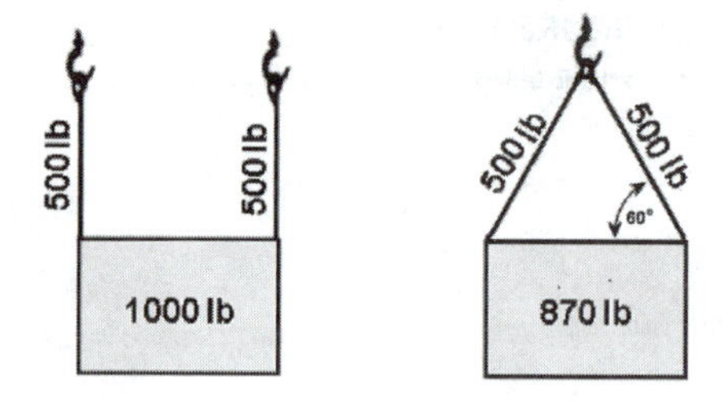

FIGURE 12.22
Straight sling versus angled sling.

Calculating the leg load of a four-leg hitch is not as simple as it appears. It would seem that each of the four legs would carry an equal portion of the load. Because of positioning, leg length, and load variation, however, it is not unusual for only three legs or even only two legs to support the entire weight. The other legs only balance the load. Be aware when using a four-leg sling that OSHA standards *require* that each leg must be sized as if it were a three-leg hitch.

So far, we have discussed the situation where the legs of the sling are all vertical. Sling legs are used more commonly at an angle if two or more legs are involved. When the sling leg is at an angle, the load capacity of each leg must be reduced. For example, Figure 12.22 shows that a leg of a double sling has a safe working load of 1,000 lb in a vertical position. If the legs are positioned to form a 60° angle, each sling leg can support only 87% of its rated safe working load, or 870 lb. If the angle is reduced to 45°, each sling leg can only support 71% of its rated capacity, or 710 lb. Reduce the angle further to 30°, and each sling leg can support only 50% of its safe working load, or 500 lb. The recommended lifting angle for sling legs is 45° or higher. The 60° angle is used quite often and is easy to determine by forming an equilateral triangle (a triangle having all sides equal). In Figure 12.23, notice that the length of each of the three sides of the triangle is the same.

The safe working load of a sling leg used with any different, unknown angle can be calculated as follows:

$$\begin{matrix}\text{Safe working load} \\ \text{at an angle}\end{matrix} = \begin{matrix}\text{safe working load} \\ \text{for a vertical lift}\end{matrix} \times \frac{\text{vertical height of sling}}{\text{length of the sling}}$$

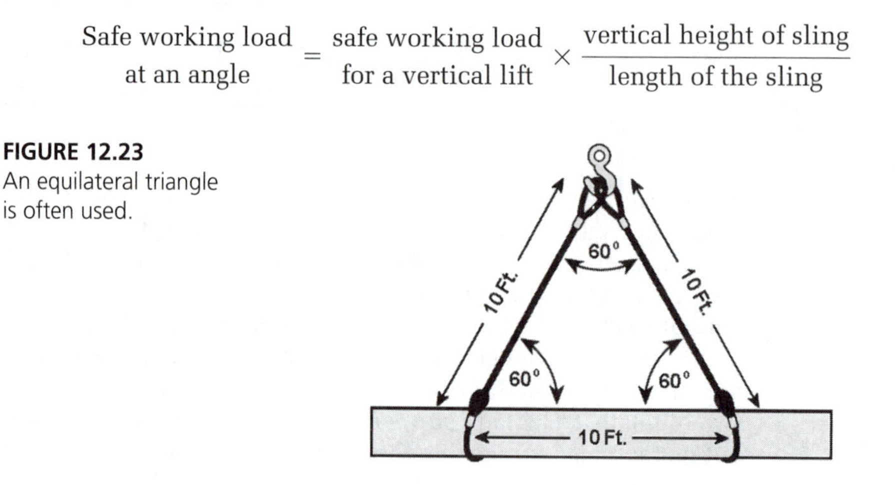

FIGURE 12.23
An equilateral triangle is often used.

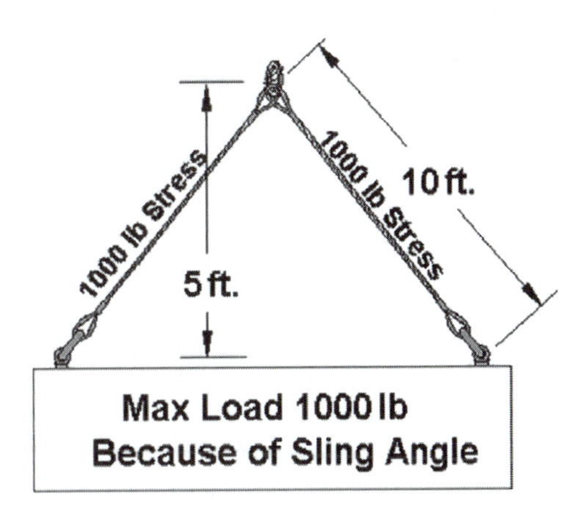

FIGURE 12.24
Maximum load is 1,000 lb because of the sling angle.

Figure 12.24 shows a two-leg hitch with a vertical height of 5 ft and a sling length of 10 ft. The safe working load for a vertical lift is 1,000 lb per leg. Applying our formula yields

$$\text{Safe working load at an angle} = 1{,}000 \text{ lb} \times \frac{5 \text{ ft}}{10 \text{ ft}}$$

$$\text{Safe working load at an angle} = 500 \text{ lb}$$

In our example, instead of each leg supporting 1,000 lb for a vertical lift, the safe working load is only one-half of the normal vertical lift, or 500 lb. The height over the length ratio of 5/10, or 1/2, indicates an angle of 30°. This is found by using the sine function (sin 30° = 1/2). Remember that the angle of lift should never be less than 30°. This means that the ratio of height over length of sling should never be less than 1/2.

Slings are used for lifting and positioning loads. Always ensure that the sling has adequate capacity and safety factors for the job to be performed. Also, determine the center of gravity to lift the load properly and safely. The load must be under control at all times when lifting and positioning; light ropes, called **tag-lines,** are attached to the load to prevent rotation and guide the load during transit and positioning. Inspection of slings for safety is the same as the inspection previously discussed for fiber ropes, wire ropes, or chains. You should always check for proper size, wear, abrasion, and corrosion. Use manufacturer's recommendations as the ultimate authority.

QUESTIONS

1. What is the correct procedure for lifting and moving material and machines?
2. Before you attempt to lift or move material or machinery, you should know its approximate ______________ and ______________.

3. A suspended object always tilts or moves so that the center of gravity is *always* ____________ the lift point.
4. When is the center of gravity located in the center of the object?
5. What is fiber rope made from?
6. What are the disadvantages of natural fiber ropes when used in rigging?
7. What is the only natural fiber rope that should be used for rigging?
8. Why is synthetic fiber rope better than natural fiber rope for rigging?
9. Before using a rope for rigging, you should check the outside of the rope for ____________, ____________, ____________, and ____________.
10. Where should you store rope?
11. What is a 6 × 19 seale-type wire rope?
12. What is meant by wire rope lay?
13. How do you measure the size of a wire rope?
14. What are the four things you should look for when inspecting a wire rope?
15. What are the advantages of chain over wire rope for rigging?
16. Where should you store chain?
17. What is meant by the rigging term *hitch*?
18. What are the three basic types of slings?
19. OSHA standards require that when using a four-leg hitch you must size each leg as if it were a ____________ leg hitch.
20. What are *tag-lines*?

CHAPTER 13

Industrial Pneumatics

13-1 INTRODUCTION TO PNEUMATICS

Pneumatic comes from a Greek word meaning "wind." Air compressors are pumps that convert mechanical energy into pneumatic energy (Figure 13.1). They do this by reducing the volume of air. This compressed air, or pneumatic energy, is a common source of power in industry. Compressed air operates mechanical devices—power tools, conveyors, and other machinery. Compressed air performs such tasks as agitating liquids, promoting chemical changes, and accelerating combustion.

You may describe pneumatic energy as fluid, gas, air, pneumatic fluid, or compressed air. The terms are interchangeable. In this unit, it is called compressed air.

FIGURE 13.1
Air compressor.

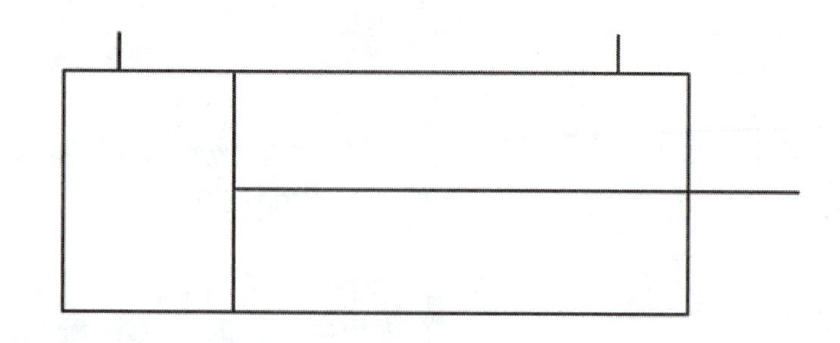

FIGURE 13.2
ANSI symbol.

Pneumatic systems and their components are represented in diagrams of systems by symbols. American National Standards Institute (ANSI) symbols (Figure 13.2) are a standardized way they are represented.

13-2 SAFETY

Working with pneumatic systems requires an awareness of the hazards they pose to people and equipment. Observing proper safety precautions reduces the risks associated with maintaining these systems.

Keep in mind that compressors will compress not only air, but everything in it. For this reason, locate system intakes well away from potentially noxious or explosive fumes. Nevertheless, someone may unknowingly park a running vehicle or set a drum of vaporous material near the intake. If drawn into a compressing system, carbon monoxide or other hazardous vapors could pose a serious threat. Be sure that large-volume compressor units are well ventilated.

Before servicing pneumatic equipment, turn off compressor drive motors. Lock and tag switches in accordance with OSHA requirements (Figure 13.3). Allow all valves, pumps, and cylinders to come to a complete stop. Before working on an individual component, be sure to isolate it from all sources of pressure. Always wear safety goggles when venting a system. The sudden release of compressed air can propel grit, filings, and liquid surprisingly far. When driven by air pressure, even fine dust can seriously damage your eyes.

FIGURE 13.3
Lockout/tagout.

During the depressurizing process, lock and tag each vent in the open position. If the vents are immediately reclosed, a rise in temperature could very quickly repressurize the system. In contrast, a temperature drop could create a vacuum. Some pneumatic components are designed to withstand only pressure. A vacuum can cause them to collapse.

Properly vent system pressure before disconnecting a line or coupling. Accidental discharge of compressed air could cause serious injury, especially if the system operates at high temperatures.

The power of compressed air is great. When confined in any sealed container, even low pressure applied over a relatively large area can exert dangerous force. Remember, high pressure concentrated in a relatively small area is dangerous too. Super-compressed air emitted through a small hole in its container can actually cut metal. Leaks in high-pressure piping are particularly hazardous to fingers, hands, and arms. As a rule, do not touch pressurized air lines or pneumatic components. Check for leaks using a brush and soapy water. Also, be aware that compressed air can cause erratic action in a system, and exercise appropriate caution.

The noise of a pneumatic system can cause permanent hearing loss. Hearing damage may occur without your feeling any particular pain or discomfort. OSHA requires hearing protection for those working around compressor systems.

Many pneumatic components are heavy. When working on them, properly secure the system so it cannot shift or fall. If you remove a heavy component, support the system with a floor jack, chain hoist, or blocks.

13-3 COMPRESSORS

A compressor draws in atmospheric air through an air intake and then compresses it (Figure 13.4). Compressors may be classified as positive or nonpositive displacement, single or multiple stage, and reciprocating or rotary types. All compressors have an intake filter element that cleans the air before it

FIGURE 13.4
Air intake.

enters the compressor. In most pneumatic systems, the air is then taken a system air-preparation package. The package removes heat and contamination from the air before the air goes into a tank for storage. The receiver is part of the air-distribution system that provides compressed air through the plant.

Most industrial compressors are fixed systems with large compressors. Often, the compressors are in a compressor room. This prevents them from adding to the noise level in the plant. Sometimes, compressors are outside the plant in a compressor house. This conserves floor space in the plant, in addition to reducing noise and heat inside the plant.

Compressors generate heat in as they operate. Most compressors have coolers that improve efficiency by cooling the air after it has been compressed.

Single-stage compressors compress atmospheric air to a usable pressure in a single operation. However, it is not efficient to produce high pressure in a single operation because of the heat generated by compression. The practical limit to the pressure in single-stage compressors is 150 psi. A single stage compressor is adequate for low-pressure systems because these systems develop little heat. For example, paint-spraying systems require only about 35 PSIG. For general plant systems (70 to 120 psig), two-stage compressors are common (Figure 13.5). For very high pressure applications (1000 to 3000 psig), five or six stages are often used. Using more stages allows enough intercooling to avoid inefficient and dangerously high temperatures.

The two-stage piston compressor is one of the most common types of multistage compressors. It consists of an intake, first-stage compressor, intercooler, and second-stage compressor. The first-stage compressor compresses atmospheric air to an intermediate pressure and expels it into the intercooler. The intercooler cools the air by radiating the heat through its fins into the surrounding air. The smaller second-stage cylinder draws the compressed air in from the intercooler and compresses it further. The second-stage cylinder expels the air into the main air line.

The air that the second-stage cylinder expels is much warmer than the atmospheric air. However, it is not nearly as hot as the air expelled by an equivalent single-stage compressor. The single-stage compressor would also use

FIGURE 13.5
Two-stage piston compressor.

more energy to do the same amount of work as the two-stage compressor because the two-stage compressor is more efficient.

A compressor that produces high pressure by using two or more compressors in a series is a multistage compressor. In a multistage compressor, the air passes from one stage (one compression device) to the next. Each stage in the series compresses the air more than the preceding stage. Between the stages, the air temperature is reduced with coolers called intercoolers. Intercoolers increase efficiency by cooling compressed air before it goes on to the next stage of compression.

Some single-stage piston compressors have two separate compression cylinders similar to two-stage compressors. In this type of unit, the two cylinders work in parallel. Both cylinders discharge air directly into the main air line. The compressed air does not flow from one cylinder to the next like the two-stage compressor.

The output of a compressor is stated as a certain volume of flow at a certain pressure. The manufacturer's nameplate on a compressor shows the rating of the compressor. For example, the output might be 6 cubic feet per minute (cfm) at 80 pounds per square inch (psi). The nameplate may also provide alternate output ratings. For example, a 6 cfm 80 psi compressor provides 12 cfm at 40 psi. (This is twice as much flow at half the pressure.) You can regulate compressors to achieve alternate output. However, the change in volume output is always in inverse proportion to the change in pressure output.

Piston or reciprocating compressors are the most common type of positive-displacement compressor used in industry. The piston moves back and forth (reciprocates) in a straight line inside a cylinder. Most piston compressors require oil to lubricate the movement of the piston. The oil also provides a seal against pressure loss. Other types of compressors are oil-less pistons with Teflon® or graphite piston rings that do not require oil.

A rotary screw compressor compresses air by meshing two rotors that have spiral surfaces. The arrangement of the spiral surfaces is helical. Helical means that each rotor has several spiral surfaces that are parallel to each other. In this type of compressor, the discharge of one lobe overlaps the discharge of the next lobe. This provides continuous, pulsation free delivery of compressed air with much less vibration than piston compressors.

The two types of screw compressors are the dry and the oil-flooded. The oil-flooded is more common. The rotors' contact surfaces are lubricated to reduce wear. The lubrication also provides a good air seal.

The dry-screw type-compressor has extremely close tolerances and does not usually need lubrication. This type of screw compressor is only efficient at high rotation speeds. It is normally limited to applications that require oil-free compressed air.

Screw compressors produce low vibration. They provide a steady flow of pressurized air without creating pulsations. Screw compressors typically provide a maximum pressure of 125 psig. They are usually less than 300 hp. The rotors in most industrial screw compressors measure about 18 to 24 in. They work better in applications where they are constant running and don't stop and start often. In typical industrial applications, the rotary screw compressor is the primary air provider with a piston compressor being the secondary unit providing additional compressed air when necessary.

13-4 INTERCOOLERS

An intercooler (Figure 13.6) is an integral part of a multistage compressor. Compressor manufacturers build intercoolers into multistage compressors to increase compression efficiency and help prevent moisture-related problems. Intercoolers on small multistage compressors often consist simply of tubes with fins. The fins radiate the heat into the surrounding air. Such an intercooler may include a fan that blows air over the fins to help remove the heat.

All intercoolers remove heat from compressed air and some also remove moisture. Intercoolers for large multistage compressors usually use water to help dissipate heat and condense moisture. One type of water-cooled intercooler has a copper water line that is wrapped around the air line. Cold water flows through the copper line to cool the compressed air as it passes through the air line. The air chamber may have a drain line that allows water to escape as it condenses. The drain contains an automatic valve that opens when condensate accumulates in the collector.

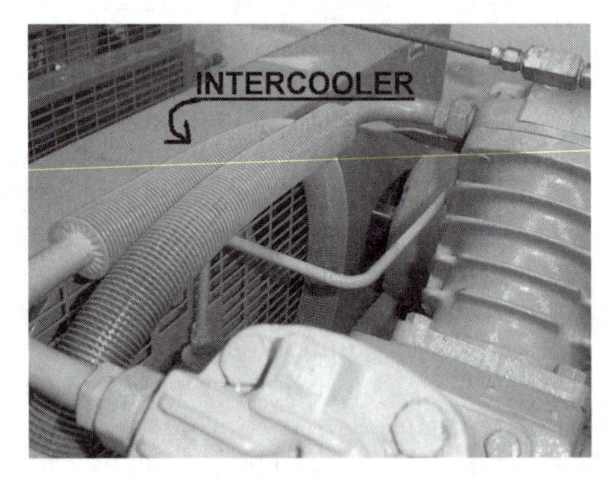

FIGURE 13.6
Multistage compressor intercooler.

13-5 AIR PREP

System air-preparation packages receive compressed air from the compressor and discharge it into the air-distribution system. The purpose of a system air-preparation package is to remove heat and contamination from compressed air. A comprehensive package removes oil, heat, moisture, and dirt from the air. Other packages perform only some of these functions. The design of the package depends on the needs of the pneumatic system.

A comprehensive package has an oil separator, cooler, dryer, and filter. However, a system air-preparation package generally includes a filter only if it provides instrument air. Instrument air is extremely clean, dry air that operates precision control devices used in industry. Instrument air must be as free of contaminants as possible to prevent damage to the instruments. If a system has a filter, it must be maintained regularly. When the filter becomes clogged, the air pressure increases on the compressor side of the filter. Less air will pass through and the compressor will work harder to maintain air flow. If left unattended the filter element may rupture allowing contaminated air into the system. Air pressure drop across the filter is a good indicator of when maintenance is required. Low air pressure drop means filter is clean, high pressure drop means filter is dirty.

Regardless of the number of components in a system air-preparation package, the components must be in a given sequence. The sequence of airflow must be from the compressor to the oil separator and then to the aftercooler. From the aftercooler, the air must flow to the dryer and, finally, to the filter.

In an existing system, you may eventually have to add components to the system air-preparation package. New instruments or other applications may require drier air, cleaner air, or oil-free air. Also, the addition of system air-preparation components can sometimes correct a recurring problem caused by moisture or dirt.

13-6 AFTERCOOLERS

The aftercooler is used to remove heat and moisture from compressed air. Although aftercoolers removes some moisture, you should use a dryer after the aftercooler to ensure moisture will not condense in air lines or equipment downstream. There are three general types of aftercoolers: air cooled, water cooled, and refrigerated.

An air-cooled aftercooler (Figure 13.7) operates on the same principle as an air-cooled intercooler. The hot compressed air passes through a tube with fins that absorb the heat. A fan blows ambient (surrounding) air over the fins, carrying the heat away from the aftercooler.

There are two types of water-cooled aftercoolers. In one, the water line runs inside a chamber and compressed air flows over it. The other design has a copper water line that wraps around the air chamber. The aftercooler operates exactly like the intercooler of the same design. Cold water flows through the copper line to cool the hot compressed air as it passes through the air tube. The air chamber has a drain that allows moisture to drain as it condenses. An automatic valve is usually used for this purpose.

FIGURE 13.7
Air-cooled aftercooler.

A refrigerated aftercooler uses a refrigeration system to cool the compressed air. In a refrigerated aftercooler, the refrigerated evaporator coil cools the air coil that carries the hot compressed air. After the moisture is removed the air enters the air warming section. The cool dry air enters a heat exchanger that pre-cools the hot compressor air entering the aftercooler. This makes the refrigerated aftercooler very efficient. Shutting down a refrigerated aftercooler can allow wet air to enter the pneumatic distribution system at the next start-up. To prevent this, refrigerated aftercoolers should run continuously including nights and weekends. Refrigeration systems require little or no maintenance and seldom break down. Refrigerated aftercoolers remove moisture more effectively than other types of aftercoolers. The use of a refrigerated aftercooler before the dryer ensures that the dryer receives the driest possible air. This enhances the efficiency of the dryer.

13-7 DRYERS

Although aftercoolers are moisture separators, they cannot remove all the moisture from the air. A desiccant dryer can remove much of the remaining moisture. A desiccant is a granular chemical compound that either absorbs the moisture or adsorbs it. Absorption is the process of a liquid filling the spaces inside a dry substance. Adsorption is the process of collecting moisture on the surface of the particles of a dry substance. Desiccant dryers operate most efficiently if most of the air's moisture has been removed before it reaches the desiccant dryer. For this reason, the air may pass through more than one aftercooler before entering a desiccant dryer.

Desiccants such as lithium chloride absorb the moisture as the air passes through a container filled with the desiccant. The desiccant must be replaced when it becomes saturated. This type of dryer usually has a paper moisture indicator. The paper is visible through a small window. When the color of the paper changes, you should replace the desiccant.

Other dryers use desiccants such as silica gel to adsorb water. This type of desiccant is reusable. After the desiccant's moisture content reaches a certain point, the dryer begins removing the moisture from the desiccant. The dryer may use either heat or dry air to dry the desiccant.

13-8 MAINTENANCE

Maintaining compressors includes cleaning and replacing worn or dirty parts, checking and adding oil, and changing oil. Basic troubleshooting includes recognizing the four most common symptoms of compressor malfunction when they occur and repairing their causes.

Before you do repairs or maintenance work on a compressor, you should turn off the drive motor. If the unit has a drive clutch, be sure to disengage it. Also, close the isolation valve that is between the compressor and the receiver. When the isolation valve is closed, no pressure can flow from the receiver back to the compressor.

Compressor parts that require maintenance most frequently include filters, screens, drive belts, and valves. Intercoolers, oil separators, aftercoolers, and dryers also require maintenance occasionally. Following a maintenance schedule will help ensure that all maintenance is done as required.

You should clean or replace filter elements on a regular basis (Figure 13.8). A filter element traps impurities in the intake air. The impurities will eventually clog the filter and prevent the compressor from getting enough air. Some filter elements are reusable and must be cleaned when they get dirty. Other elements are disposable. The manufacturer of the compressor specifies how frequently you should clean or replace the filter element. However, conditions at the location of each particular intake often determine the frequency of necessary filter maintenance.

Inspect the filters regularly and frequently to see if they need changing. Also, check the pressure in the intake to be sure it is at least 97% of atmospheric

FIGURE 13.8
Intake filter element.

pressure. If it drops below this minimum pressure, you may need to clean or replace the element. Proper filter maintenance is cost-effective. In positive displacement compressors, contaminants wear down the moving parts. Studies have shown that regular filter maintenance is less expensive than maintaining and replacing compressor components.

You should check compressor drive belts regularly. Be sure to keep the motor and compressor aligned. Improper alignment causes uneven belt wear and shortens the life of the belt. If a belt is shiny, it is slipping and should be adjusted or replaced. If a belt is cracked or worn, replace it.

You should inspect safety valves routinely as part of your scheduled maintenance. If a safety valve becomes stuck in the open position, the compressor cannot build pressure, and you will know immediately there is a problem. However, the valve can also become stuck in the closed position. You may not discover the problem until excess pressure has damaged the equipment.

Clean the outside of an intercooler when dirt begins to collect on it. Dirt traps the heat that the intercooler is supposed to dissipate. If an intercooler develops a leak, repair it if possible.

The oil that an oil separator takes out of the air is usually recirculated to lubricate the compressor. There are three key maintenance activities you must perform on oil separators to ensure proper lubrication of the compressor. You must check the oil, change it, and clean or replace the oil strainer. You should check the oil level in an oil separator regularly. Follow the compressor manufacturer's recommendations on how many hours of operation are allowed before the oil must be changed. The recommendations may give two specifications; one for operation under normal conditions and one for operation under severe conditions. Severe conditions include unusually high operating temperatures and unusually dirty intake air. When you change the oil, remove and clean or replace the oil strainer.

Aftercoolers have either automatic or manual water drains. If the drain is automatic, verify that it is operating properly. The automatic drain should blow out water regularly (Figure 13.9). Check the automatic drain at least once each day. This check is especially important on refrigerated aftercoolers because they remove such large quantities of water from the air.

FIGURE 13.9
Automatic drain.

If the aftercooler's drain is manual, open it briefly to allow the compressed air to blow out the water that has accumulated. Do this at least once each day. If you allow too much water to collect in an aftercooler, the water will enter the air system.

Keep the outside of an aftercooler clean. When dirt begins to collect on it the heat will not dissipate and the aftercooler will not operate efficiently.

In some types of dryers, the desiccant is disposable. Other types of dryers have reusable desiccant, which must be dried and reused. When the desiccant will no longer remove moisture, it must be replaced. If the desiccant is disposable, check the moisture indicator regularly to determine if the desiccant needs to be changed. You should change the desiccant as needed.

13-9 FRLS

Some air-preparation components are placed close to the machine using them and are often placed close together in a single package. The filter, pressure regulator, and lubricator (FRL unit) often come together as a triple assembly (Figure 13.10). The first stage of air preparation is always filtration. The air-line filter cleans and sometimes dehumidifies air immediately before it enters pneumatic equipment. An air-line filter has two sections: the element and the bowl. The filter element catches any particles or moisture remaining in the air. A good filter will extend the life of air-operated tools and equipment, and reduce maintenance cost. There are two types of filters; an absolute filter has holes that are all exactly the same size, a nominal filter has holes that are approximately the same size.

Some filters are chemically treated to remove vapor also. These filters are either coalescent or absorbent. Coalescent filters cause oil and water aerosols to combine into droplets that fall to the bottom of the bowl. Absorbent filters extract vapor and retain it by absorption. All filters have a drain for removal of the condensate that collects in the bowl. You should check and drain the bowl

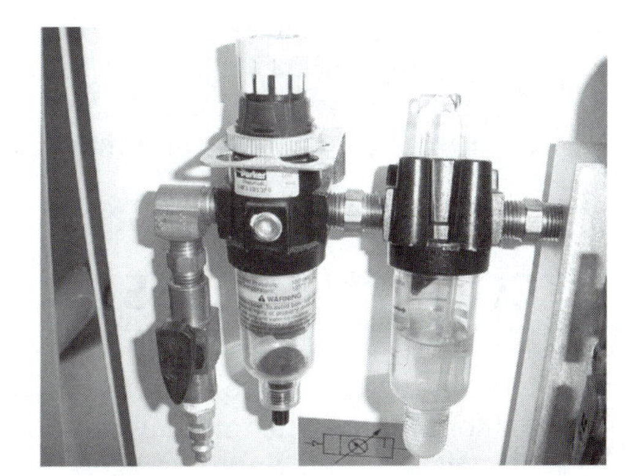

FIGURE 13.10
FRL assembly.

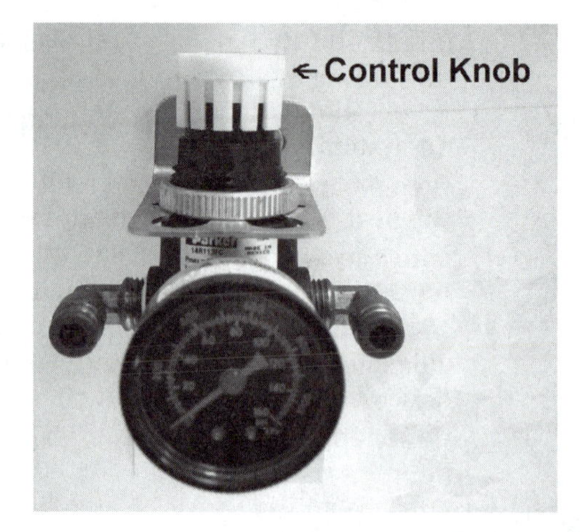

FIGURE 13.11
Pressure regulator.

regularly. The water level must not rise above the lower baffle, as it will block the filter element. Most drains are manual, but some filters have automatic drains.

Filter bowls can be plastic, plastic with metal guards, or metal. Metal guards are used for external protection of plastic bowls. Should a bowl become fractured, the guard will contain all fragments for safety. Metal bowls are used where filters are exposed to higher pressures and temperatures. Metal bowls will withstand rough usage and are less susceptible to physical damage.

Different machines require different air pressures to operate efficiently. After the filter purifies the air, a pressure regulator (Figure 13.11) is used to adjust the air pressure to make it suitable for each machine or tool. Turning the control knob on the pressure regulators clockwise will increase the air pressure and counterclockwise will decrease the air pressure.

The third stage of air preparation is lubrication. Lubrication is always after filtration and pressure regulation. The compressed air at this point should be clean, dry, and at the correct pressure for the equipment using it. The lubricator then injects or mists the compressed air with oil. The oil-treated air then passes through the mechanism it serves, keeping all internal parts lubricated. Not all machines require lubrication and follow manufacturer's recommendation for each situation.

You should always adjust the lubricator while air is flowing. The air pressure will affect the drip rate, so set the regulator pressure first. Time the drip rate by watching it through the sight glass, adjust the flow-control knob to increase or decrease it. It may take several tries to get proper lubrication.

13-10 CONTROL VALVES

Control valves are used to change the path of air flow or to shut off air flow completely. The internal moving part of a control valve will connect or disconnect internal flow passages that results in changing the air flow direction.

Pneumatic control valves are usually of two main types: the poppet-type and the sliding or shear-action-type valve.

The poppet valve has a movable element (the poppet) that is used to direct the flow of air through the valve body. The poppet inside is moved through a stem that pushes the poppet off its seat allowing a flow path (in the case of a two-way, normally closed valve), or closing off a flow path by pushing the poppet onto a seat (in the case of a two-way, normally open valve). The stem is moved by some sort of actuator which may be manual, pilot- or solenoid-operated.

Poppet valves are usually very tolerant of air-line contaminants and solids when used in compressed air service. This type of valve construction is typically characterized as being a high flow, fast-acting design due to the fact that large flow paths through the body that can be opened quickly. You can think of a poppet valve like a plug in a bath tub drain. When the plug is pulled, the water will flow quickly and the drain exposed is quite large. The large opening of a poppet allows trash to pass through the valve easily without valve damage.

Some advantages of poppet valves are the following:

1. Reliable design
2. Easily manufactured
3. Fast action
4. High air flows
5. Tolerant of contaminants
6. Requires little or no lubrication

The most common sliding action valve is the spool valve. Spool valves are used extensively for pneumatic service. A spool valve consists of a cylindrical spool that slides back and forth inside the valve body to either allow air flow or block air flow. The spool has a land that is a larger diameter portion of the spool that blocks air flow by sliding to cover a port.

Directional control valves are manually actuated or automatically actuated. Automatic operation can be controlled by mechanical, pneumatic, or electrical actuation. Directional control valves are generally classified in three areas:

1. The number of *ways*
2. The number of positions
3. The type of actuator

The term *way* is used to describe the flow path through the valve body. Two-way, three-way, and four-way valves are the most common industrial pneumatic valves. Two-way valves have two ports and allow air flow in both directions. They are typically used as common on-off valves. Three-way valves have three ports. The third port is the exhaust port; it allows the pressure in the valve to bleed out. Think of a garden hose with a spray nozzle attached to a faucet. If the faucet is turned off pressure still remains in the garden hose, when a three-way valve is turned off the remaining pressure is allowed to bleed out the third port—the exhaust port. Both two-way and three-way can be either

FIGURE 13.12
Four-way control valve.

normally closed or normally open. A normally closed valve has the pressure or flow blocked when the valve is in the nonactuated position. A normally open valve permits air flow when the valve is in the nonactuated position. Normally closed valves are more common in industry than normally open valves.

Four-way valves (Figure 13.12) are commonly used to control double-acting cylinders and bi-directional air motors. Four-way valves may have four or five ports: one pressure port, two cylinder ports, and two exhaust ports that are sometimes combined into one port. The simplest four-way valves have two positions used to extend or retract a cylinder, the cylinder is always pressurized in one position or the other and there is no "off" position. With a three position valve, the middle position can be used as an "off" position. The "off" or center position is commonly found with two types.

The closed center has all ports blocked with no air flow whatsoever. This effectively locks the cylinder rod allowing no movement. The float center has a closed pressure port but allows the cylinder ports to be connected to the exhaust ports allowing the cylinder rod free movement in either direction. Three position valves usually have springs to return them to the center position when not activated.

There are many ways to actuate or operate a valve. Each has its own method and symbol. Sometimes the manual symbol is used instead of giving the more specific symbol such as the button, lever, or foot pedal that would operate the valve.

13-11 ACTUATORS

Actuators or pneumatic cylinders may be single-acting or double-acting. Single-acting cylinders are pressurized in one direction only and usually on the out or push stroke. A spring, gravity or the load is used to return the cylinder to its

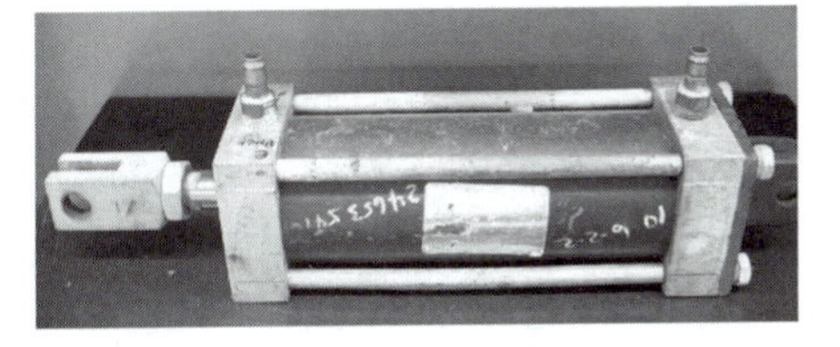

FIGURE 13.13
Double-acting actuator.

original position when the work is completed. A single-acting cylinder has only one port and is usually activated with a three-way valve. The double-acting cylinder (Figure 13.13) can be pressurized in either direction. In order to do this it must have two ports, one to pressurize the cylinder to extend and the other to pressurize the cylinder to retract. A four-way valve is required to operate a double-acting cylinder. Most standard air cylinders are double-acting. Most cylinders are constructed with aluminum or other nonferrous materials to transfer heat and reduce problems with corrosion from water and contaminants in the air lines.

Air cylinders may be cushioned with polyurethane shock pads placed inside the cylinder between the piston and the cylinder end caps. These pads reduce cylinder shock and noise that would occur if the piston rod was allowed to slam into the end of the cylinder. Shock pads increase the length of the cylinder by $\frac{1}{4}$ inch per pad. Sometimes cylinder cushioning is accomplished by adding center spear or extension. When the piston approaches the end of its stroke, the spear or sleeve enters cushion port blocking the exhaust. With the exhaust closed, the remaining air is forced through an adjustable small built-in needle valves metering the air exhaust in the last half inch of stroke. One or both end caps may be cushioned.

QUESTIONS

1. Who is responsible for standardizing pneumatic symbols?
2. Why should vents be locked in the open position when working on a pressurized system?
3. What is the practical pressure limit for single-stage compressors?
4. Is the single-stage or two-stage compressor more efficient?
5. What is the most common type of positive-displacement compressor used in industry?
6. Which are more common: dry or oil-flooded rotary screw compressors?
7. What are the three types of aftercoolers?
8. What is a "FRL"?
9. When should metal bowls be used on filters and lubricators?
10. What are the two main types of pneumatic control valves?
11. What type of control valve is required to operate a double-acting cylinder?
12. Why do some air cylinders have cushions?

APPENDIX

DECIMAL AND MILLIMETER EQUIVALENTS OF FRACTIONS

Inches		Milli-meters	Inches		Milli-meters	Inches		Milli-meters
Fractions	Decimals		Fractions	Decimals		Fractions	Decimals	
1/64	.015625	.397	11/32	.34375	8.731	11/16	.6875	17.463
1/32	.03125	.794	23/64	.359375	9.128	45/64	.703125	17.859
3/64	.046875	1.191	3/8	.375	9.525	23/32	.71875	18.256
1/16	.0625	1.588	25/64	.390625	9.922	47/64	.734375	18.653
5/64	.078125	1.984	13/32	.40625	10.319	3/4	.750	19.050
3/32	.09375	2.381	27/64	.421875	10.716	49/64	.765625	19.447
7/64	.109375	2.778	7/16	.4375	11.113	25/32	.78125	19.844
1/8	.125	3.175	29/64	.453125	11.509	51/64	.796875	20.241
9/64	.140625	3.572	15/32	.46875	11.906	13/16	.8125	20.638
5/32	.15625	3.969	31/64	.484375	12.303	53/64	.828125	21.034
11/64	.171875	4.366	1/2	.500	12.700	27/32	.84375	21.431
3/16	.1875	4.763	33/64	.515625	13.097	55/64	.859375	21.828
13/64	.203125	5.159	17/32	.53125	13.494	7/8	.875	22.225
7/32	.21875	5.556	35/64	.546875	13.891	57/64	.890625	22.622
15/64	.234375	5.953	9/16	.5625	14.288	29/32	.90625	23.019
1/4	.250	6.350	37/64	.578125	14.684	59/64	.921875	23.416
17/64	.265625	6.747	19/32	.59375	15.081	15/16	.9375	23.813
9/32	.28125	7.144	39/64	.609375	15.478	61/64	.953125	24.209
19/64	.296875	7.541	5/8	.625	15.875	31/32	.96875	24.606
5/16	.3125	7.938	41/64	.640625	16.272	63/64	.984375	25.003
21/64	.328125	8.334	21/32	.65625	16.669	1	1.000	25.400
			43/64	.671875	17.066			

MILLIMETER-INCH EQUIVALENTS

Milli-meter	Decimal	Milli-meter	Decimal	Milli-meter	Decimal	Milli-meter	Decimal	Milli-meter	Decimal
1	.03937	52	2.04724	103	4.05511	154	6.06299	205	8.07086
2	.07874	53	2.08661	104	4.09448	155	6.10236	206	8.11023
3	.11811	54	2.12598	105	4.13385	156	6.14173	207	8.14960
4	.15748	55	2.16535	106	4.17322	157	6.18110	208	8.18897
5	.19685	56	2.20472	107	4.21259	158	6.22047	209	8.22834
6	.23622	57	2.24409	108	4.25196	159	6.25984	210	8.26771
7	.27559	58	2.28346	109	4.29133	160	6.29921	211	8.30708
8	.31496	59	2.32283	110	4.33070	161	6.33858	212	8.34645
9	.35433	60	2.36220	111	4.37007	162	6.37795	213	8.38582
10	.39370	61	2.40157	112	4.40944	163	6.41732	214	8.42519
11	.43307	62	2.44094	113	4.44881	164	6.45669	215	8.46456
12	.47244	63	2.48031	114	4.48818	165	6.49606	216	8.50393
13	.51181	64	2.51968	115	4.52755	166	6.53543	217	8.54330
14	.55118	65	2.55905	116	4.56692	167	6.57480	218	8.58267
15	.59055	66	2.59842	117	4.60629	168	6.61417	219	8.62204
16	.62992	67	2.63779	118	4.64566	169	6.65354	220	8.66141
17	.66929	68	2.67716	119	4.68503	170	6.69291	221	8.70078
18	.70866	69	2.71653	120	4.72440	171	6.73228	222	8.74015
19	.74803	70	2.75590	121	4.76378	172	6.77165	223	8.77952
20	.78740	71	2.79527	122	4.80315	173	6.81102	224	8.81889
21	.82677	72	2.83464	123	4.84252	174	6.85039	225	8.85826
22	.86614	73	2.87401	124	4.88189	175	6.88976	226	8.89763
23	.90551	74	2.91338	125	4.92126	176	6.92913	227	8.93700
24	.94488	75	2.95275	126	4.96063	177	6.96850	228	8.97637
25	.98425	76	2.99212	127	5.00000	178	7.00787	229	9.01574
26	1.02362	77	3.03149	128	5.03937	179	7.04724	230	9.05511
27	1.06299	78	3.07086	129	5.07874	180	7.08661	231	9.09448
28	1.10236	79	3.11023	130	5.11811	181	7.12598	232	9.13385
29	1.14173	80	3.14960	131	5.15748	182	7.16535	233	9.17322
30	1.18110	81	3.18897	132	5.19685	183	7.20472	234	9.21259
31	1.22047	82	3.22834	133	5.23622	184	7.24409	235	9.25196
32	1.25984	83	3.26771	134	5.27559	185	7.28346	236	9.29133
33	1.29921	84	3.30708	135	5.31496	186	7.32283	237	9.33070
34	1.33858	85	3.34645	136	5.35433	187	7.36220	238	9.37007
35	1.37795	86	3.38582	137	5.39370	188	7.40157	239	9.40944
36	1.41732	87	3.42519	138	5.43307	189	7.44094	240	9.44881
37	1.45669	88	3.46456	139	5.47244	190	7.48031	241	9.48818
38	1.49606	89	3.50393	140	5.51181	191	7.51968	242	9.52755
39	1.53543	90	3.54330	141	5.55118	192	7.55905	243	9.56692
40	1.57480	91	3.58267	142	5.59055	193	7.59842	244	9.60629
41	1.61417	92	3.62204	143	5.62992	194	7.63779	245	9.64566
42	1.65354	93	3.66141	144	5.66929	195	7.67716	246	9.68503
43	1.69291	94	3.70078	145	5.70866	196	7.71653	247	9.72440
44	1.73228	95	3.74015	146	5.74803	197	7.75590	248	9.76378
45	1.77165	96	3.77952	147	5.78740	198	7.79527	249	9.80315
46	1.81102	97	3.81889	148	5.82677	199	7.83464	250	9.84252
47	1.85039	98	3.85826	149	5.86614	200	7.87401	251	9.88189
48	1.88976	99	3.89763	150	5.90551	201	7.91338	252	9.92126
49	1.92913	100	3.93700	151	5.94488	202	7.95275	253	9.96063
50	1.96850	101	3.97637	152	5.98425	203	7.99212	254	10.00000
51	2.00787	102	4.01574	153	6.02362	204	8.03149	...	....

TORQUE AND HORSEPOWER EQUIVALENTS

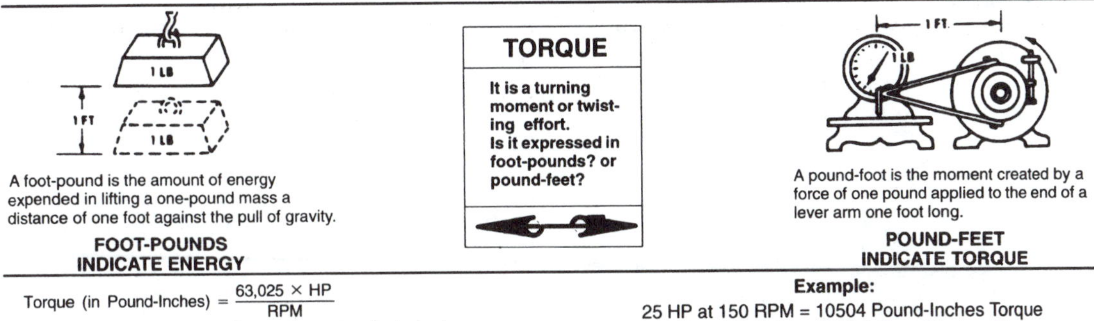

A foot-pound is the amount of energy expended in lifting a one-pound mass a distance of one foot against the pull of gravity.

FOOT-POUNDS INDICATE ENERGY

A pound-foot is the moment created by a force of one pound applied to the end of a lever arm one foot long.

POUND-FEET INDICATE TORQUE

$$\text{Torque (in Pound-Inches)} = \frac{63{,}025 \times HP}{RPM}$$
$$= \text{Force} \times \text{Lever Arm (In Inches)}$$

$$\text{Torque (in Pound-Feet)} = \frac{5{,}252 \times HP}{RPM}$$
$$= \text{Force X Lever Arm (In Feet)}$$

Force = Working Load in Pounds.
FPM = Feet Per Minute.
RPM = Revolutions Per Minute.
Lever Arm = Distance from the Force to the center of rotation in Inches or Feet.

Example:

25 HP at 150 RPM = 10504 Pound-Inches Torque
2.5 HP at 150 RPM = 1050.4 Pound-Inches Torque

For other values of RPM move decimal point in RPM values to the left or right as desired, and in torque values move to the right or left (opposite way) the same number of places.

Example:

25 HP at 150 RPM = 10504 Pound-Inches Torque
25 HP at 1.50 RPM = 1050400 Pound-Inches Torque
2.5 HP at 1.50 RPM = 105040 Pound-Inches Torque

HORSEPOWER

Common Unit of Mechanical power. (HP) One HP is the rate of work required to raise 33,000 pounds one foot in one minute.

ONE FOOT PER MINUTE
33,000 LBS.

$$HP = \frac{\text{Force} \times RPM}{33{,}000}$$

$$HP = \frac{\text{Torque (in Pound-Inches)} \times RPM}{63{,}025}$$

$$HP = \frac{\text{Torque (in Pound-Feet)} \times RPM}{5{,}252}$$

Overhung Loads

An overhung load is a bending force imposed on a shaft due to the torque transmitted by V-drives, chain drives, and other power transmission devices, other than flexible couplings.

Most motor and reducer manufacturers list the maximum values allowable for overhung loads. It is desirable that these figures be compared with the load actually imposed by the connected drive.

Overhung loads may be calculated as follows:

$$\text{O.H.L.} = \frac{63.000 \times HP \times F}{N \times R}$$

where HP = Transmitted hp X service factor
N = RPM of shaft
R = Radius of sprocket, pulley, etc.
F = Factor

Weights of the drive components are usually negligible. The formula is based on the assumption that the load is applied at a point equal to one shaft diameter from the bearing face. Factor F depends on the type of drive used:

F =
- 1.00 for single chain drives
- 1.3 for TIMING belt drives and HTD belt Drives
- 1.25 for spur or helical gear or double chain drives
- 1.50 for V-belt drives
- 2.50 for flat belt drives

Example: Find the overhung load imposed on a reducer by a double chain drive transmitting 7 hp @ 30 RPM. The pitch diameter of the sprocket is 10″; service factor is 1.3.

Solution:

$$\text{O.H.L.} = \frac{(63{,}000)(7 \times 1.3)(1.25)}{(30)(5)} = 4{,}780 \text{ lbs.}$$

Mathematical Equations

To find circumference of a circle, multiply diameter by 3.1416.
To find diameter of a circle, multiply circumference by .31831.
To find area of a circle, multiply square of diameter by .7854.
To find area of a rectangle, multiply length by breadth.
To find area of a triangle, multiply base by 1/2 perpendicular height.
To find area of ellipse, multiply product of both diameters by .7854.
To find area of parallelogram, multiply base by altitude.
To find side of an inscribed square, multiply diameter by 0.7071 or multiply circumference by 0.2251 or divide circumference by 4.4428.
To find side of inscribed cube, multiply radius of sphere by 1.1547.
To find side of an equal square, multiply diameter by .8862.
To find the surface of a sphere, square the diameter and multiply by 3.1416.
To find the volume of a sphere, cube the diameter and multiply by .5236.
A side of a square multiplied by 1.4142 equals diameter of its circumscribing circle.
A side of a square multiplied by 4.443 equals circumference of its circumscribing circle.
A side of a square multiplied by 1.128 equals diameter of an equal circle.
A side of a square multiplied by 3.547 equals circumference of an equal circle.
To find gallon capacity of tanks (given dimensions of a cylinder in inches): square the diameter of the cylinder, multiply by the length and by .0034.

V-BELT DRIVE FORMULAS

V-belt tensioning – In cases where tensioning of a drive effects belt pull and bearing loads, the following formulas may be used.

$$T_1 - T_2 = 33{,}000 \left(\frac{HP}{V}\right)$$

where
- T_1 = tight side tension, pounds
- T_2 = slack side tension, pounds
- HP = design horsepower
- V = belt speed, feet per minute

$$T_1 + T_2 = 33{,}000\ (2.5\text{–}G)\left(\frac{HP}{GV}\right)$$

where
- T_1 = tight side tension, pounds
- T_2 = slack side tension, pounds
- HP = design horsepower
- V = belt speed, feet per minute
- G = arc of contact correction factor*

$$T_1/T_2 = \frac{1}{1\text{–}0.8G} \quad (\text{also } T_1/T_2 = eK\emptyset)$$

where
- T_1 = tight side tension, pounds
- T_2 = slack side tension, pounds
- G = arc of contact correction factor*
- e = base of natural logarithms
- K = .51230, a constant for V-belt drive design
- Ø = arc of contact in radians

$$T_1 = 41{,}250 \left(\frac{HP}{GV}\right)$$

where
- T_1 = tight side tension, pounds
- HP = design horsepower
- V = belt speed, feet per minute
- G = arc of contact correction factor*

$$T_2 = 33{,}000\ (1.25\text{–}G)\left(\frac{HP}{GV}\right)$$

where
- T_2 = slack side tension, pounds
- HP = design horsepower
- V = belt speed, feet per minute
- G = arc of contact correction factor*

Belt Speed

$$V = \frac{(PD)\ (rpm)}{3.82} = (PD\ (rpm)\ (.262))$$

where
- V = belt speed, feet per minute
- PD = pitch diameter of sheave or pulley
- rpm = revolutions per minute of the same sheave or pulley

*See Table A1, at left.

Table A1—Arc of Contact Correction Factors G and R

$\frac{D-d}{C}$	Small Sheave Arc of Contact	Factor G	Factor R	$\frac{D-d}{C}$	Small Sheave Arc of Contact	Factor G	Factor R
.00	180°	1.00	1.000	.80	133°	.87	.917
.10	174°	.99	.999	.90	127°	.85	.893
.20	169°	.97	.995	1.00	120°	.82	.866
.30	163°	.96	.989	1.10	113°	.80	.835
.40	157°	.94	.980	1.20	106°	.77	.800
.50	151°	.93	.968	1.30	99°	.73	.760
.60	145°	.91	.954	1.40	91°	.70	.714
.70	139°	.89	.937	1.50	83°	.65	.661

D = Diam. of large sheave. d = Diam. of small sheave.
C = Center distance.

Table A2—Allowable Sheave Rim Speed

Sheave Material	Rim Speed in Feet per Minute
Cast Iron	6,500
Ductile Iron	8,000
Steel	10,000

Note: Above rim speed values are maximum for normal considerations. In some cases these values may be exceeded. Consult factory and include complete details of proposed application.

Bearing Load Calculations

To find actual bearing loads it is necessary to know machine component weights and values of all other forces contributing to the load. Sometimes it becomes desirable to know the bearing load imposed by the V-belt drive alone. This can be done if you know bearing spacing with respect to the sheave center and shaft load and apply it to the following formulas:

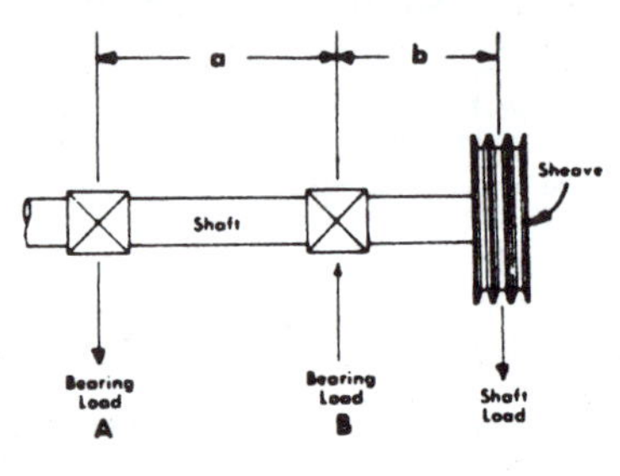

Overhung Sheave

$$\text{Load at B, lbs.} = \frac{\text{Shaft Load} \times (a + b)}{a}$$

$$\text{Load at A, lbs.} = \text{Shaft Load} \times \frac{b}{a}$$

Where: a and b = Spacing, inches

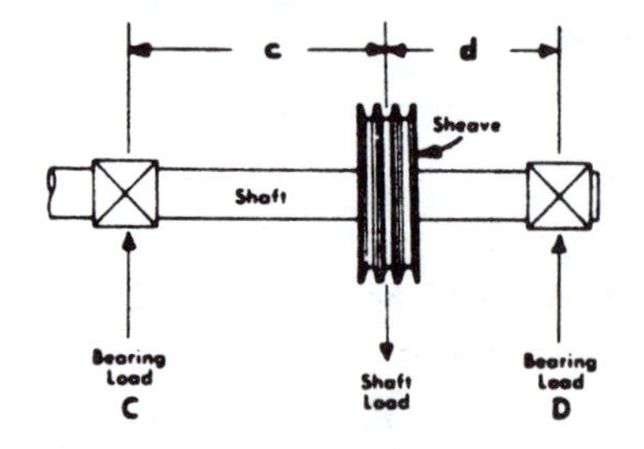

Sheave Between Bearings

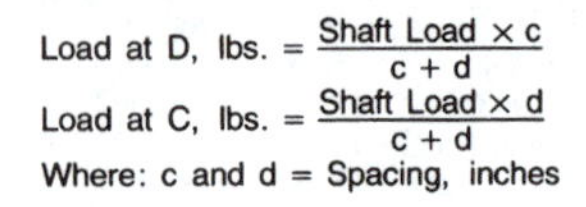

$$\text{Load at D, lbs.} = \frac{\text{Shaft Load} \times c}{c + d}$$

$$\text{Load at C, lbs.} = \frac{\text{Shaft Load} \times d}{c + d}$$

Where: c and d = Spacing, inches

ACCELERATION

COLUMN A

To Convert From...	To...	Multiply Col. A by
feet per second per second (ft per sec ^{2})	meters per second per second (m per sec^2)	0.3048
m per sec^2	ft per sec^2	3.281
revolutions per minute per second (rpm per sec)	radians per second per second (rad per sec^2)	0.1047
rad per sec^2	rpm per sec	9.55

PRESSURE

To Convert From...	To...	Multiply Col. A by
pascals (Pa)	pounds per square inch (psi)	0.0001450
	pounds per square foot (lb per ft^2)	0.02089
	newtons per square meter	1
pounds per square inch (psi)	atmospheres, std. (atm)	0.0680
	pounds per square foot (lb per ft^2)	144
	pascals (Pa)	6894.8
	foot of water (ft of H_20) 60F	2.301
atmospheres (atm), standard	psi	14.70
	lb per ft^2	2116.8
	Pa	101325
inch of water, 60F (in of H_20)	psi	0.03609
	lb per ft^2	5.197
	Pa	248.84
foot of water, 60F (ft of H_20)	psi	0.4331
	lb per ft^2	62.36
	Pa	2985.9

VELOCITY

To Convert From...	To...	Multiply Col. A by
centimeters per second (cm per sec)	feet per second (fps or ft per sec)	0.3281
feet per second (fps)	centimeters per second (cm per sec)	30.48
	meters per second (m per sec)	0.3048
	kilometers per hour (km per hr)	1.097
	miles per hour (mph)	0.6818
kilometers per hour (km per hr)	knots (kn)	0.5396
	feet per second (fps)	1.467
	kilometers per hour (km per hr)	1.609
	feet per minute (ft per min)	88
knots (kn)	miles per hour (mph)	1.152
	kilometers per hour (km per hr)	1.853
radians per second (rad per sec)	revolutions per minute (rpm)	9.55
	degrees per minute (deg per min)	3437.7
revolutions per minute (rpm)	radians per second (rad per sec)	0.1047
	degrees per minute (deg per min)	360

VOLUMETRIC FLOW RATES

To Convert From...	To...	Multiply Col. A by
gallons per minute, US (gpm)	liters per second (l per sec)	0.008434
	cubic feet per minute (cfm)	0.1337
	cubic feet per hour (cu ft per hr)	8.022
gallons per minute, UK or Canadian (gpm)	liters per second (l per sec)	0.0101
	cubic feet per minute (cfm)	0.1606
	cubic feet per hour (cu ft per hr)	9.634
cubic feet per second (cfs)	gpm (UK or Canadian)	373.77
	gpm (US)	448.86
	liters per second (l per sec)	1699.2
liters per second (l per sec)	cubic feet per minute (cfm)	2.119
	gpm (UK or Canadian)	13.20
	gpm (US)	15.85
millions of gallons per day, US (MGD)	liters per second (l per sec)	43.81
	cubic feet per minute (cfm)	92.85
	gallons per minute, US (gpm)	694.44

WEIGHT, MASS, INERTIA

To convert from	To	Multiply by
pounds (lb)*	kilograms (kg)	0.4536
	ounces (oz)	16
kilograms (kg)	pounds (lb)	2.205
	ounces (oz)	35.27
tons (short)	metric tons	0.9072
	kilograms (kg)	907.2
	pounds (lb)	2000
metric tons	tons (short)	1.102
	kilograms	1000
	pounds	2205
pounds, weight (lb)	slugs, mass (lb-sec^2 per ft)	0.03106
pound-foot^2 (lb-ft^2)	kilogram-meters^2 (kg-m^2)	0.04214

*pounds and ounces are avoirdupois

FORCE AND TORQUE

To convert from	To	Multiply by
pounds (lb)	newtons(N)	4.448
newtons (N)	pounds (lb)	0.2248
newton-meters (N-m)	pound-feet (lb-ft)	0.7376
	pound-inches (lb-in)	8.851
	ounce-inches (oz-in)	141.60
ounce-inches (oz-in)	lb-ft	0.005208
	N-m	0.007062
	lb-in	0.0625
pound-inches (lb-in)	lb-ft	0.0833
	N-m	0.1298
	oz-In	16
pound-feet (lb-ft)	N-m	1.356
	lb -in	12
	oz-In	192

POWER

To convert from	To	Multiply by
horsepower (hp)	kilowatts (kW)	0.7457
	foot-pounds per second (ft-lb per sec)	550
	foot-pounds per minute (ft-lb per min)	33000
kilowatts (kW)	horsepower (hp)	1.341

TEMPERATURE

To convert from	To	Use This Relationship
degrees Fahrenheit (F)	degrees Celsius (C)	C =5/9 (F−32)
degrees Celsius (C)	degrees Fahrenheit (F)	F=9/5C+32
degrees Fahrenheit (F)	degrees Rankine (R)	R =F+459.69
degrees Celsius (C)	degrees Kelvin (K)	K=C+273.16

DODGE®

ENGINEERING/TECHNICAL

Electrical Tables

Table A3—Electrical Formulas

To Find	Alternating Current	
	Single-Phase	Three-Phase
Amperes when horsepower is known	$\frac{Hp \times 746}{E \times Eff \times pf}$	$\frac{Hp \times 746}{1.73 \times E \times Eff \times pf}$
Amperes when kilowatts are known	$\frac{Kw \times 1000}{E \times pf}$	$\frac{Kw \times 1000}{1.73 \times E \times pf}$
Amperes when Kva are known	$\frac{Kva \times 1000}{E}$	$\frac{Kva \times 1000}{1.73 \times E}$
Kilowatts	$\frac{I \times E \times pf}{1000}$	$\frac{1.73 \times I \times E \times pf}{1000}$
Kva	$\frac{I \times E}{1000}$	$\frac{1.73 \times I \times E}{1000}$
Horsepower = (Output)	$\frac{I \times E \times Eff \times pf}{746}$	$\frac{1.73 \times I \times E \times Eff \times pf}{746}$

I - Amperes; E = Volts; Eff = Efficiency; pf = power factor; Kva = Kilovolt amperes; Kw = Kilowatts; R = Ohms.

To Find	Alternating or Direct Current
Amperes when voltage and resistance is known	$\frac{E}{R}$
Voltage when resistance and current are known	IR
Resistance when voltage and current are known	$\frac{E}{I}$

General Information (Approximation)

All Values At 100% Load:
- At 1800 rpm, a motor develops 36 lb.-in per hp
- At 1200 rpm, a motor develops 54 lb.-in per hp
- At 575 volts, a 3-phase motor draws 1 amp per hp
- At 460 volts, a 3-phase motor draws 1.25 amp per hp
- At 230 volts, a 3-phase motor draws 2.5 amp per hp
- At 230 volts, a single-phase motor draws 5 amp per hp
- At 115 volts, a single-phase motor draws 10 amp per hp

Temperature Conversion:
Deg C = (Deg F - 32) × 5/9
Deg F = (Deg C × 9/5) + 32

Table A4—AC Motor Recommended Wire Size

Volts	Motor Horsepower																				
	1-3	5	$7^1/_2$	10	15	20	25	30	40	50	60	75	100	125	150	200	250	300	350	400	450	500
230	14	12	10	8	6	4	3	1	0	000	000	300	500	...	...	...	...	...	...	...	...	...
460	14	14	14	12	10	8	6	6	4	3	2	0	000	0000	300	500	700	900	1500	600*	750*	900*
575	14	14	14	14	12	10	8	6	6	4	3	2	0	000	0000	250	500	600	800	1000	1500	600*

Insure that the requirements of the National Electric Code are fully met in all installations. This table is included as a guide only and is based on 3 phase, continuous duty, design B, standard efficiency motors using 6oo volt Insulation, Type THW, with individual cooper conductors run in rigid conduit as defined in the 1987 NEC.

Table A5—Motor Amps @ Full Load†

HP	Alternating Current		DC	HP	Alternating Current		DC	HP	Alternating Current		DC	HP	Alternating Current		DC
	Single-phase	3-phase			Single-phase	3-phase			Single-phase	3-phase			Single-phase	3-phase	
1/2	4.9	2.0	2.7	5	28	14.4	20	25	....	60	92	75	....	180	268
1	8.0	3.4	4.8	7-1/2	40	21.0	29	30	....	75	110	100	....	240	355
1-1/2	10.0	4.8	6.6	10	50	26.0	38	40	....	100	146	125	....	300	443
2	12.0	6.2	8.5	15	....	38.0	56	50	....	120	180	150	....	360	534
3	17.0	8.6	12.5	20	....	50.0	74	60	....	150	215	200	....	480	712

† Values are for all speeds and frequencies @ 230 volts. Amperage other than 230 volts can be figured:

$$A = \frac{230 \times \text{Amp from Table}}{\text{New Voltage}}$$

Example:

For 60 hp, 3 phase @ 550 volts : $\frac{(230 \times 150)}{550} = 62$ amps.

Power Factor estimated @ 80% for most motors. Efficiency is usually 80–90%.

Table A6—NEMA Electrical Enclosure Types

Type	Description	Type	Description
NEMA Type 1 (General Purposes)	For indoor use wherever oil, dust or water is not a problem.	NEMA Type 5 Dust Tight (Non-Hazardous)	Used for excluding dust. (All NEMA 12 enclosures are usually suitable for NEMA 5 use.)
NEMA Type 2 (Driptight)	Used indoors to exclude falling moisture and dirt.	NEMA Type 9 Dust Tight (Hazardous)‡	For locations where combustible dusts are present.
NEMA Type 3 (Weatherproof)	Provides protection against rain, sleet and snow.		
NEMA Type 4 (Watertight) ◆	Needed when subject to great amounts of water from any angle—such as areas which are repeatedly hosed down.	NEMA Type 12 (Industrial Use)	Used for excluding oil, coolant, flying dust, lint, etc.

◆ Not designed to be submerged.

‡ Class II Groups E, F and G.

ENGINEERING/TECHNICAL

A-C Motor Information Tables

Table A7—Frame Assignments

HP	Motor Speed, rpm				HP	Motor Speed, rpm			
	3600	1800	1200	900		3600	1800	1200	900
1/8-1/3	....	48	....	....	15	215T, 256U	254T, 284U	284T, 324U	286T, 326U
1/8-1/2	48	....	56	....	20	254T, 284U	256T, 286U	286T, 326U	324T, 364U
1/6	....	....	48	....	25	256T, 286U	284T, 324U	324T, 364U	326T, 365U
1/3-1	....	56	....	....	30	284TS, 324S	286T, 326U	326T, 365U	364T, 404U
3/4-1	56	....	....	....	40	286TS, 326S	324T, 364U	364T, 404U	365T, 405U
1/2	....	....	....	143T	50	324TS, 364US	326T, 365U, 365US	365T, 405U	404T, 444U
3/4	....	....	143T	145T	60	326TS, 365US	364TS▲, 404U, 404US	404T, 444U	405T, 445U
1	....	143T	145T	182T	75	364TS, 404US	365TS▲, 405U, 405US	405T, 445U	444T
1-1/2	143T	145T	182T	184T	100	365TS, 405US	404TS▲, 444US	444T	445T
2	145T	145T	184T	213T	125	404TS, 444US	405TS▲, 445US	445T	....
3	145T	182T	213T	215T, 254U	150	405TS, 445US	444TS▲	....	....
5	182T	184T	215T, 254U	254T, 256U	200	444TS	445TS▲	....	....
7-1/2	184T	213T, 254U	254T, 256U	256T, 284U	250	445TS	....	....	....
10	213T, 254U	215T, 256U	256T, 284U	284T, 286U	...	....	....	....	....

Table A8—Motor Frame Dimensions

Frame Size	D	E	2F	H Dia. (4) Holes	U Dia.	BA	V Min.	Key
48	3	2-1/8	2-3/4	11/32	1/2	2-1/2	...	3/64 Flat
56	3-1/2	2-7/16	3	11/32	5/8	2-3/4	...	3/16×3/16 ×1-3/8
143T	3-1/2	2-3/4	4	11/32	7/8	2-1/4	2	3/16×3/16 ×1-3/8
145T	3-1/2	2-3/4	5	11/32	7/8	2-1/4	2	3/16×3/16 ×1-3/8
182T	4-1/2	3-3/4	4-1/2	13/32	1-1/8	2-3/4	2-1/2	1/4 ×1/4 ×1-3/4
184T	4-1/2	3-3/4	5-1/2	13/32	1-1/8	2-3/4	2-1/2	1/4 ×1/4 ×1-3/4
213T	5-1/4	4-1/4	5-1/2	13/32	1-3/8	3-1/2	3-1/8	5/16×5/16×2-3/8
215T	5-1/4	4-1/4	7	13/32	1-3/8	3-1/2	3-1/8	5/16×5/16×2-3/8
254U	6-1/4	5	8-1/4	17/32	1-3/8	4-1/4	3-1/2	5/16×5/16×2-3/4
254T	6-1/4	5	8-1/4	17/32	1-5/8	4-1/4	3-3/4	3/8 ×3/8 ×2-7/8
256U	6-1/4	5	10	17/32	1-3/8	4-1/4	3-1/2	5/16×5/16×2-3/4
256T	6-1/4	5	10	17/32	1-5/8	4-1/4	3-3/4	3/8 ×3/8 ×2-7/8
284U	7	5-1/2	9-1/2	17/32	1-5/8	4-3/4	4-5/8	3/8 ×3/8 ×3-3/4
284T	7	5-1/2	9-1/2	17/32	1-7/8	4-3/4	4-3/8	1/2 ×1/2 ×3-1/4
284TS	7	5-1/2	9-1/2	17/32	1-5/8	4-3/4	3	3/8 ×3/8 ×1-7/8
286U	7	5-1/2	11	17/32	1-5/8	4-3/4	4-5/8	3/8 ×3/8 ×3-3/4
286T	7	5-1/2	11	17/32	1-7/8	4-3/4	4-3/8	1/2 ×1/2 ×3-1/4
286TS	7	5-1/2	11	17/32	1-5/8	4-3/4	3	3/8 ×3/8 ×1-7/8
324U	8	6-1/4	10-1/2	21/32	1-7/8	5-1/4	5-3/8	1/2 ×1/2 ×4-1/4
324T	8	6-1/4	10-1/2	21/32	2-1/8	5-1/4	5	1/2 ×1/2 ×3-7/8
324TS	8	6-1/4	10-1/2	21/32	1-7/8	5-1/4	3-1/2	1/2 ×1/2 ×2
326U	8	6-1/4	12	21/32	1-7/8	5-1/4	5-3/8	1/2 ×1/2 ×4-1/4
326T	8	6-1/4	12	21/32	2-1/8	5-1/4	5	1/2 ×1/2 ×3-7/8
326TS	8	6-1/4	12	21/32	1-7/8	5-1/4	3-1/2	1/2 ×1/2 ×2
364U	9	7	11-1/4	21/32	2-1/8	5-7/8	6-1/8	1/2 ×1/2 ×5
364US	9	7	11-1/4	21/32	1-7/8	5-7/8	3-1/2	1/2 ×1/2 ×2
364T	9	7	11-1/4	21/32	2-3/8	5-7/8	5-5/8	5/8 ×5/8 ×4-1/4
364TS	9	7	11-1/4	21/32	1-7/8	5-7/8	3-1/2	1/2 ×1/2 ×2
365U	9	7	12-1/4	21/32	2-1/8	5-7/8	6-1/8	1/2 ×1/2 ×5
365US	9	7	12-1/4	21/32	1-7/8	5-7/8	3-1/2	1/2 ×1/2 ×2
365T	9	7	12-1/4	21/32	2-3/8	5-7/8	5-5/8	5/8 ×5/8 ×4-1/4
365TS	9	7	12-1/4	21/32	1-7/8	5-7/8	3-1/2	1/2 ×1/2 ×2
404U	10	8	12-1/4	13/16	2-3/8	6-5/8	6-7/8	5/8 ×5/8 ×5-1/2
404US	10	8	12-1/4	13/16	2-1/8	6-5/8	4	1/2 ×1/2 ×2-3/4
404T	10	8	12-1/4	13/16	2-7/8	6-5/8	7	3/4 ×3/4 ×5-5/8
404TS	10	8	12-1/4	13/16	2-1/8	6-5/8	4	1/2 ×1/2 ×2-3/4
405U	10	8	13-3/4	13/16	2-3/8	6-5/8	6-7/8	5/8 ×5/8 ×5-1/2
405US	10	8	13-3/4	13/16	2-1/8	6-5/8	4	1/2 ×1/2 ×2-3/4
405T	10	8	13-3/4	13/16	2-7/8	6-5/8	7	3/4 ×3/4 ×5-5/8
405TS	10	8	13-3/4	13/16	2-1/8	6-5/8	4	1/2 ×1/2 ×2-3/4
444U	11	9	14-1/2	13/16	2-7/8	7-1/2	8-3/8	3/4 ×3/4 ×7
444US	11	9	14-1/2	13/16	2-1/8	7-1/2	4	1/2 ×1/2 x2-3/4
444T	11	9	14-1/2	13/16	3-3/8	7-1/2	8-1/4	7/8 ×7/8 ×6-7/8
444TS	11	9	14-1/2	13/16	2-3/8	7-1/2	4-1/2	5/8 ×5/8 ×3
445U	11	9	16-1/2	13/16	2-7/8	7-1/2	8-3/8	3/4 ×3/4 ×7
445US	11	9	16-1/2	13/16	2-1/8	7-1/2	4	1/2 ×1/2 ×2-3/4
445T	11	9	16-1/2	13/16	3-3/8	7-1/2	8-1/4	7/8 ×7/8 ×6-7/8
445TS	11	9	16-1/2	13/16	2-3/8	7-1/2	4-1/2	5/8 ×5/8 ×3

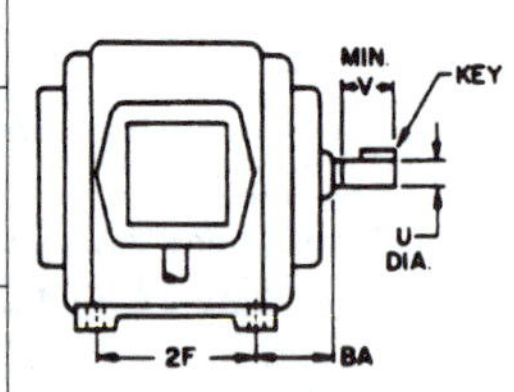

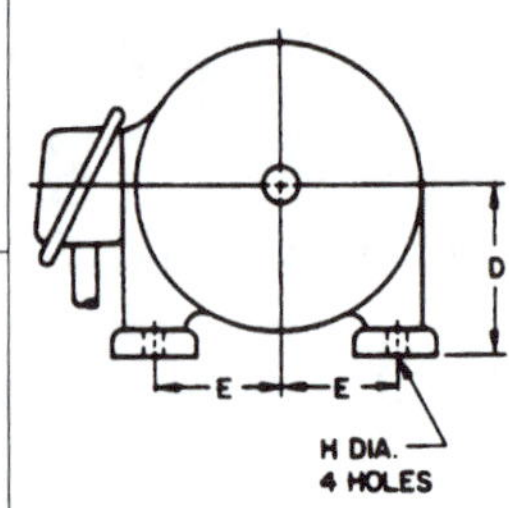

ENGLISH STANDARD MEASURES

Long Measure

1 mile = 1760 yards = 5280 feet.
1 yard = 3 feet = 36 inches.
1 foot = 12 inches.

Surveyor's Measure

1 mile = 8 furlongs = 80 chains.
1 furlong = 10 chains = 220 yards.
1 chain = 4 rods = 22 yards = 66 feet = 100 links.
1 link = 7.92 inches.

Square Measure

1 square mile = 640 acres = 6400 square chains.
1 acre = 10 square chains = 4840 square yards = 43,560 square feet.
1 square chain = 16 square rods = 484 square yards = 4356 square feet.
1 square rod = 30.25 square yards = 272.25 square feet = 625 square links.
1 square yard = 9 square feet.
1 square foot = 144 square inches.
An acre is equal to a square, the side of which is 208.7 feet.

Dry Measure

1 bushel (U.S. or Winchester struck bushel) = 1.2445 cubic foot = 2150.42 cubic inches.
1 bushel = 4 pecks = 32 quarts = 64 pints.
1 peck = 8 quarts = 16 pints.
1 quart = 2 pints.
1 heaped bushel = 1 1/4 struck bushel.
1 cubic foot = 0.8036 struck bushel.
1 British Imperial bushel = 8 Imperial gallons = 1.2837 cubic foot = 2218.19 cubic inches.

Liquid Measure

1 U.S. gallon = 0.1337 cubic foot = 231 cubic inches = 4 quarts = 8 pints.
1 quart = 2 pints = 8 gills.
1 pint = 4 gills.
1 British Imperial gallon = 1.2003 U.S. gallon = 277.27 cubic inches.
1 cubic foot = 7.48 U.S. gallons.

Circular and Angular Measure

60 seconds (″)	=	1 minute (′).
60 minutes	=	1 degree (°).
360 degrees	=	1 circumference (C).
57.3 degrees	=	1 radian.
2 π radians	=	1 circumference (C).

Specific Gravity

The specific gravity of a substance is its weight as compared with the weight of an equal bulk of pure water.
For making specific gravity determinations the temperature of the water is usually taken at 62° F when 1 cubic foot of water weighs 62.355 lbs. Water is at its greatest density at 39.20° F or 4° C.

Temperature

The following equation will be found convenient for transforming temperature from one system to another:

Let F = degrees Fahrenheit; C = degrees Centigrade; R = degrees Reamur.

$$\frac{F-32}{180} = \frac{C}{100} = \frac{R}{80}$$

Avoirdupois or Commercial Weight

1 gross or long ton = 2240 pounds.
1 net or short ton = 2000 pounds.
1 pound = 16 ounces = 7000 grains.
1 ounce = 16 drams = 437.5 grains.

Measures of Pressure

1 pound per square inch = 144 pounds per square foot = 0.068 atmosphere = 2.042 inches of mercury at 62 degrees F = 27.7 inches of water at 62 degrees F = 2.31 feet of water at 62 degrees F.
1 atmosphere = 30 inches of mercury at 62 degrees F = 14.7 pounds per square inch = 2116.3 pounds per square foot = 33.95 feet of water at 62 degrees F.
1 foot of water at 62 degrees F = 62.355 pounds per square foot = 0.433 pound per square inch.
1 inch of mercury at 62 degrees F = 1.132 foot of water = 13.58 inches of water = 0.491 pound per square inch.
Column of water 12 in. high, 1 in. dia. = .341 lbs.

Cubic Measure

1 cubic yard = 27 cubic feet.
1 cubic foot = 1728 cubic inches.
The following measures are also used for wood and masonry:
1 cord of wood = 4 × 4 × 8 feet = 128 cubic feet.
1 perch of masonry = 16-1/2 × 1-1/2 × 1 foot = 24-3/4 cubic feet.

Shipping Measure

For measuring entire internal capacity of a vessel: 1 register ton = 100 cubic feet.
For measurement of cargo:
1 U.S. shipping ton = 40 cubic feet = 32.143 U.S. bushels = 31.16 Imperial bushels.
British shipping ton = 42 cubic feet = 33.75 U.S. bushels = 32.72 Imperial bushels.

Troy Weight, Used for Weighing Gold and Silver

1 pound = 12 ounces = 5760 grains.
1 ounce = 20 pennyweights = 480 grains.
1 pennyweight = 24 grains.
1 carat (used in weighing diamonds) = 3.086 grains.
1 grain Troy = 1 grain avoirdupois = 1 grain apothecaries' weight.

Measure Used for Diameters and Areas of Electric Wires

1 circular inch = area of circle 1 inch in diameter = 0.7854 square inch.
1 circular inch = 1,000,000 circular mils.
1 square inch = 1.2732 circular inch = 1,273,239 circular mils.
A circular mil is the area of a circle 0.001 inch in diameter.

Board Measure

One foot board measure is a piece of wood 12 inches square by 1 inch thick, or 144 cubic inches. 1 cubic foot therefore equals 12 feet board measure.

METRIC SYSTEM OF MEASUREMENTS

Measures of Length

10 millimeters (mm.)	=	1 centimeter (cm.)
10 centimeters	=	1 decimeter (dm.)
10 decimeters	=	1 meter (m.)
1000 meter	=	1 kilometer (km.)

Measures of Weight

10 milligrams (mg.)	=	1 centigram (cg.)
10 centigrams	=	1 decigram (dg.)
10 decigrams	=	1 gram (g.)
10 grams	=	1 decagram (Dg.)
10 decagrams	=	1 hectogram (Hg.)
10 hectograms	=	1 Kilogram (Kg.)
1000 kilograms	=	1 (metric) ton (T.)

Surveyor's Square Measure

100 square meters ($m.^2$)	=	1 acre (ar.)
100 acres	=	1 hectare (har.)
100 hectares	=	1 sq. kilometer ($Km.^2$)

Square Measure

100 sq. millimeters ($mm.^2$)	=	1 sq. centimeter ($cm.^2$)
100 sq. centimeters	=	1 sq. decimeter ($dm.^2$)
100 sq. decimeters	=	1 sq. meter ($m.^2$)

Cubic Measure

1000 cu. millimeters ($mm.^3$)	=	1 cu. centimeter ($cm.^3$)
1000 cu. centimeters	=	1 cu. decimeter ($dm.^3$)
1000 cu. decimeters	=	1 cu. meter ($m.^3$)

Dry and Liquid Measure

10 milliliters (ml.)	=	1 centiliter (cl.)
10 centiliters	=	1 deciliter (dl.)
10 deciliters	=	1 liter (l.)
100 liters	=	1 hectoliter (Hl.)

1 liter = 1 cubic decimeter = the volume of 1 kilogram of pure water at a temperature of 39.2 degrees F.

Length Conversion Constants for Metric and U.S. Units

Millimeters × .039370 = inches.
Meters × 39.370 = inches.
Meters × 3.2808 = feet.
Meters × 1.09361 = yards.
Kilometers × 3,280.8 = feet.
Kilometers × .62137 = Statute Miles.
Kilometers × .53959 = Nautical Miles.

Inches × 25.4001 = millimeters.
Inches × .0254 = meters.
Feet × .30480 = meters.
Yards × .91440 = meters.
Feet × .0003048 = kilometers.
Statute Miles × 1.60935 = kilometers.
Nautical Miles × 1.85325 = kilometers.

Weight Conversion Constants for Metric and U.S. Units

Grams × 981 = dynes.
Grams × 15.432 = grains.
Grams × .03527 = ounces (Avd.).
Grams × .033818 = fluid ounces (water).
Kilograms × 35.27 = ounces (Avd.).
Kilograms × 2.20462 = pounds (Avd.).
Metric Tons (1000 Kg.) × 1.10231 = Net Ton (2000 lbs.).
Metric Tons (1000 Kg.) × .98421 = Gross Ton (2240 lbs.).

Dynes × .0010193 = grams.
Grains × .0648 = grams.
Ounces (Avd.) × 28.35 = grams.
Fluid Ounces (Water) × 29.57 = grams.
Ounces (Avd.) × .02835 = kilograms.
Pounds (Avd.) × .45359 = kilograms.
Net Ton (2000 lbs.) × .90719 = Metric Tons (1000 Kg.).
Gross Ton (2240 lbs.) × 1.01605 = Metric Tons (1000 Kg.).

Area Conversion Constants for Metric and U.S. Units

Square Millimeters × .00155 = square inches.
Square centimeters × .155 = square inches.
Square Meters × 10.76387 = square feet.
Square Meters × 1.19599 = square yards.
Hectares × 2.47104 = acres.
Square Kilometers × 247.104 = acres.
Square Kilometers × .3861 = square miles.

Square Inches × 645.163 = square millimeters.
Square Inches × 6.45163 = square centimeters.
Square Feet × .0929 = square meters.
Square Yards × .83613 = square meters.
Acres × .40469 = hectares.
Acres × .0040469 = square kilometers.
Square Miles × 2.5899 = square kilometers.

Volume Conversion Constants for Metric and U.S. Units

Cubic centimeters × .033818 = fluid ounces.
Cubic centimeters × .061023 = cubic inches.
Cubic centimeters × .271 = fluid drams.
Liters × 61.023 = cubic inches.
Liters × 1.05668 = quarts.
Liters × .26417 = gallons.
Liters × .035317 = cubic feet.
Hectoliters × 26.417 = gallons.
Hectoliters × 3.5317 = cubic feet.
Hectoliters × 2.83794 = bushel (2150.42 cu. in.).
Hectoliters × .1308 = cubic yards.
Cubic Meters × 264.17 = gallons.
Cubic Meters × 35.317 = cubic feet.
Cubic Meters × 1.308 = cubic yards.

Fluid Ounces × 29.57 = cubic centimeters.
Cubic Inches × 16.387 = cubic centimeters.
Fluid Drams × 3.69 = cubic centimeters.
Cubic Inches × .016387 = liters.
Quarts × .94636 = liters.
Gallons × 3.78543 = liters.
Cubic Feet × 28.316 = liters.
Gallons × .0378543 = hectoliters.
Cubic Feet × .28316 = hectoliters.
Bushels (2150.42 cu. in.) × .352379 = hectoliters.
Cubic Yards × 7.645 = hectoliters.
Gallons × .00378543 = cubic meters.
Cubic Feet × .028316 = cubic meters.
Cubic Yards × .7645 = cubic meters.

Power and Heat Conversion Constants for Metric and U.S. Units

Calorie × 0.003968 = B.T.U.
Joules × .7373 = pound-feet.
Newton-Meters × 8.851 = pound-inches
Cheval Vapeur × .9863 = Horsepower.
Kilowatts × 1.34 = Horsepower.
Kilowatt Hours × 3415 = B.T.U.
(Degrees Cent. × 1.8) +32 = degrees Fahr.
(Degrees Reamur × 2.25) + 32 = degrees Fahr.

B.T.U. × 252 = calories.
Pound-Feet × 1.3563 = joules.
Pound-inches × .11298 = Newton-meters.
Horsepower × 1.014 = Cheval Vapeur.
Horsepower × .746 = kilowatts.
B.T.U. × .00029282 = kilowatt hours.
(Degrees Fahr. – 32) × .555 = degrees Cent.
(Degrees Fahr. – 32) × .444 = degrees Reamur.

VISCOSITY CLASSIFICATION EQUIVALENTS

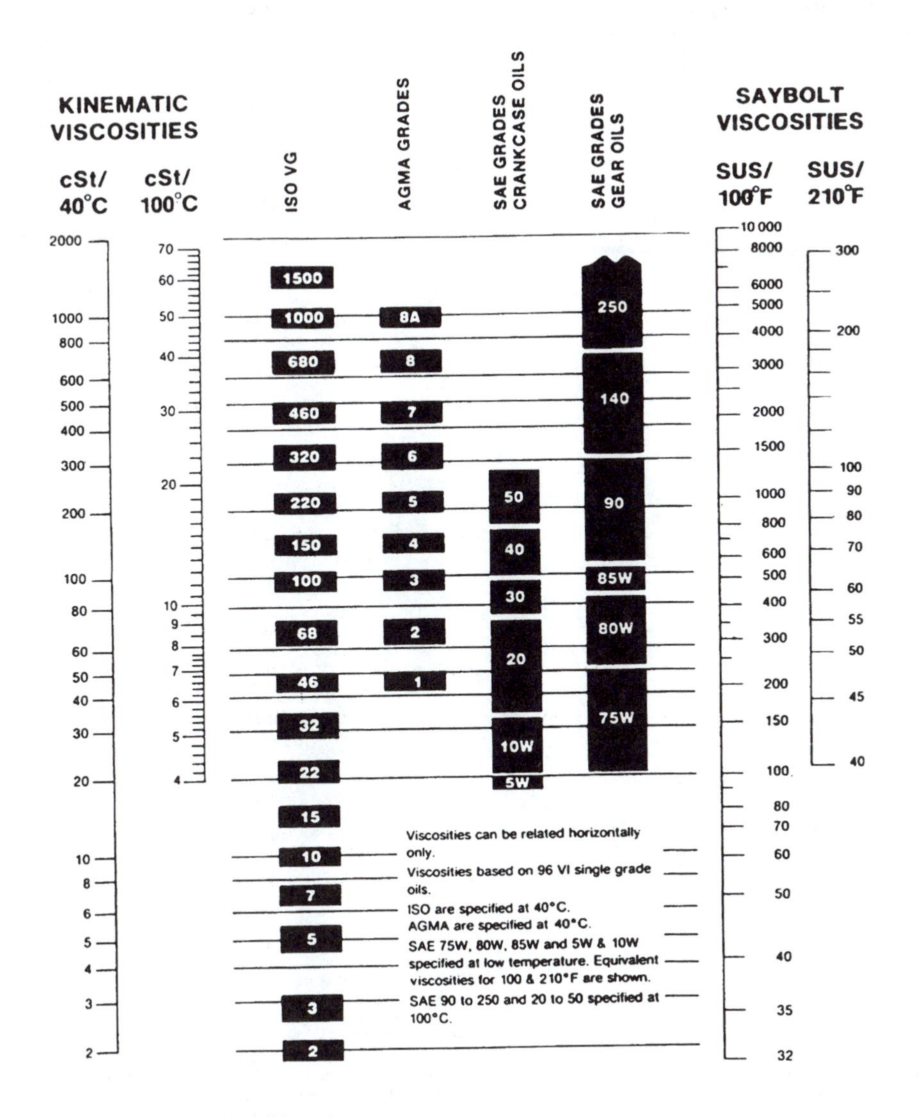

ISO Viscosity Classification System

All industrial oils are graded according to the ISO Viscosity Classification System, approved by the International Standards Organizations (ISO). Each ISO viscosity grade number corresponds to the mid-point of viscosity range expressed in centistokes (cSt) at 40°C. For example, a lubricant with an ISO grade of 32 has a viscosity within the range of 28.80-35.2, the midpoint of which is 32.

Rule-of-Thumb: The comparable ISO grade of a competitive product whose viscosity in SUS at 100°F is known can be determined by using the following conversion formula:

SUS @ 100°F ÷ 5 = cSt @ 40°C

GLOSSARY

actual mechanical advantage **(*AMA*)** A measurement that takes into account efficiency and friction losses.

addendum The height of the tooth above the pitch circle.

allowances How loosely or tightly fasteners fit with their mating parts.

antifoam additives Antifoaming agents that prevent oil from foaming.

axial, or thrust, loads Force exerted parallel to the shaft axis.

axial thrust Force, or push, along the axis of the shaft. Also called *end thrust.*

axial-type friction clutches Clutches or brakes that have contact pressure applied perpendicular to the rotating shaft. The disc brake and clutch are common examples of axial-type friction brakes and clutches.

babbitt Lead- or tin-based alloys with copper and antimony. Babbitt is used in bearings for moisture or chemical environments.

backlash The amount of side clearance of a gear tooth when two teeth mesh. More specifically, it is the space between meshing teeth measured at the pitch circle.

bearing The part of any machine that is used to reduce the friction of a rotating shaft or between two moving surfaces.

belt pitch length Circumference of the tensile cord of the V-belt located just above the center of the cross section.

boundary film lubrication Lubrication occurring when only an extremely thin film of lubricant is present to separate bearing surfaces.

cape chisel A chisel with a narrow pointed edge. The area behind the edge is ground smaller to avoid binding when you use it for working in a slot. The cape chisel is used for cleaning keyway slots or in any application that requires a sharp inside corner.

carbon bearings These bearings have good chemical and moisture resistance and may be lubricated or be used as self-lubricating bearings.

cast-iron bearings These are plain bearings used primarily for slow, light applications. Their major advantages are their low cost, long life, and low maintenance requirement.

center distance The shortest distance between the axes of two mating gears.

center of gravity The point at which a suspended object balances. The weight of the object is concentrated at that point.

chamfer Beveling the edge of a fastener or rod by 30° to 60° to make it easier to start cutting threads.

chordal action The vibratory motion caused by the rise and fall of the chain as it goes over a small sprocket.

chordal thickness The thickness of a tooth measured at the pitch circle.

circular pitch The distance from a point on one tooth to a corresponding point on the next tooth measured on the pitch circle.

clearance The distance between the top of the tooth of one gear and the bottom of the meshing space.

clearance fit Fit in which the coupling hub slides onto the shaft without forcing or heating.

coefficient of friction Resistance to sliding expressed as a ratio of the force holding the surfaces together divided by the force that resists sliding.

cold chisel A general-purpose chisel with a straight edge that is used to cut metal that has not been heated. It can be used to cut metals of all kinds except hardened steel.

combination sheave A sheave that can be used with A- or B-section belts.

cone clutches A cone-shaped combination of radial- and axial-type clutches. Because of the wedging action of the cone inside a matching drum, the force required for activation is less than purely radial or axial clutches.

crest The top portion of an individual thread.

dead-blow hammer A hammer with a hollow head that is loaded with lead shot that absorbs most bounce, or rebound. The face of this hammer is a tough, modern plastic that can be used on sharp edges without damage.

dedendum Distance from the pitch circle to the bottom of the tooth.

deflecting-beam torque wrench The most basic type of torque wrench. When you pull the handle, a pointer on a scale indicates the amount of torque applied.

detergent additive A substance incorporated in lubricating oils which gives them the property of keeping insoluble material in suspension.

dial-indicating torque wrench A torque wrench with a precision dial indicator built into it. It is used when more accurate torque settings are required.

diametral pitch Ratio of the number of teeth for each inch of pitch diameter of the gear.

diamond-point chisel A chisel ground at an angle across the corners that gives its cutting edge a diamond-shaped face. This sharp-pointed chisel can cut V-grooves, slot a joint for welding, and start slots for cutting with a cold chisel.

dispersant additive A substance incorporated in lubricating oils to assist the detergent additives to maintain dispersed particles in suspension.

dynamic load rating The life expectancy of a bearing rotating under load; the bearing radial load that gives a basic rating life of one million revolutions.

dynamic seals Designed to seal moving surfaces. Rotating or reciprocating shafts require some type of dynamic seal. Lip seals and mechanical seals are two widely used dynamic seals.

elastomer Flexible polymer used for seals and O-rings.

end-float The inclination of shafts to move back and forth across their bearings. Also called *end-play.*

equilibrium Special state when all forces are balanced.

extreme-pressure additives This chemical compound increases the load-carrying ability of the lubricant to cushion the shock and rubbing action associated with gearboxes. They are most effective in reducing friction and wear at high temperatures.

fiber The smallest component of a rope.

fire point The minimum temperature at which a lubricant will continuously burn.

flanks The sloped surfaces joining the crests and roots.

flash point The minimum temperature required for the lubricant to give off a vapor that will flash into flames. The lubricant itself is not hot enough to burn—only the vapor or flames.

garter spring Used to provide additional force to help maintain lip contact with the shaft and the lip seal.

hard-laid rope Rope twisted tightly. Hard-laid rope is stiffer and has better resistance to abrasion and moisture penetration than soft-laid rope.

helix The curving path that a point would follow if it were to travel in an even spiral around a cylinder and in line with the axis of that cylinder.

hitch The total configuration of a sling or slings and all the attachments connected to the load.

hydrodynamic The state in which a rotating shaft is supported by a layer or wedge of oil so that no surface contact occurs.

hydromatic The state in which a rotating shaft becomes fully supported by a full oil film due to hydroplaning action.

hydrostatic The state in which the lubricant pressure required to separate moving metal surfaces is supplied from an external source (an oil pump is an example).

interference fit Fit in which the coupling hub must be heated so it will expand enough to slide onto the shaft.

involute The curved line produced by a point of a stretched string when it is unwrapped from a given cylinder.

ISO International Standards Organization.

kinetic energy Energy in motion.

kinetic friction The resistance to continued movement after an object is moving.

labyrinth seal Seal that does not touch the shaft but has a series of teeth with a small clearance between the teeth and the shaft.

lantern gears An early type of gear consisting of wooden discs fitted with pegs.

lantern ring A special-shaped spacer (usually made of metal) that allows the introduction of an outside fluid in a stuffing box.

linear pitch Distance from a point on one tooth to the corresponding point on the next tooth of a gear rack.

line shafting Rigidly coupled shafts used to transmit mechanical power.

major diameter The maximum diameter measured from crest to crest of a bolt. The major diameter of a bolt is measured to determine the bolt's *nominal size.*

mallet A soft-faced hammer. Soft faces may be made of rawhide, rubber, plastic, brass, copper, or lead.

match numbers Used to match exactly V-belts used in a multiple V-belt drive. A match number of 50 means that the V-belt is exactly the length it was designed to be.

micrometer torque wrench A wrench where a micrometer-type handle is used to preset a specified torque. When this preset torque is reached, you will hear a loud click and feel a slight jump.

minimum plain bore A sprocket with no keyway and with the smallest bore available.

minor diameter The minimum diameter measured from the thread valleys.

module The pitch diameter of a gear divided by the number of teeth. An actual dimension, unlike diametral pitch, which is a ratio of the number of teeth to the pitch diameter.

overhung load Force applied at right angles at the end of the input or output shaft. An example is the force applied to a shaft because of the tension of a V-belt drive.

oxidation inhibitors Additives that slow the rate of a lubricant's natural tendency to oxidize.

pedestal rings Rings used to give U-cups proper back support and side support to prevent the lips from collapsing until fluid pressure becomes high enough to provide hydraulic support. Pedestal rings are typically made of nylon or Teflon.

pintle chain An offset-type chain usually made of malleable iron.

pitch The distance from the center of one crest of a thread to the center of the next one.

pitch diameter The imaginary circle between the major and minor diameters where the force applied by a screw thread is centered. The gear diameter defined by the point on the teeth where force is applied to rotate the gear. Diameter of the pitch circle.

plain, or journal, bearing Cylindrical sleeve that supports a rotating or sliding shaft. It uses a sliding action rather than a rolling action.

plastic bearings Class III bearings. These bearings require no lubrication, have good wear resistance, and have long life.

positive belt drive A type of belt drive that does not rely on friction for the driving force. It transmits power by the positive engagement of belt teeth with pulley teeth, much like a chain drive. Because of its construction, there is no slippage or speed variations.

potential energy Stored energy.

pour point The point at which the lubricant becomes so thick that it no longer flows.

pressure angle The angle of contact of a gear tooth. It is usually 14.5° or 20°.

QD bushing Type of tapered bushing with a characteristic flange that a taper lock does not have. Instead of set screws, QD bushings use hex-head bolts to hold the sheave and squeeze the shaft.

radial loads A situation in which force is exerted perpendicular to the shaft axis.

radial-type friction clutches Clutches or brakes that have contact pressure applied to the peripheral of a drum or rim.

radian (rad) A unit of angle measure of a circle. It can be directly converted to degrees and is equal to 57.3°.

rolling-element bearing A cylinder containing a moving inner ring of steel balls or rollers. It is also called an *antifriction bearing*, because the friction created by this bearing is rolling friction rather than the sliding friction created by the plain bearing.

root The bottom section or valley of a thread. It is the opposite of the crest.

round-nose chisel A chisel with a round cross section at the cutting edge. The edge is ground at an angle of 60° with the chisel axis. You can use round-nose chisels to clean up round slots, damaged bolt holes, and rounded inside corners or to cut oil grooves in plain bearings or bushings.

runout Amount of wobble caused by loose machinery bearings, bent shafts, or poorly machined coupling hubs.

rust inhibitors Additives that improve a lubricant's ability to stick to metal surfaces. This coating action protects the metal surfaces from oxidation and rust formation by preventing moisture from penetrating the protective film.

Saybolt seconds universal (SSU) Unit of measure for industrial lubricant viscosity. It may also be referred to as *Saybolt universal seconds (SUS).*

scalars Force with magnitude but no direction.

sintered bronze bearings Class II bearings. They are made from a powdered bronze composite and impregnated with oil.

soft foot The condition that exists when the feet of a machine do not equally support its weight. It can be compared to a chair with one short leg, causing the chair to rock.

soft-laid rope Rope twisted loosely. Soft-laid rope is more flexible and has better tolerance to shock loading than hard-laid rope. It stretches and rebounds rather than breaking.

solid bronze bearings These bearings have high hardness and good fatigue resistance. They are harder than babbitt and can also damage shafts if not properly lubricated.

static friction The resistance to movement between two contacting surfaces.

static load rating The life expectancy of a bearing stationary or slow-moving loads.

static seals Designed for service where there is little or no movement between mating surfaces. Gaskets and O-rings are the most common examples of static seals.

strand Several yarns twisted together.

tag-lines Light ropes attached to a load to prevent rotation and to guide the load during transit and positioning.

taper-lock bushing Type of bushing that is tapered and has no flange. It is assembled with set screws that cause the bushing to squeeze the shaft as they are tightened.

tapping The process of cutting an internal thread for a fastener.

theoretical mechanical advantage (TMA) Ideal measurement that does not consider any losses due to friction or efficiency.

thermal rating The maximum torque or power that a gearbox can transmit continuously, based on its ability to dissipate heat generated by friction.

threading Using a die to cut external threads on a rod or fastener.

threads per inch This is determined by using a ruler and physically counting the number of threads in 1 in. of bolt. Bolts with more threads per inch are referred to as *fine threads*, and bolts with fewer threads per inch are called *coarse threads*.

tolerances The amount of size difference allowed from the exact size of a fastener. For metric screw threads, tolerance is referred to as *fit*.

tooth thickness Thickness of the tooth measured on the pitch circle.

torque Forces produced in a twisting or rotating motion.

vector A force with a magnitude and direction.

viscometer Instrument used to measure viscosity of a lubricant. Also called *viscosimeter*.

viscosity Property of a fluid, semifluid, or semisolid substance that causes it to resist flow.

viscosity index Measure of the rate at which the viscosity of a lubricant will change as the temperature changes. The higher the viscosity index, the more resistant the lubricant will be to thinning out with increased temperature.

whole depth Full depth of the tooth.

yarns A bundle of fibers twisted together.

INDEX